Dietary Strategies for Prevention of Geriatric Diseases and Exploring the Mechanism of Aging

Dietary Strategies for Prevention of Geriatric Diseases and Exploring the Mechanism of Aging

Guest Editors

Xiaoyu Wang

Raquel Sanchez-Varo

MDPI

Basel • Beijing • Wuhan • Barcelona • Belgrade • Novi Sad • Cluj • Manchester

Guest Editors
Xiaoyu Wang
College of Food Science and Nutritional Engineering
China Agricultural University
Beijing
China

Raquel Sanchez-Varo
Departamento Fisiologia Humana
Universidad de Malaga
Malaga
Spain

Editorial Office
MDPI AG
Grosspeteranlage 5
4052 Basel, Switzerland

This is a reprint of the Special Issue, published open access by the journal *Nutrients* (ISSN 2072-6643), freely accessible at: https://www.mdpi.com/journal/nutrients/special_issues/86R23COS2T.

For citation purposes, cite each article independently as indicated on the article page online and as indicated below:

Lastname, A.A.; Lastname, B.B. Article Title. *Journal Name* **Year**, *Volume Number*, Page Range.

ISBN 978-3-7258-5077-8 (Hbk)
ISBN 978-3-7258-5078-5 (PDF)
https://doi.org/10.3390/books978-3-7258-5078-5

Contents

nutrients

MDPI

Editorial

Frontiers in Age-Related Diseases and Nutritional Interventions

Caiziyu Yu [1], Lili Qiu [1] and Xiaoyu Wang [1,2,*]

[1] College of Food Science & Nutritional Engineering, China Agricultural University, Beijing 100083, China; yuczy@cau.edu.cn (C.Y.); qiulily980829@126.com (L.Q.)
[2] Key Laboratory of Precision Nutrition and Food Quality, Department of Nutrition and Health, China Agricultural University, Beijing 100193, China
* Correspondence: xy.wang@cau.edu.cn

1. Introduction

The aging population presents one of the greatest challenges to sustainable development, progressing at an unprecedented rate in recent years. Aging involves an irreversible, progressive decline in physiological functions, ultimately leading to a range of age-related diseases, including neurodegenerative, cardiovascular, gastrointestinal, and immune system diseases. Over the years, extensive efforts have been devoted to investigating the biology of aging and the mechanisms underlying age-associated degenerative diseases, with the goal of developing strategies to alleviate age-related health issues. A growing body of research has highlighted the significant role of diet in regulating aging and the development of age-associated diseases. This Special Issue of *Nutrients*, titled "Dietary Strategies for Prevention of Geriatric Diseases and Exploring the Mechanism of Aging", aims to present dietary and nutritional strategies for alleviating aging and age-related degenerative diseases in the elderly population, while also advancing our knowledge of the underlying mechanisms of and age-related pathology. The studies featured in this Special Issue can be categorized into the following two main areas: (i) Aging and Age-related Diseases, and (ii) Nutritional Interventions for the Elderly. Together, these two themes offer new insights and practical approaches to understanding and managing age-related health issues, paving the way for more effective strategies to improve the quality of life among the elderly and promote healthy aging.

2. Aging and Age-Related Diseases

One article explored the factors influencing the cellular and molecular hallmarks of aging. Reduced telomere length and various epigenetic alterations (DNA methylation, histone modifications, and chromatin remodeling) are among the key cellular and molecular markers of aging. Additionally, S. Liu et al. investigated the causal relationship between the intake of different types of meat and biological aging [1]. Their findings revealed a positive causal relationship between the intake of meat and accelerated DNA methylation PhenoAge. Moreover, processed meat consumption was negatively associated with telomere length, suggesting that increased meat intake may accelerate the biological aging process.

In addition, several studies in this Special Issue focused on age-related degenerative diseases. Although each study addressed a different topic, their findings collectively provide further insight into the mechanisms and contributing factors of such diseases. C. Chang et al. identified a significant association between nutritional risk and dynapenia, specifically in elderly people aged 75 years and above, suggesting that age may moderate the relationship between nutritional status and dynapenia in older populations [2]. A comprehensive review authored by W. Yu et al. summarized the alterations in the immune system associated with aging and elucidated the mechanisms of immunosenescence [3]. Aging impairs both innate and adaptive immunity; this decline in immunity not only exacerbates the aging process but also accelerates the onset of infectious diseases, autoimmune disorders, and cancer. The study by L. Gaussens et al. examined the associations between vitality components (appetite loss and weight loss) and other domains of Intrinsic Capacity (IC) in the elderly, such as cognition, vision, hearing, psychology and locomotion [4]. Appetite and weight loss, both common among the elderly, were found to be associated with other potential IC impairments, i.e., psychological and locomotor functions. In another study, X. Sang et al. explored the changes in physicochemical properties of the colonic mucus layer with increasing age and the underlying mechanisms [5]. In aging mice, changes in the content and structure of mucin were found to be linked to impaired mucus barrier function. Additionally, decreased expression of the genes that typically regulate goblet cell formation resulted in a reduction in the proportion of goblet cells and mucin content, while the low expression of mucin glycosylase led to shorter O-glycan chains, affecting mucus microstructure and thus reducing mucus barrier function with aging. Together, these studies provide valuable insights into the mechanisms related to aging by exploring the diverse physiological changes in the body during the aging process and the various factors that trigger age-related diseases.

3. Nutritional Interventions for the Elderly

Dietary interventions have been proven to be among the most powerful and consistent approaches to mitigating age-related diseases and dysfunction. The prevention and treatment of aging-induced diseases can be achieved through proper dietary habits, nutritional supplements, and adequate nutrient intake.

Two papers in this Issue provide significant insights into dietary intervention strategies. Through latent class trajectory analysis, C. Zhang et al. discovered that Chinese older adults with a dietary variety score (DVS) change trajectory characterized as "Moderate–Slow decline–Slow growth" were at an increased risk of frailty [6]. Their findings suggest that maintaining a high dietary diversity trajectory could reduce the risk of frailty in older Chinese adults, while the optimal time for intervention is in the early stages of aging. In addition, H. -Y. Lee et al. presented a review on the interplay between dietary restriction-related molecular signaling proteins, signaling pathways, lipid metabolism, and aging. The review examined diet-induced modifications in cell membrane lipids and commensal bacteria-driven shifts in lipid metabolism, ultimately highlighting the pivotal role of lipid metabolism in the aging process [7].

Among the articles related to specific nutrient-based intervention strategies for aging, Q. Chen et al. conducted the first comprehensive assessment of the relationship between the intake of methyl donor nutrients (MDNs) and cognitive function in the elderly. They found that MDNs may have the potential to enhance cognitive function by regulating microbial homeostasis [8]. In addition, C. Cochet et al. conducted a systematic review of the existing evidence on the efficacy of nutritional supplementation in supporting muscle and mitochondrial health in elderly individuals with sarcopenia or malnutrition [9]. The review concluded that branched-chain amino acids, in association with vitamin D, may help in treating sarcopenia and enhancing mitochondrial bioenergetics and redox activity. Based on these findings, the authors proposed dietary recommendations for treating sarcopenia in older people. In a related study, O. Ezaki summarized three potential extracellular signals that could improve muscle mass and function through the supplementation of medium-chain triglycerides (MCT; C8:0 and C10:0) during meals in frail older adults. They provide guidance for future clinical trials evaluating the use of MCT in treating primary (age-related) sarcopenia [10]. S. Wang et al. evaluated the efficacy and safety of exogenous nucleotides as aging supplements for the elderly [11], finding that nucleotide intervention may be a promising strategy due to its anti-aging effects and ability to enhance the quality of life in older adults. In another review, W. Yu et al. emphasized the importance of counteracting immunosenescence through the use of specific nutrients, such as macronutrients including proteins, fatty acids (e.g., short-chain fatty acids), and micronutrients including vitamins (A, D, E), minerals (e.g., zinc), as well as probiotics and prebiotics [3]. A review by N.A. Rivero-Segura et al. explored the potential of nutraceuticals—such as vitamins, polyphenols, flavonoids, saponins, fatty acids, probiotics, and plant extracts—in preventing age-related diseases by targeting biological markers of aging (such as telomere shortening, epigenetic changes, mitochondrial dysfunction, etc.) and emphasized their role as natural anti-aging agents [12]. Overall, these articles present viable strategies for alleviating aging and age-associated diseases through specific supplements and nutrients.

4. Conclusions

This Special Issue of *Nutrients* sheds light on the underlying mechanisms related to aging and age-associated pathology, while also exploring optimal nutritional strategies for the prevention and mitigation of geriatric diseases. The aging process and associated decline in physiological functions pose substantial challenges to the health and quality of life of the elderly. Several studies included here emphasize the potential of dietary and nutritional interventions to counteract these age-related alterations. Collectively, the papers published in this Special Issue underscore the significant role of dietary and nutritional interventions in addressing the multifaceted challenges of aging and age-related diseases. In summary, this Special Issue serves as a platform to enhance our understanding of the biological factors and pathogenic mechanisms involved in aging and age-related diseases, offering meaningful insights into dietary approaches for the prevention and management of geriatric diseases.

Author Contributions: Conceptualization, X.W.; writing—original draft preparation, C.Y. and L.Q.; writing—review and editing X.W. and C.Y. All authors have read and agreed to the published version of the manuscript.

Funding: This research was funded by Young Elite Scientists Sponsorship Program by CAST (NO. 2022QNRC001).

Conflicts of Interest: The authors declare no conflicts of interest.

References

1. Liu, S.; Deng, Y.; Liu, H.; Fu, Z.; Wang, Y.; Zhou, M.; Feng, Z. Causal Relationship between Meat Intake and Biological Aging: Evidence from Mendelian Randomization Analysis. *Nutrients* **2024**, *16*, 2433. [CrossRef] [PubMed]
2. Chang, C.-C.; Lai, T.-F.; Chen, J.; Liao, Y.; Park, J.-H.; Chang, Y.-J. Age Difference in the Association Between Nutritional Status and Dynapenia in Older Adults. *Nutrients* **2025**, *17*, 734. [CrossRef] [PubMed]
3. Yu, W.; Yu, Y.; Sun, S.; Lu, C.; Zhai, J.; Lei, Y.; Bai, F.; Wang, R.; Chen, J. Immune Alterations with Aging: Mechanisms and Intervention Strategies. *Nutrients* **2024**, *16*, 3830. [CrossRef] [PubMed]
4. Gaussens, L.; González-Bautista, E.; Bonnefoy, M.; Briand, M.; Tavassoli, N.; De Souto Barreto, P.; Rolland, Y.; on behalf of the GEGN Group. Associations between Vitality/Nutrition and the Other Domains of Intrinsic Capacity Based on Data from the INSPIRE ICOPE-Care Program. *Nutrients* **2023**, *15*, 1567. [CrossRef] [PubMed]
5. Sang, X.; Wang, Q.; Ning, Y.; Wang, H.; Zhang, R.; Li, Y.; Fang, B.; Lv, C.; Zhang, Y.; Wang, X.; et al. Age-Related Mucus Barrier Dysfunction in Mice Is Related to the Changes in Muc2 Mucin in the Colon. *Nutrients* **2023**, *15*, 1830. [CrossRef] [PubMed]

6. Zhao, C.; Wang, Y.; Jia, X.; Fan, J.; Wang, N.; Yang, Y.; Shi, X. Associations of Dietary Diversity Trajectories with Frailty among Chinese Older Adults: A Latent Class Trajectory Analysis Based on a CLHLS Cohort. *Nutrients* **2024**, *16*, 1445. [CrossRef] [PubMed]
7. Lee, H.-Y.; Min, K.-J. Dietary Restriction and Lipid Metabolism: Unveiling Pathways to Extended Healthspan. *Nutrients* **2024**, *16*, 4424. [CrossRef] [PubMed]
8. Chen, Q.; Fan, R.; Song, L.; Wang, S.; You, M.; Cai, M.; Wu, Y.; Li, Y.; Xu, M. Association of Methyl Donor Nutrients' Dietary Intake and Cognitive Impairment in the Elderly Based on the Intestinal Microbiome. *Nutrients* **2024**, *16*, 2061. [CrossRef] [PubMed]
9. Cochet, C.; Belloni, G.; Buondonno, I.; Chiara, F.; D'Amelio, P. The Role of Nutrition in the Treatment of Sarcopenia in Old Patients: From Restoration of Mitochondrial Activity to Improvement of Muscle Performance, a Systematic Review. *Nutrients* **2023**, *15*, 3703. [CrossRef] [PubMed]
10. Ezaki, O. Possible Extracellular Signals to Ameliorate Sarcopenia in Response to Medium-Chain Triglycerides (8:0 and 10:0) in Frail Older Adults. *Nutrients* **2024**, *16*, 2606. [CrossRef] [PubMed]
11. Wang, S.; Song, L.; Fan, R.; Chen, Q.; You, M.; Cai, M.; Wu, Y.; Li, Y.; Xu, M. Targeting Aging and Longevity with Exogenous Nucleotides (TALENTs): Rationale, Design, and Baseline Characteristics from a Randomized Controlled Trial in Older Adults. *Nutrients* **2024**, *16*, 1343. [CrossRef] [PubMed]
12. Rivero-Segura, N.A.; Zepeda-Arzate, E.A.; Castillo-Vazquez, S.K.; Fleischmann-delaParra, P.; Hernández-Pineda, J.; Flores-Soto, E.; García-delaTorre, P.; Estrella-Parra, E.A.; Gomez-Verjan, J.C. Exploring the Geroprotective Potential of Nutraceuticals. *Nutrients* **2024**, *16*, 2835. [CrossRef] [PubMed]

nutrients

MDPI

Article

Age Difference in the Association Between Nutritional Status and Dynapenia in Older Adults

Chih-Ching Chang [1], Ting-Fu Lai [2,3], Jiaren Chen [2], Yung Liao [2], Jong-Hwan Park [3,4,5,*] and Yen-Jung Chang [1,*]

1 Department of Health Promotion and Health Education, College of Education, National Taiwan Normal University, Taipei 106308, Taiwan; 81305010e@ntnu.edu.tw
2 Graduate Institute of Sport, Leisure and Hospitality Management, National Taiwan Normal University, Taipei 106308, Taiwan; 80905010e@ntnu.edu.tw (T.-F.L.); 80931005a@ntnu.edu.tw (J.C.); liaoyung@ntnu.edu.tw (Y.L.)
3 Department of Convergence Medicine, Pusan National University School of Medicine, Yangsan 50612, Republic of Korea
4 Department of Clinical Bio-Convergence, Graduate School of Convergence in Biomedical Science, Pusan National University School of Medicine, Yangsan 50612, Republic of Korea
5 Convergence Medical Institute of Technology, Pusan National University Hospital, Busan 49241, Republic of Korea
* Correspondence: parkj@pusan.ac.kr (J.-H.P.); yjchang@ntnu.edu.tw (Y.-J.C.)

Abstract: Background: Although nutritional status plays a critical role in maintaining muscle strength, limited evidence exists regarding its association with dynapenia. Objectives: We aimed to investigate the association between different nutritional statuses and dynapenia among Taiwanese older adults, and assessed whether age modifies this relationship. Methods: In this study, we enrolled individuals aged 65 years and older living in community settings through convenience sampling from 2020 to 2021, following a cross-sectional design. The Mini-Nutritional Assessment Short Form (MNA-SF) was used to assess whether the participants were at nutritional risk. Standardized assessments measured muscle strength (handgrip measurement), physical performance (6 m walking test), and muscle mass (bioelectrical impedance analysis) to confirm dynapenia classifications. The interaction terms were tested using likelihood ratio tests to examine for dynapenia between nutritional status and age. For overall sample and subgroup analyses, binary logistic regression was employed. Results: Among 211 participants (mean age: 80.7 $\pm$ 7.1 years), after adjusting for potential confounders, those at nutritional risk (OR: 3.11; 95% CI: 1.31–7.36) were positively associated with dynapenia, whereas higher MNA-SF scores (OR: 0.73; 95% CI: 0.57–0.93) were negatively associated. Interactions regarding dynapenia were observed between nutritional status and age group ($p = 0.014$), with nutritional risk significantly associated with dynapenia only in the old–old group ($\geq$75 years) (OR = 4.11, 95% CI: 1.39–12.15). Conclusions: Age is a potential moderator of nutritional status and dynapenia among older populations. Nutritional status appeared to be more profound in the old–old group in terms of the risk of dynapenia. These findings offer insights for monitoring nutritional status and implementing targeted interventions to prevent dynapenia in those aged over 75 years. Future studies using prospective designs should explore the underlying mechanisms linking nutritional status to dynapenia and assess the effectiveness of nutritional interventions in preventing muscle strength decline.

Keywords: malnutrition; undernutrition; handgrip strength; physical function; sarcopenia; aging

1. Introduction

The rapidly growing aging population is contributing to an increased occurrence of age-related health outcomes [1], leading to the global burden of disability and morbidity among older populations [2]. Research indicates that disability and mortality in older adults shows a stronger association with reduced muscle strength than with muscle mass [3–5]. Declining muscle strength not only elevates the likelihood of frailty, but also increases the risk of cardiovascular disease and all-cause mortality [6–8], underscoring its critical role in maintaining physical abilities in older adults. Dynapenia is characterized by age-associated reductions in muscle strength [9,10] and could be a preliminary indicator for screening sarcopenia [11]. Furthermore, dynapenia has a stronger association with physical ability limitations than that of sarcopenia [12] and correlates with a higher risk of mortality [3,13]. The prevalence of dynapenia varies worldwide, ranging from 10% to 84.6% [14–17]. In Taiwan, it is approximately 28.6% to 31.3% [18,19], and a previous 6-year follow-up study reported that nearly 18.5% of community-dwelling older adults transitioned from a healthy state to dynapenia, suggesting that deterioration in muscle function could be more common [20]. Given these findings, a better comprehension of the modifiable factors associated with dynapenia is crucial, not only to facilitate early identification, but also to reduce its impact on physical ability outcomes among older adults.

Nutritional status, a modifiable lifestyle factor, and particularly adequate protein and essential nutrient intake is critical for promoting muscle protein synthesis among older populations [10,21,22]. However, malnutrition is prevalent in this population, largely because of several age-related changes that elevate the risk of nutritional deficits [23,24]. Such nutritional deficits can contribute to decreased physical functionality, poor quality of life, and increased mortality [25–27]. Although previous research has established nutritional status as a key factor influencing muscle strength [28,29] and physical performance [23,30], current evidence on the correlation between nutritional status and dynapenia remains limited and is primarily focused on specific populations, such as individuals with Parkinson's disease [31], post-COVID-19 [32], and postmenopausal women [33]. This focus on specific populations highlights a research gap in understanding this relationship in the community-dwelling older population, a broader at-risk population. Moreover, prior research [20] has indicated that dynapenia has gradually become a health concern in the aging population of Taiwan. Despite this concerning trend, a deficiency of localized research exists exploring the relationship between nutritional status and dynapenia among Taiwanese older adults in community settings. Addressing this gap is essential for developing targeted nutritional interventions aimed at mitigating the impact of dynapenia.

Furthermore, prior research has identified age as a crucial factor of malnutrition among older populations, with the prevalence rates increasing as the age advances due to physiological changes, decreased appetite, and reduced nutrient absorption [34–36]. Additionally, muscle strength declines progressively with aging [37], and this decline accelerates beyond the age of 60 at a rate of approximately 3% per year [38]. These age-related physiological changes may further contribute to muscle loss and exacerbate the negative impact of malnutrition on muscle function. Given this age-related risk, we hypothesized that the association between nutritional status and dynapenia differs between the young–old group (comprising older adults aged 65–74 years) and the old–old group (comprising those aged $\geq$ 75 years) [39,40]. Therefore, the purpose of this study was to explore the relationship between different nutritional statuses (normal/at risk) and dynapenia among Taiwanese older adults living in community settings, as well as to assess whether age serves as a moderator of this relationship.

2. Materials and Methods

2.1. Research Design and Study Samples

In this study, we employed a cross-sectional design and recruited participants aged over 65 years living in community settings using convenience sampling from the Department of Geriatrics and Gerontology at National Taiwan University Hospital (NTNU). From September 2020 to December 2021, participants were recruited through two methods. One method involved the preliminary screening of potentially eligible individuals who had completed periodic physical examinations, whereas the other method identified eligible individuals attending the outpatient clinic through referrals from outpatient physicians.

The participants first completed a series of questionnaires through in-person interviews to report their sociodemographic information, health behaviors, and health status, with the Mini-Nutritional Assessment Short Form (MNA-SF) utilized to assess whether they were at nutritional risk. Next, standardized procedures were employed to measure handgrip strength, gait speed, and muscle mass to assess whether the participants met the dynapenia criteria outlined by the 2019 Asian Working Group for Sarcopenia (AWGS) guidelines. Finally, to objectively record moderate-to-vigorous physical activity (MVPA), participants were instructed to wear a triaxial accelerometer for at least 10 h/day over 7 consecutive days.

The minimum sample size for a binary logistic regression analysis was estimated to be 131, with an alpha level (α) of 0.05 and a statistical power of 0.8, calculated using G*Power version 3.1.9.7 [41]. Given the possibility of missing data, we initially recruited 299 participants; however, eight individuals declined to participate, resulting in 291 participants being enrolled. Subsequently, the participants were screened according to the following exclusion criteria: (1) inability to complete the questionnaires, failure to follow the standard measurement procedures for dynapenia, or noncompliance with wearing the triaxial accelerometer (n = 40); (2) incomplete or missing data from the questionnaires or MVPA records (n = 33); and (3) being classified as potentially having sarcopenia (n = 7). In the final analysis, 211 participants were included (Figure 1).

The rights of all participants were safeguarded through the acquisition of signed informed consent forms before the study's initiation. At the completion of the study, each participant was given a US$7 convenience store gift voucher in appreciation. The approval for the study protocol was granted by the Research Ethics Committee of National Taiwan University Hospital (REC number: 202008046RINC).

2.2. Measurements

2.2.1. Nutritional Status

This study applied the MNA-SF, a tool developed for older adults aged over 65 years, to evaluate the participants' nutritional status [42–44]. The MNA-SF assesses an individual's changes over the past three months in degrees of weight loss, food intake, physical activity level, body mass index (BMI), and presence of psychiatric problems or psychological stress [43]. The total MNA-SF score ranges from 0 to 14, where higher scores indicate a lower nutritional risk. The MNA-SF defines scores of >11 as normal and $\leq$11 as at nutritional risk [43].

2.2.2. Dynapenia

The definition of dynapenia was low muscle function, indicated by reduced muscle strength and/or poor physical performance, while maintaining normal muscle mass [45]. Muscle strength was assessed by handgrip measurements, physical performance by a 6 m walking test, and appendicular skeletal muscle mass (ASM) by bioelectrical impedance analysis (BIA). The cutoffs for these measurements were based on the 2019 updated AWGS guidelines [11]. Therefore, the participants were classified as having dynapenia if they exhibited reduced handgrip strength (<28.0 kg for men; <18.0 kg for women) or a gait speed of <1.0 m/s (for both sexes), while maintaining normal ASM (≥7.0 kg/m^2 for men; ≥5.7 kg/m^2 for women). The participants with normal ASM and muscle function were classified as not having dynapenia. Additionally, to allow a clear comparison between the participants with dynapenia and without (healthy individuals), those with decreased ASM (<7.0 kg/m^2 for men; <5.7 kg/m^2 for women), who may be classified as potentially having sarcopenia [11], were excluded.

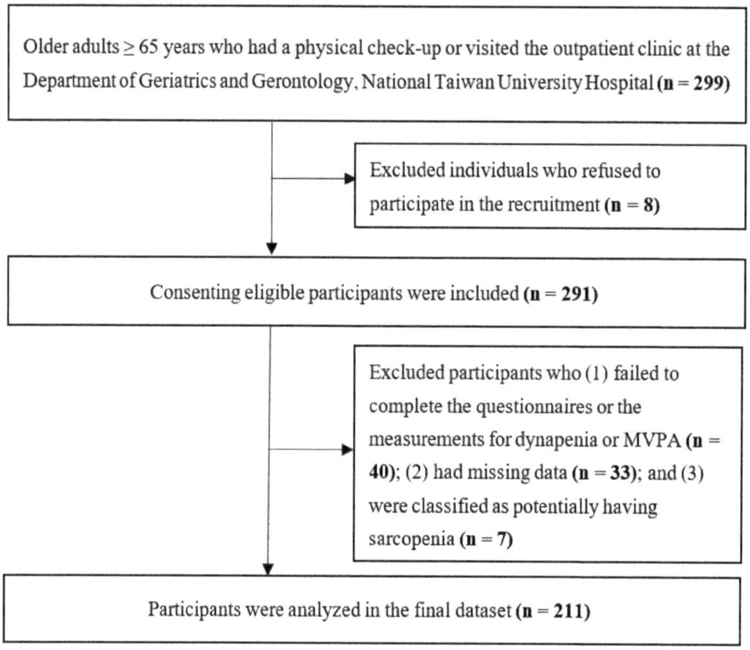

Figure 1. Study process flowchart.

2.2.3. Muscle Strength

To evaluate muscle strength, a hydraulic handheld dynamometer (Jamar Plus+ Digital Hand Dynamometer; Lafayette Instrument Company, Lafayette, IN, USA) was utilized for the handgrip measurements. The participants were asked to perform at maximum strength to grip the device with their dominant hand, and the highest handgrip strength recorded from the two trials was used for the analysis.

2.2.4. Physical Performance

Physical performance was measured as the walking speed of the participants using a 6 m walking test. Each participant completed the walk twice under safe conditions, and the shorter of the two was recorded for the analysis.

2.2.5. Muscle Mass

A bioelectrical impedance analyzer (DC-430MA; TANITA, Tokyo, Japan) was utilized to measure ASM. We instructed the participants to stand barefoot on the analyzer and maintain a firm grip on the device handles throughout the measurement process to ensure accurate results.

2.2.6. Covariates

We collected sociodemographic data including age (categorized as 65 to 74 years or 75 years and above), sex (men or women), educational level (less than university level or university-educated and above), and living status (living alone or with others). The assessment of health behaviors focused on documenting the participants' smoking and alcohol consumption habits over the past year, with the responses categorized as either "yes" or "no". The health status parameters, including BMI, were calculated as weight (in kilograms) divided by height (in meters) squared (kg/m^2). MVPA was objectively measured using a triaxial accelerometer (GT3X+ ActiGraph; Pensacola, FL, USA) worn over 7 consecutive days to accurately assess the average weekly minutes of MVPA [46]. The threshold for MVPA was set based on the World Health Organization (WHO) recommendations, advising that older populations should achieve a minimum of 150 min weekly of MVPA [47].

2.3. Statistical Analysis

The basic characteristics of the participants were presented using descriptive statistics. We conducted chi-square tests and t-tests to assess dynapenia in relation to categorical and continuous variables, respectively, and the significant covariates were adjusted in the logistic regression analyses. Binary logistic regression analyses, including both unadjusted and adjusted models, were performed to examine the relationship between different nutritional statuses and dynapenia in the overall sample. Likelihood ratio tests were performed to determine whether interaction terms existed for dynapenia between the nutritional status and the characteristic variables, particularly the age groups. Upon confirming a significant interaction, the sample was stratified into two age groups: old–old (75+ years) and young–old (65–74 years) [39,40]. Finally, binary logistic regression analyses were performed for the two age groups as subgroup analyses to examine the relationship between different nutritional statuses and dynapenia. The results were presented as odds ratios (ORs) along with 95% confidence intervals (CIs). IBM SPSS (version 23.0, Chicago, IL, USA) was applied, and the statistical significance level was established at $p < 0.05$.

3. Results

3.1. Sample Characteristics

Table 1 summarizes an overview of the participants' sociodemographic, health behavior, and health status variables. Among the 211 participants, the mean age was 80.7 $\pm$ 7.1 years; 46.0% were men, 51.2% had an education level below university, and 90.5% lived with others. Regarding health behaviors, most participants did not use tobacco (91.9%) or alcohol (89.6%). Of the participants, 31.3% engaged in a minimum of 150 min of MVPA weekly. In terms of health status, the mean BMI was 24.4 $\pm$ 3.7 kg/m^2 for all participants. The mean MNA-SF scores were 12.6 $\pm$ 1.7, and 43 (20.4%) participants were at nutritional risk. There were 123 (58.3%) participants who were classified as having dynapenia. The results of the chi-square and t-tests showed that nutritional status, educational level, BMI, smoking, and MVPA were significantly associated with dynapenia. Moreover, a greater proportion of the participants at nutritional risk were distributed within the dynapenia group (Figure 2).

Table 1. Descriptive statistics of categorical and continuous variables for all participants.

Variables	Total (n = 211)	Non-Dynapenia (n = 88)	Dynapenia (n = 123)	p-Value
Categorical Variables [a]	N (%)	N (%)	N (%)	
Age group (years)				
65–74	53 (25.1)	22 (25.0)	31 (25.2)	0.973
75+	158 (74.9)	66 (75.0)	92 (74.8)	
Sex				
Men	97 (46.0)	44 (50.0)	53 (43.1)	0.321
Women	114 (54.0)	44 (50.0)	70 (56.9)	
Educational level				
Lower than university	108 (51.2)	31 (35.2)	77 (62.6)	<0.001 *
University and over	103 (48.8)	57 (64.8)	46 (37.4)	
Living status				
Alone	20 (9.5)	8 (9.1)	12 (9.8)	0.871
With others	191 (90.5)	80 (90.9)	111 (90.2)	
Smoking				
No	194 (91.9)	85 (96.6)	109 (88.6)	0.036 *
Yes	17 (8.1)	3 (3.4)	14 (11.4)	
Alcohol drinking				
No	189 (89.6)	83 (94.3)	106 (86.2)	0.056
Yes	22 (10.4)	5 (5.7)	17 (13.8)	
MVPA (mins/per week)				
<150	145 (68.7)	50 (56.8)	95 (77.2)	0.002 *
≥150	66 (31.3)	38 (43.2)	28 (22.8)	
Nutritional status				
Normal	168 (79.6)	77 (87.5)	91 (74.0)	0.016 *
At risk	43 (20.4)	11 (12.5)	32 (26.0)	
Continuous variables [b]	Mean (SD)	Mean (SD)	Mean (SD)	p-value
Age (years)	80.69 (7.06)	80.25 (7.16)	81.00 (7.00)	0.448
BMI (kg/m^2)	24.39 (3.67)	23.64 (3.15)	24.93 (3.92)	0.009 *
MNA-SF (scores)	12.62 (1.68)	12.90 (1.43)	12.42 (1.81)	0.035 *
MVPA (mins/per week)	104.17 (167.85)	173.27 (219.32)	54.74 (91.04)	<0.001 *

[a] Chi-square test for associations with dynapenia; * $p < 0.05$, significant. [b] T-test for associations with dynapenia; $p < 0.05$, significant. Abbreviations: SD, standard deviation; BMI, body mass index; MNA-SF, Mini-Nutritional Assessment Short Form; MVPA, moderate-to-vigorous physical activity.

3.2. Associations Between MNA-SF, Nutritional Status, and Dynapenia Among All Participants

Table 2 shows the associations between nutritional status (normal or at risk), MNA-SF scores, and dynapenia across the two logistic regression models for the entire sample. Model 1 represented a crude model. Model 2 accounted for adjustments related to age, sex, educational level, BMI, smoking status, and MVPA. Both the nutritional status at risk (model 1: OR: 2.46; 95% CI: 1.16–5.21; $p = 0.018$, model 2: OR: 3.11; 95% CI: 1.31–7.36; $p = 0.009$) and the MNA-SF scores (model 1: OR: 0.83; 95% CI: 0.70–1.00; $p = 0.045$, model 2: OR: 0.73; 95% CI: 0.57–0.93; $p = 0.011$) demonstrated a significant association with dynapenia in the unadjusted and adjusted models. Figure 3 illustrates the results from model 2, showing the OR and 95% CI for the associations between the nutritional status at risk, MNA-SF scores, and dynapenia.

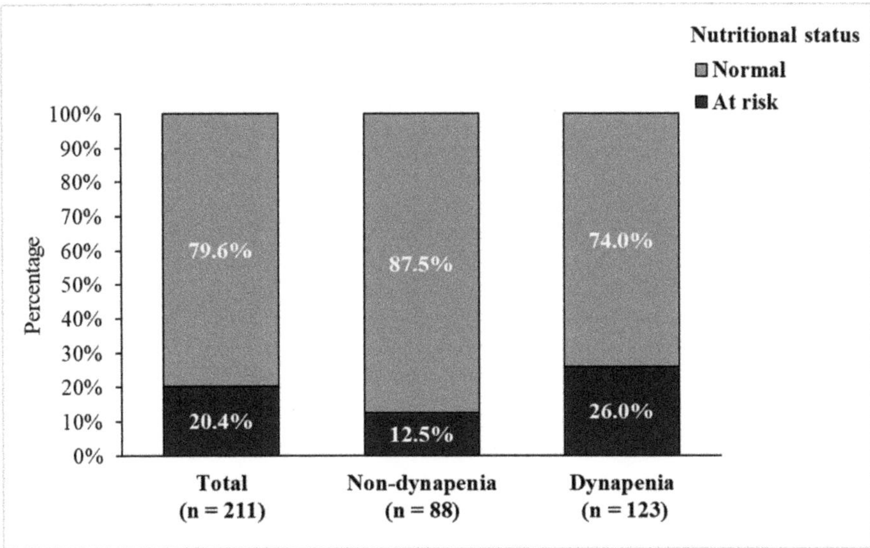

Figure 2. Distribution of nutritional status among total participant, non-dynapenia, and dynapenia groups.

Table 2. A binary logistic regression analysis of the relationship between MNA-SF, nutritional status, and dynapenia in the total sample (n = 211).

Variables	Model 1		Model 2	
	OR (95%CI)	*p*-Value	OR (95%CI)	*p*-Value
Nutritional status at risk (Ref. Normal)	2.46 (1.16, 5.21)	0.018 *	3.11 (1.31, 7.36)	0.009 **
MNA-SF (scores)	0.83 (0.70, 1.00)	0.045 *	0.73 (0.57, 0.93)	0.011 *

Model 1: unadjusted; model 2: adjusted for age, sex, educational level, BMI, smoking status, and MVPA. Abbreviations: Ref., reference; OR, odds ratio; CI, confidence interval. * $p < 0.05$. ** $p < 0.01$.

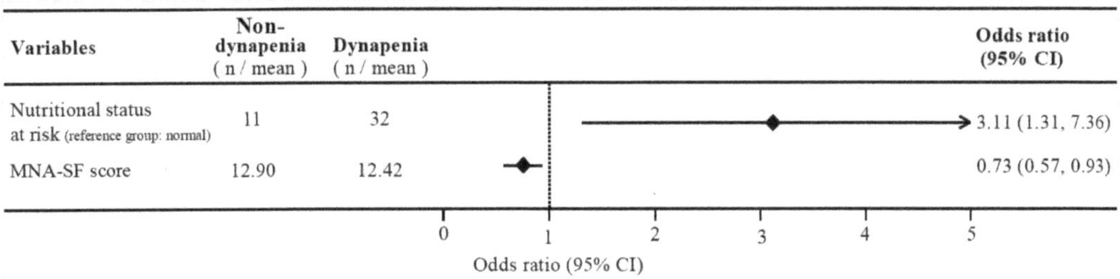

Figure 3. Adjusted odds ratios for association between nutritional status, MNA-SF scores, and dynapenia.

3.3. Interactions Regarding Dynapenia Between Nutritional Status and Sociodemographic Variants

Table 3 shows the significant interactions regarding dynapenia between nutritional status and age ($p = 0.014$).

Table 3. Significance of interactions between nutritional status and characteristic variables by binary logistic regression model.

Variables	*p*-Value for Interaction Termwith Nutritional Status
	Dynapenia (*p*-Value)
Age	0.014 *
Sex	0.125
Educational level	0.863
Living status	0.675
Smoking	0.999
Alcohol drinking	0.999

* $p < 0.05$.

3.4. *Associations Between Nutritional Status and Dynapenia in Different Age Groups*

Table 4 shows the associations between the nutritional status at risk and dynapenia in different age groups. Model 1 represented a crude model, whereas Model 2 was an adjusted model (adjusted for sex, educational level, BMI, smoking status, and MVPA). In the aged 65–74 group (n = 53), nutritional status was not significantly associated with dynapenia in both models, whereas in the aged 75+ group (n = 158), the nutritional status at risk was significantly associated with dynapenia, with an OR of 3.15 (95% CI: 1.27–7.80; $p = 0.013$) in model 1 and 4.11 (95% CI: 1.39–12.15; $p = 0.010$) in model 2. Figure 4 illustrates the results from model 2, presenting the OR and 95% CI for the association between the nutritional status at risk and dynapenia, stratified by age groups.

Table 4. Age-stratified binary logistic regression models of nutritional status and dynapenia.

Variables	Model 1		Model 2	
	OR (95%CI)	***p*-Value**	**OR (95%CI)**	***p*-Value**
Aged 65–74 (n = 53)				
Nutritional status at risk (Ref. Normal)	1.31 (0.33, 5.18)	0.698	1.81 (0.35, 9.40)	0.480
Aged 75+ (n = 158)				
Nutritional status at risk (Ref. Normal)	3.15 (1.27, 7.80)	0.013 *	4.11 (1.39, 12.15)	0.010 *

Model 1: unadjusted; model 2: adjusted for sex, educational level, BMI, smoking status, and MVPA. Abbreviations: Ref., reference; OR, odds ratio; CI, confidence interval. * $p < 0.05$.

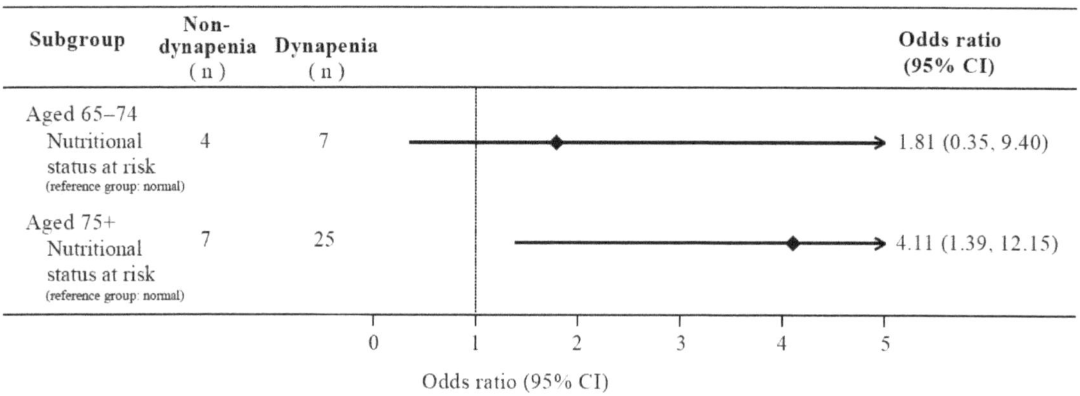

Figure 4. Adjusted odds ratios for association between nutritional status at risk and dynapenia across age groups.

4. Discussion

The present study is the first to explore the relationship between different nutritional statuses and dynapenia, based on the 2019 AWGS criteria, among older adults from community settings in Taiwan and to assess whether age is a potential modifier of this relationship. The primary findings indicate that, in the overall sample, being at nutritional risk is correlated with a higher risk of dynapenia. Furthermore, the study provides evidence of age as a potential moderator of these associations, showing a significant correlation in the old–old group, but not in the young–old group. Regarding the goal of dynapenia prevention, our results indicate that public health and nutrition practitioners should develop an age-specific intervention strategy to enhance nutritional status, especially in the old–old population.

Consistent with previous research [31–33], our findings reveal an association between nutritional risk and dynapenia. Several studies from different regions of the world have also explored this relationship in distinct populations. While our study examines community-dwelling older adults, a Brazilian study indicated that in patients with Parkinson's disease, dynapenia is associated with a reduced calf circumference and an increased body fat percentage [31]. In line with this, an Italian study highlighted a high prevalence of malnutrition and dynapenia in post-COVID-19 patients [32]. Moreover, a Canadian study emphasized the nutritional aspect, particularly protein intake, as a key modifiable factor influencing dynapenia among postmenopausal women [33]. Our findings align with these international studies; however, more importantly, we emphasize a community-dwelling population, which contrasts with the studies focusing on individuals with specific diseases. This distinction underscores the broader impact of nutritional status on dynapenia, potentially independent of underlying medical conditions. Several possible explanations exist for these findings. First, malnutrition in older adults can result from inadequate dietary protein intake [28,36] due to age-associated alterations in sensory function and metabolism [28,48], which influence muscle protein synthesis pathways and impair muscle growth and repair [49], ultimately declining muscle strength and functional capacity in aging populations [22,50,51]. Second, older adults at malnutrition risk may be more likely to experience vitamin D deficiency [49,52], which plays a critical role in muscle function by promoting calcium absorption and facilitating protein synthesis [53]. Vitamin D deficiency often presents with marked proximal muscle weakness [54], and some studies have accordingly established an association between this deficiency and reduced muscle strength [55,56]. Third, the risk of malnutrition could exacerbate chronic inflammation in older adults [25,36], with elevated inflammatory markers potentially promoting muscle breakdown, inhibiting muscle synthesis, and further decreasing muscle strength in older adults [57–59]. Given that poor nutritional status may elevate the risk of dynapenia through age-related physiological factors, future research is suggested to understand these mechanisms and develop targeted interventions to prevent and manage dynapenia in aging populations.

Age moderates the relationship between nutritional status and dynapenia, with a significant association observed in the old–old ($\geq$75 years), but not in the young–old (65–74 years) population. These findings are similar to previous studies [39,60–62]. A study conducted in Korea found that old–old individuals exhibited poorer gait function compared to young–old individuals [39]. Similarly, a study from Japan demonstrated that declines in handgrip strength and walking speed became more pronounced after the age of 70 [62]. Additionally, a cross-sectional study on hospitalized older adults revealed that individuals aged 75 and older had significantly poorer nutritional status than those younger than 75 [61]. Furthermore, a study from New Zealand reported that the odds of nutritional risk increased with each additional year of age and identified a significant association between lower nutritional risk and better physical performance [60]. These findings indicate that old–old individuals are disproportionately affected by the combined effects of malnutrition and muscle function decline compared to young–old individuals. Considering that aging is a complex process that affects physical, mental, and social health dimensions [28,63], there are likely underlying reasons for this relationship. First, from a physiological perspective, aging accelerates muscle fiber atrophy and loss, significantly reducing muscle mass and strength [37] with decreases in muscle strength often preceding losses in muscle mass [20,64]. As the age advances, reduced gastrointestinal function and diminished appetite often occur due to sensory changes, potentially leading to inadequate nutrient intake and a heightened risk of malnutrition [23,25]. In addition, aging correlates with an elevated risk of chronic diseases and persistent low-grade inflammation, both of which pose challenges to nutrient absorption [36] and heighten the risk of dynapenia [10]. From a psychosocial perspective, aging presents emotional and social challenges, including increased risks of depression, anxiety [24,65], and reduced social participation [66]. These factors collectively contribute to a decline in appetite [28,67] as well as in physical activity levels [68,69]. Therefore, early identification and intervention for malnutrition risk, especially in those aged over 75, may help prevent dynapenia.

This study applied the latest 2019 updated AWGS criteria to classify dynapenia, specifically tailored to older Asian populations. This approach enhanced the study's validity by ensuring that the classification standards aligned with region-specific health characteristics, thereby improving the applicability of the findings. Additionally, the use of these criteria provides a foundation for international comparisons, enabling future research to examine dynapenia prevalence and characteristics across different Asian countries. Certain limitations were present in this study. First, the recruitment of participants was conducted in a highly urbanized region in Taiwan, limiting the generalizability of our findings to the entire Taiwanese population. Second, the cross-sectional methodology restricted the ability to infer causal relationships. Third, the MNA-SF depends primarily on self-reported data and lacks objective measurements of food intake, which may affect the accuracy of the results.

5. Conclusions

Age is a potential moderator of the relationship between nutritional status and dynapenia in Taiwanese older adults living in community settings. Old–old individuals (over 75 years) who were at nutritional risk had about four times the likelihood of being at risk of dynapenia, whereas this relationship was not observed in young–old individuals (65 to 74 years). Given that the aging population is expanding globally, addressing the intersection of nutritional status and dynapenia has become a pressing public health priority. Our findings offer insights for monitoring nutritional status and implementing targeted interventions to prevent dynapenia in older adults, particularly those over 75 years old. Future studies should explore the underlying physiological mechanisms linking malnutrition to muscle strength decline and evaluate the effectiveness of targeted nutritional strategies.

Author Contributions: Conceptualization, C.-C.C. and Y.L.; Formal analysis, C.-C.C., T.-F.L. and J.C.; Funding acquisition, J.-H.P. and Y.L.; Supervision, Y.L., J.-H.P. and Y.-J.C.; Writing—original draft, C.-C.C.; Writing—review and editing, T.-F.L., J.C., Y.L., J.-H.P. and Y.-J.C. All authors have read and agreed to the published version of the manuscript.

Funding: This research was funded by Yung Liao through a personal grant from the Ministry of Science and Technology of Taiwan, grant number: MOST 111-2628-H-003-006-MY3. The National Taiwan Normal University (NTNU) also contributed financial support under the framework of the Higher Education Sprout Project of the Ministry of Education (MOE) in Taiwan. This work was also supported by Institute of Information & communications Technology Planning & Evaluation (IITP) grant funded by the Korea government(MSIT) (No.2022-0-00501, Development of Wearable Device Assessment technology).

Institutional Review Board Statement: Informed consent was obtained from all subjects involved in the study. The study was conducted in accordance with the Declaration of Helsinki, and the protocol was approved by the Research Ethics Committee of NTUH (REC number: 202008046RINC) on 9 September 2020.

Informed Consent Statement: Informed consent was obtained from all participants involved in the study.

Data Availability Statement: The original contributions presented in the study are included in the article, further inquiries can be directed to the corresponding author.

Acknowledgments: The authors appreciate all participants in this study.

Conflicts of Interest: The authors declare no conflicts of interest.

References

1. DESA/Population Division. *World Population Prospects 2024: Summary of Results*, 9th ed.; United Nations: New York, NY, USA, 2024.
2. GBD 2019 Ageing Collaborators. Global, regional, and national burden of diseases and injuries for adults 70 years and older: Systematic analysis for the Global Burden of Disease 2019 Study. *BMJ* **2022**, *376*, e068208. [CrossRef]
3. Manini, T.M.; Clark, B.C. Dynapenia and aging: An update. *J. Gerontol. Ser. A Biol. Sci. Med. Sci.* **2012**, *67*, 28–40. [CrossRef] [PubMed]
4. Seene, T.; Kaasik, P. Muscle weakness in the elderly: Role of sarcopenia, dynapenia, and possibilities for rehabilitation. *Eur. Rev. Aging Phys. Act.* **2012**, *9*, 109–117. [CrossRef]
5. Mitchell, W.K.; Williams, J.; Atherton, P.; Larvin, M.; Lund, J.; Narici, M. Sarcopenia, dynapenia, and the impact of advancing age on human skeletal muscle size and strength; a quantitative review. *Front. Physiol.* **2012**, *3*, 260. [CrossRef] [PubMed]
6. Leong, D.P.; Teo, K.K.; Rangarajan, S.; Lopez-Jaramillo, P.; Avezum, A.; Orlandini, A.; Seron, P.; Ahmed, S.H.; Rosengren, A.; Kelishadi, R.; et al. Prognostic value of grip strength: Findings from the Prospective Urban Rural Epidemiology (PURE) study. *Lancet* **2015**, *386*, 266–273. [CrossRef] [PubMed]
7. Riviati, N.; Indra, B. Relationship between muscle mass and muscle strength with physical performance in older adults: A systematic review. *SAGE Open Med.* **2023**, *11*, 20503121231214650. [CrossRef]
8. Burbank, C.M.; Branscum, A.; Bovbjerg, M.L.; Hooker, K.; Smit, E. Muscle power predicts frailty status over four years: A retrospective cohort study of the National Health and Aging Trends Study. *J. Frailty Sarcopenia Falls* **2023**, *8*, 1–8. [CrossRef] [PubMed]
9. Clark, B.C.; Manini, T.M. Sarcopenia =/= dynapenia. *J. Gerontol. Ser. A Biol. Sci. Med. Sci.* **2008**, *63*, 829–834. [CrossRef] [PubMed]
10. Clark, B.C.; Manini, T.M. What is dynapenia? *Nutrition* **2012**, *28*, 495–503. [CrossRef] [PubMed]
11. Chen, L.K.; Woo, J.; Assantachai, P.; Auyeung, T.W.; Chou, M.Y.; Iijima, K.; Jang, H.C.; Kang, L.; Kim, M.; Kim, S.; et al. Asian Working Group for Sarcopenia: 2019 Consensus Update on Sarcopenia Diagnosis and Treatment. *J. Am. Med. Dir. Assoc.* **2020**, *21*, 300–307.E2. [CrossRef]
12. Iwamura, M.; Kanauchi, M. A cross-sectional study of the association between dynapenia and higher-level functional capacity in daily living in community-dwelling older adults in Japan. *BMC Geriatr.* **2017**, *17*, 1. [CrossRef] [PubMed]
13. Alexandre Tda, S.; Duarte, Y.A.; Santos, J.L.; Wong, R.; Lebrão, M.L. Sarcopenia according to the European Working Group on Sarcopenia in Older People (EWGSOP) versus dynapenia as a risk factor for mortality in the elderly. *J. Nutr. Health Aging* **2014**, *18*, 751–756. [CrossRef]
14. Kobayashi, K.; Imagama, S.; Ando, K.; Nakashima, H.; Machino, M.; Morozumi, M.; Kanbara, S.; Ito, S.; Inoue, T.; Yamaguchi, H.; et al. Dynapenia and physical performance in community-dwelling elderly people in Japan. *Nagoya J. Med. Sci.* **2020**, *82*, 415–424. [CrossRef] [PubMed]
15. Zhou, W.; Tong, J.; Wen, Z.; Mao, M.; Wei, Y.; Li, X.; Zhou, M.; Wan, H. Prevalence and factors associated with dynapenia among middle-aged and elderly people in rural southern China. *Prev. Med. Rep.* **2024**, *38*, 102630. [CrossRef] [PubMed]
16. Neves, T.; Ferriolli, E.; Lopes, M.B.M.; Souza, M.G.C.; Fett, C.A.; Fett, W.C.R. Prevalence and factors associated with sarcopenia and dynapenia in elderly people. *J. Frailty Sarcopenia Falls* **2018**, *3*, 194–202. [CrossRef]
17. Benjumea, A.M.; Curcio, C.L.; Duque, G.; Gómez, F. Dynapenia and Sarcopenia as a Risk Factor for Disability in a Falls and Fractures Clinic in Older Persons. *Open Access Maced. J. Med. Sci.* **2018**, *6*, 344–349. [CrossRef] [PubMed]

18. Kao, T.W.; Peng, T.C.; Chen, W.L.; Han, D.S.; Chen, C.L.; Yang, W.S. Impact of adiposity on muscle function and clinical events among elders with dynapenia, presarcopenia and sarcopenia: A community-based cross-sectional study. *Aging* **2021**, *13*, 7247–7258. [CrossRef] [PubMed]
19. Huang, C.-Y.; Hwang, A.-C.; Liu, L.-K.; Lee, W.-J.; Chen, L.-Y.; Peng, L.-N.; Lin, M.-H.; Chen, L.-K. Association of Dynapenia, Sarcopenia, and Cognitive Impairment Among Community-Dwelling Older Taiwanese. *Rejuvenation Res.* **2016**, *19*, 71–78. [CrossRef] [PubMed]
20. Chao, Y.P.; Fang, W.H.; Chen, W.L.; Peng, T.C.; Yang, W.S.; Kao, T.W. Exploring Muscle Health Deterioration and Its Determinants Among Community-Dwelling Older Adults. *Front. Nutr.* **2022**, *9*, 817044. [CrossRef] [PubMed]
21. Chen, L.K.; Arai, H.; Assantachai, P.; Akishita, M.; Chew, S.T.H.; Dumlao, L.C.; Duque, G.; Woo, J. Roles of nutrition in muscle health of community-dwelling older adults: Evidence-based expert consensus from Asian Working Group for Sarcopenia. *J. Cachex-Sarcopenia Muscle* **2022**, *13*, 1653–1672. [CrossRef] [PubMed]
22. Deer, R.R.; Volpi, E. Protein intake and muscle function in older adults. *Curr. Opin. Clin. Nutr. Metab. Care* **2015**, *18*, 248–253. [CrossRef]
23. Ahmed, T.; Haboubi, N. Assessment and management of nutrition in older people and its importance to health. *Clin. Interv. Aging* **2010**, *5*, 207–216. [CrossRef] [PubMed]
24. World Health Organization. *World Report on Ageing and Health*; World Health Organization: Geneva, Switzerland, 2015.
25. Dent, E.; Wright, O.R.L.; Woo, J.; Hoogendijk, E.O. Malnutrition in older adults. *Lancet* **2023**, *401*, 951–966. [CrossRef]
26. Söderström, L.; Rosenblad, A.; Thors Adolfsson, E.; Bergkvist, L. Malnutrition is associated with increased mortality in older adults regardless of the cause of death. *Br. J. Nutr.* **2017**, *117*, 532–540. [CrossRef]
27. Wei, K.; Nyunt, M.S.Z.; Gao, Q.; Wee, S.L.; Ng, T.P. Long-term changes in nutritional status are associated with functional and mortality outcomes among community-living older adults. *Nutrition* **2019**, *66*, 180–186. [CrossRef]
28. Amarya, S.; Singh, K.; Sabharwal, M. Changes during aging and their association with malnutrition. *J. Clin. Gerontol. Geriatr.* **2015**, *6*, 78–84. [CrossRef]
29. Robinson, S.; Granic, A.; Sayer, A.A. Nutrition and Muscle Strength, As the Key Component of Sarcopenia: An Overview of Current Evidence. *Nutrients* **2019**, *11*, 2942. [CrossRef] [PubMed]
30. Kramer, C.S.; Groenendijk, I.; Beers, S.; Wijnen, H.H.; van de Rest, O.; de Groot, L.C. The Association between Malnutrition and Physical Performance in Older Adults: A Systematic Review and Meta-Analysis of Observational Studies. *Curr. Dev. Nutr.* **2022**, *6*, nzac007. [CrossRef]
31. de Medeiros, L.C.; da Luz, M.C.L.; da Costa Pereira, J.P.; Bezerra, G.K.A.; Cabral, P.C. Nutritional status and dynapenia in people living with Parkinson's disease: A cross-sectional study. *Neurol. Sci.* **2022**, *43*, 2509–2517. [CrossRef] [PubMed]
32. de Blasio, F.; Scalfi, L.; Castellucci, B.; Sacco, A.M.; Berlingieri, G.M.; Capitelli, L.; Alicante, P.; Sanduzzi, A.; Bocchino, M. Poor Nutritional Status and Dynapenia Are Highly Prevalent in Post-Acute COVID-19. *Front. Nutr.* **2022**, *9*, 888485. [CrossRef] [PubMed]
33. Filion, M.E.; Barbat-Artigas, S.; Dupontgand, S.; Fex, A.; Karelis, A.D.; Aubertin-Leheudre, M. Relationship between protein intake and dynapenia in postmenopausal women. *J. Nutr. Health Aging* **2012**, *16*, 616–619. [CrossRef] [PubMed]
34. Elia, M.; Stratton, R.J. An analytic appraisal of nutrition screening tools supported by original data with particular reference to age. *Nutrition* **2012**, *28*, 477–494. [CrossRef] [PubMed]
35. Leij-Halfwerk, S.; Verwijs, M.H.; van Houdt, S.; Borkent, J.W.; Guaitoli, P.R.; Pelgrim, T.; Heymans, M.W.; Power, L.; Visser, M.; Corish, C.A.; et al. Prevalence of protein-energy malnutrition risk in European older adults in community, residential and hospital settings, according to 22 malnutrition screening tools validated for use in adults $\geq$ 65 years: A systematic review and meta-analysis. *Maturitas* **2019**, *126*, 80–89. [CrossRef] [PubMed]
36. Norman, K.; Haß, U.; Pirlich, M. Malnutrition in Older Adults-Recent Advances and Remaining Challenges. *Nutrients* **2021**, *13*, 2764. [CrossRef] [PubMed]
37. Wilkinson, D.J.; Piasecki, M.; Atherton, P.J. The age-related loss of skeletal muscle mass and function: Measurement and physiology of muscle fibre atrophy and muscle fibre loss in humans. *Ageing Res. Rev.* **2018**, *47*, 123–132. [CrossRef] [PubMed]
38. Reid, K.F.; Pasha, E.; Doros, G.; Clark, D.J.; Patten, C.; Phillips, E.M.; Frontera, W.R.; Fielding, R.A. Longitudinal decline of lower extremity muscle power in healthy and mobility-limited older adults: Influence of muscle mass, strength, composition, neuromuscular activation and single fiber contractile properties. *Eur. J. Appl. Physiol.* **2014**, *114*, 29–39. [CrossRef] [PubMed]
39. Chung, E.; Lee, S.-H.; Lee, H.-J.; Kim, Y.-H. Comparative study of young-old and old-old people using functional evaluation, gait characteristics, and cardiopulmonary metabolic energy consumption. *BMC Geriatr.* **2023**, *23*, 400. [CrossRef]
40. Ouchi, Y.; Rakugi, H.; Arai, H.; Akishita, M.; Ito, H.; Toba, K.; Kai, I.; Joint Committee of Japan Gerontological Society (JGLS); Japan Geriatrics Society (JGS). Redefining the elderly as aged 75 years and older: Proposal from the Joint Committee of Japan Gerontological Society and the Japan Geriatrics Society. *Geriatr. Gerontol. Int.* **2017**, *17*, 1045–1047. [CrossRef]
41. Faul, F.; Erdfelder, E.; Buchner, A.; Lang, A.G. Statistical power analyses using G*Power 3.1: Tests for correlation and regression analyses. *Behav. Res. Methods* **2009**, *41*, 1149–1160. [CrossRef]

42. Vellas, B.; Guigoz, Y.; Garry, P.J.; Nourhashemi, F.; Bennahum, D.; Lauque, S.; Albarede, J.L. The Mini Nutritional Assessment (MNA) and its use in grading the nutritional state of elderly patients. *Nutrition* **1999**, *15*, 116–122. [CrossRef] [PubMed]
43. Rubenstein, L.Z.; Harker, J.O.; Salvà, A.; Guigoz, Y.; Vellas, B. Screening for undernutrition in geriatric practice: Developing the short-form mini-nutritional assessment (MNA-SF). *J. Gerontol. Ser. A Biol. Sci. Med. Sci.* **2001**, *56*, M366–M372. [CrossRef]
44. Tsai, A.C.; Yang, S.F.; Wang, J.Y. Validation of population-specific Mini-Nutritional Assessment with its long-term mortality-predicting ability: Results of a population-based longitudinal 4-year study in Taiwan. *Br. J. Nutr.* **2010**, *104*, 93–99. [CrossRef]
45. Yamada, M.; Kimura, Y.; Ishiyama, D.; Nishio, N.; Abe, Y.; Kakehi, T.; Fujimoto, J.; Tanaka, T.; Ohji, S.; Otobe, Y.; et al. Differential Characteristics of Skeletal Muscle in Community-Dwelling Older Adults. *J. Am. Med Dir. Assoc.* **2017**, *18*, 807.e9–807.e16. [CrossRef] [PubMed]
46. Troiano, R.P.; Berrigan, D.; Dodd, K.W.; Mâsse, L.C.; Tilert, T.; McDowell, M. Physical activity in the United States measured by accelerometer. *Med. Sci. Sports Exerc.* **2008**, *40*, 181–188. [CrossRef]
47. World Health Organization. *WHO Guidelines on Physical Activity and Sedentary Behaviour: At a Glance*; World Health Organization: Geneva, Switzerland, 2020.
48. Deutz, N.E.; Bauer, J.M.; Barazzoni, R.; Biolo, G.; Boirie, Y.; Bosy-Westphal, A.; Cederholm, T.; Cruz-Jentoft, A.; Krznariç, Z.; Nair, K.S.; et al. Protein intake and exercise for optimal muscle function with aging: Recommendations from the ESPEN Expert Group. *Clin. Nutr.* **2014**, *33*, 929–936. [CrossRef] [PubMed]
49. Boucher, B.J. The problems of vitamin d insufficiency in older people. *Aging Dis* **2012**, *3*, 313–329. [PubMed]
50. Campbell, W.W.; Deutz, N.E.P.; Volpi, E.; Apovian, C.M. Nutritional Interventions: Dietary Protein Needs and Influences on Skeletal Muscle of Older Adults. *J. Gerontol. Ser. A Biol. Sci. Med. Sci.* **2023**, *78*, 67–72. [CrossRef]
51. Coelho-Júnior, H.J.; Calvani, R.; Tosato, M.; Landi, F.; Picca, A.; Marzetti, E. Protein intake and physical function in older adults: A systematic review and meta-analysis. *Ageing Res. Rev.* **2022**, *81*, 101731. [CrossRef]
52. Giustina, A.; Bouillon, R.; Dawson-Hughes, B.; Ebeling, P.R.; Lazaretti-Castro, M.; Lips, P.; Marcocci, C.; Bilezikian, J.P. Vitamin D in the older population: A consensus statement. *Endocrine* **2023**, *79*, 31–44. [CrossRef]
53. Sahota, O. Understanding vitamin D deficiency. *Age Ageing* **2014**, *43*, 589–591. [CrossRef] [PubMed]
54. Holick, M.F. Vitamin D deficiency. *N. Engl. J. Med.* **2007**, *357*, 266–281. [CrossRef]
55. Delinocente, M.L.B.; Luiz, M.M.; de Oliveira, D.C.; de Souza, A.F.; Ramírez, P.C.; de Oliveira Máximo, R.; Soares, N.C.; Steptoe, A.; de Oliveira, C.; da Silva Alexandre, T. Are Serum 25-Hydroxyvitamin D Deficiency and Insufficiency Risk Factors for the Incidence of Dynapenia? *Calcif. Tissue Int.* **2022**, *111*, 571–579. [CrossRef] [PubMed]
56. Kitsu, T.; Kabasawa, K.; Ito, Y.; Kitamura, K.; Watanabe, Y.; Tanaka, J.; Nakamura, K.; Narita, I. Low serum 25-hydroxyvitamin D is associated with low grip strength in an older Japanese population. *J. Bone Miner. Metab.* **2020**, *38*, 198–204. [CrossRef] [PubMed]
57. Granic, A.; Davies, K.; Martin-Ruiz, C.; Jagger, C.; Kirkwood, T.B.L.; von Zglinicki, T.; Aihie Sayer, A. Grip strength and inflammatory biomarker profiles in very old adults. *Age Ageing* **2017**, *46*, 976–982. [CrossRef]
58. Haeri, N.S.; Perera, S.; Nadkarni, N.K.; Greenspan, S.L. Association of inflammatory markers with muscle and cognitive function in early and late-aging older adults. *J. Nutr. Health Aging* **2024**, *28*, 100207. [CrossRef] [PubMed]
59. Schaap, L.A.; Pluijm, S.M.; Deeg, D.J.; Harris, T.B.; Kritchevsky, S.B.; Newman, A.B.; Colbert, L.H.; Pahor, M.; Rubin, S.M.; Tylavsky, F.A.; et al. Higher inflammatory marker levels in older persons: Associations with 5-year change in muscle mass and muscle strength. *J. Gerontol. Ser. A Biol. Sci. Med. Sci.* **2009**, *64*, 1183–1189. [CrossRef]
60. Chatindiara, I.; Williams, V.; Sycamore, E.; Richter, M.; Allen, J.; Wham, C. Associations between nutrition risk status, body composition and physical performance among community-dwelling older adults. *Aust. N. Z. J. Public Health* **2019**, *43*, 56–62. [CrossRef] [PubMed]
61. Forster, S.; Gariballa, S. Age as a determinant of nutritional status: A cross sectional study. *Nutr. J.* **2005**, *4*, 28. [CrossRef]
62. Makizako, H.; Shimada, H.; Doi, T.; Tsutsumimoto, K.; Lee, S.; Lee, S.C.; Harada, K.; Hotta, R.; Nakakubo, S.; Bae, S.; et al. Age-dependent changes in physical performance and body composition in community-dwelling Japanese older adults. *J. Cachex-Sarcopenia Muscle* **2017**, *8*, 607–614. [CrossRef] [PubMed]
63. Tieland, M.; Trouwborst, I.; Clark, B.C. Skeletal muscle performance and ageing. *J. Cachexia Sarcopenia Muscle* **2018**, *9*, 3–19. [CrossRef] [PubMed]
64. Hughes, V.A.; Frontera, W.R.; Wood, M.; Evans, W.J.; Dallal, G.E.; Roubenoff, R.; Singh, M.A.F. Longitudinal Muscle Strength Changes in Older Adults: Influence of Muscle Mass, Physical Activity, and Health. *J. Gerontol. Ser. A* **2001**, *56*, B209–B217. [CrossRef]
65. Donovan, N.J.; Blazer, D. Social Isolation and Loneliness in Older Adults: Review and Commentary of a National Academies Report. *Am. J. Geriatr. Psychiatry* **2020**, *28*, 1233–1244. [CrossRef] [PubMed]
66. Rosso, A.L.; Taylor, J.A.; Tabb, L.P.; Michael, Y.L. Mobility, disability, and social engagement in older adults. *J. Aging Health* **2013**, *25*, 617–637. [CrossRef] [PubMed]

67. Simmons, W.K.; Burrows, K.; Avery, J.A.; Kerr, K.L.; Bodurka, J.; Savage, C.R.; Drevets, W.C. Depression-Related Increases and Decreases in Appetite: Dissociable Patterns of Aberrant Activity in Reward and Interoceptive Neurocircuitry. *Am. J. Psychiatry* **2016**, *173*, 418–428. [CrossRef] [PubMed]
68. Lindsay Smith, G.; Banting, L.; Eime, R.; O'Sullivan, G.; van Uffelen, J.G.Z. The association between social support and physical activity in older adults: A systematic review. *Int. J. Behav. Nutr. Phys. Act.* **2017**, *14*, 56. [CrossRef] [PubMed]
69. Luo, M.; Ding, D.; Bauman, A.; Negin, J.; Phongsavan, P. Social engagement pattern, health behaviors and subjective well-being of older adults: An international perspective using WHO-SAGE survey data. *BMC Public Health* **2020**, *20*, 99. [CrossRef] [PubMed]

nutrients

MDPI

Article

Causal Relationship between Meat Intake and Biological Aging: Evidence from Mendelian Randomization Analysis

Shupeng Liu †, Yinyun Deng †, Hui Liu, Zhengzheng Fu, Yinghui Wang, Meijuan Zhou * and Zhijun Feng *

Department of Radiation medicine, Guangdong Provincial Key Laboratory of Tropical Disease Research, School of Public Health, Southern Medical University, Guangzhou 510515, China; 20220062@smu.edu.cn (S.L.); yinyun9845@foxmail.com (Y.D.); lhuidoc@163.com (H.L.); smufuzz@163.com (Z.F.); wyh045@smu.edu.cn (Y.W.)
* Correspondence: lkzmj@smu.edu.cn (M.Z.); fengzhj18@lzu.edu.cn (Z.F.)
† These authors contributed equally to this work.

Abstract: Existing research indicates that different types of meat have varying effects on health and aging, but the specific causal relationships remain unclear. This study aimed to explore the causal relationship between different types of meat intake and aging-related phenotypes. This study employed Mendelian randomization (MR) to select genetic variants associated with meat intake from large genomic databases, ensuring the independence and pleiotropy-free nature of these instrumental variables (IVs), and calculated the F-statistic to evaluate the strength of the IVs. The validity of causal estimates was assessed through sensitivity analyses and various MR methods (MR-Egger, weighted median, inverse-variance weighted (IVW), simple mode, and weighted mode), with the MR-Egger regression intercept used to test for pleiotropy bias and Cochran's Q test employed to evaluate the heterogeneity of the results. The findings reveal a positive causal relationship between meat consumers and DNA methylation PhenoAge acceleration, suggesting that increased meat intake may accelerate the biological aging process. Specifically, lamb intake is found to have a positive causal effect on mitochondrial DNA copy number, while processed meat consumption shows a negative causal effect on telomere length. No significant causal relationships were observed for other types of meat intake. This study highlights the significant impact that processing and cooking methods have on meat's role in health and aging, enhancing our understanding of how specific types of meat and their preparation affect the aging process, providing a theoretical basis for dietary strategies aimed at delaying aging and enhancing quality of life.

Keywords: causal relationship; meat consumption; biological aging; Mendelian randomization; red meat; white meat; processed meat

1. Introduction

Changes and trends in global meat consumption have had profound impacts on public health and the ecological environment [1]. In recent years, especially in rapidly developing economies, meat consumption levels have significantly increased [2,3]. This trend has sparked widespread attention and research from health scholars regarding its potential health effects [4,5]. Studies indicate that high meat consumption, particularly of red and processed meats, is closely associated with the occurrence of various chronic diseases, including cardiovascular diseases [6–8], cancer [9–12], and metabolic syndrome [13–15]. For example, several studies indicated that higher intake of red and processed meats was linked to increased mortality from cardiovascular diseases and cancer [12,16,17]. In contrast, the intake of white meat (such as poultry and fish) is generally associated with lower health risks [18–20]. These findings underscore the importance of distinguishing between red and white meat in dietary intake concerning health risks, providing a scientific basis for public health policy formulation.

Aging is a complex biological process involving a series of physiological and functional changes [21,22]. Studying biomarkers of aging-related phenotypes is crucial for understanding the mechanisms of aging and promoting healthy aging [23,24]. Commonly used aging biomarkers include telomere length [25,26], mitochondrial DNA copy number [27–29], and DNA methylation levels [30–32]. Telomere length is an important marker of cellular aging, shortening with cell division, and is thus used as an indicator of an individual's biological age [33–35]. Changes in mitochondrial DNA copy number are also closely related to the aging process [36,37]. In previous analyses, we identified a negative causal relationship between telomere length shortening and mitochondrial DNA copy number, indicating that as telomeres shorten, aging accelerates, and mitochondrial DNA copy number significantly decreases [38]. Similar results have been reported in other studies [39–42]. In recent years, with advancements in molecular biology techniques, assessing and predicting biological age through detecting DNA methylation levels at specific sites has become a hotspot in aging research, using methods such as the Horvath clock [43–48]. In daily life, aging is often manifested as declines in cognitive function and physical fitness, which significantly affect an individual's quality of life and independence. The role of diet structure in the aging process has also gradually become a research focus [49–52]. Meat intake, as a primary source of animal protein in daily life, plays an important role in ensuring healthy life activities, yet its potential health risks still need to be thoroughly explored. Currently, there is no definitive conclusion on whether meat intake causes or accelerates aging-related phenotypes.

The Mendelian randomization (MR) method uses genetic variations as instrumental variables to study the causal relationship between exposure factors (such as meat consumers) and outcomes (such as aging-related phenotypes) [53,54]. The significant advantage of the MR method is its ability to reduce the impact of confounding biases and reverse causation common in observational studies, thereby providing more reliable causal inferences [55]. Genetic variations are randomly assigned during fertilization, a natural randomization process akin to the random grouping in randomized controlled trials, giving the MR method a unique advantage in causal inference [56,57]. This study used the MR method, selecting genetic variants associated with meat intake as instrumental variables (IVs) to analyze their association with aging-related phenotypes. Through large-sample data analysis, we aimed to clarify the mechanisms by which meat consumption affects the aging process and verify the causal relationship between different types of meat intake and aging-related phenotypes. Specifically, this study aimed to identify genetic variants associated with meat consumers and use them as IVs; analyze the causal relationship between these genetic variants and various aging-related phenotypes (such as telomere length, mitochondrial DNA copy number, DNA methylation levels related to human age); infer the causal relationship between meat consumers and aging-related phenotypes through the MR method; and explore the different impacts of red, white, and processed meat intakes on the aging process. Through these research steps, we aimed to reveal the causal relationship of meat intake on the aging process, provide the public with more scientific dietary recommendations, and promote healthy aging.

2. Material and Methods

2.1. Study Design and Data Source

This study followed the STROBE-MR guidelines (Strengthening the Reporting of Observational Studies in Epidemiology Using Mendelian Randomization) [58] and employed a two-sample MR approach. Eleven GWAS datasets related to meat intake were retrieved and downloaded from the openGWAS database (https://gwas.mrcieu.ac.uk/, accessed on 10 June 2024). These datasets include one for overall meat consumers [59], three for red meat (including pork, lamb, and beef) intake, three for white meat (including poultry, oily fish, and non-oily fish) intake [60], and four for processed meat (including overall processed meat consumption, ham, bacon, and sausage) intake. Additionally, four aging-related phenotype datasets were selected as outcome variables, including telomere length [61], mitochondrial DNA copy number [62], PhenoAge (a DNA methylation predictor trained on 42 clinical markers and age), and Hannum age (intrinsic epigenetic age acceleration) [46].

Detailed information on these datasets is provided in Supplementary Table S1. Based on these data, a comprehensive MR analysis framework (Figure 1) was constructed to explore the potential causal relationships between meat intake and aging-related phenotypes.

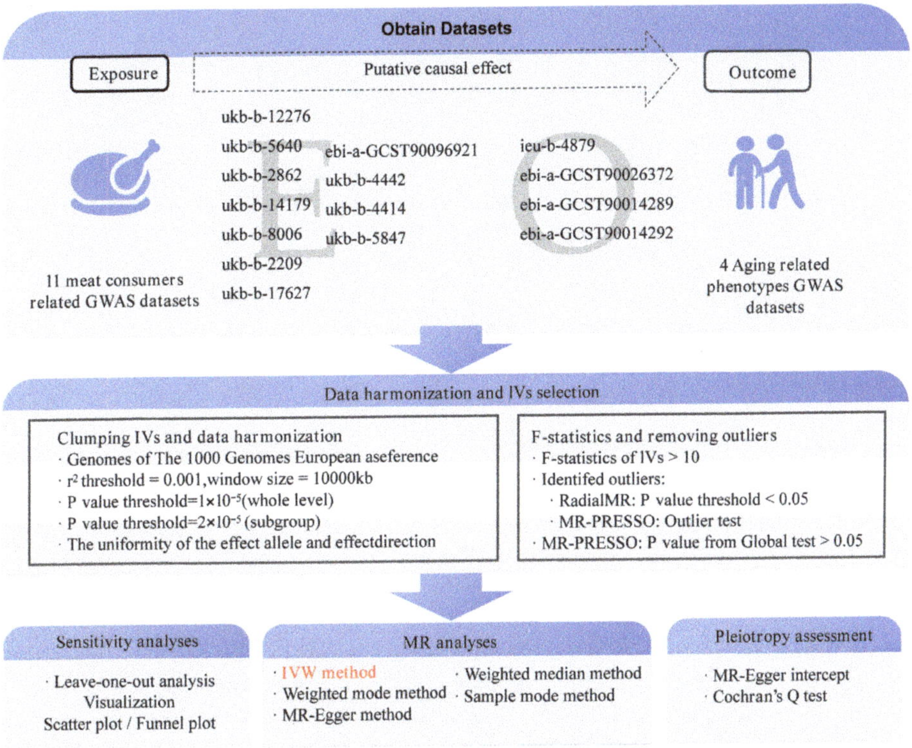

Figure 1. Flow chart showing the design of the study and the steps involved in the data analysis. E represents exposure variable, and the text above it represents the identification of genome-wide association study (GWAS) datasets; O represents outcome variable, and the text above it represents the identification of genome-wide association study (GWAS) datasets; MR, Mendelian randomization; IVW, inverse-variance weighted.

To ensure the robustness of the MR analysis, three key assumptions regarding IVs were established [63,64]: (1) IVs must be significantly associated with the exposure; (2) IVs must not be related to any confounders that might simultaneously affect the exposure and the outcome; (3) IVs must influence the outcome solely through the exposure and not via any other pathways. Since all data used in this study were obtained from publicly available databases, no additional ethical approval or informed consent was required. Figure 1 provides a detailed description of the dataset selection, IV screening, and main steps of the MR analysis.

2.2. IVs Selection and Data Cleaning

According to Figure 1, the "TwoSampleMR" R package was utilized to obtain IVs related to meat consumers and meat intake [65], ensuring sufficient sample sizes and appropriate effect allele frequencies (EAFs). Missing EAFs were supplemented using data from the 1000 Genomes Project (https://www.internationalgenome.org/home, accessed on 10 June 2024) [66]. Only single nucleotide polymorphisms (SNPs) with an F-statistic ($F = \beta^2/se^2$) greater than 10 were included in the analysis [67,68]. Data cleaning involved removing potential confounding IVs identified through the "LDtrait" database [69]. Additionally, IVs with p-values less than 0.05 were considered outliers and excluded using the "RadialMR" R package (https://github.com/WSpiller/RadialMR/, accessed on 10 June 2024) [70], with further assessment of IV pleiotropy conducted via Global t-tests [71,72].

2.3. MR Analysis

Five MR analysis methods (MR-Egger [73], weighted median [74], inverse-variance weighted (IVW) [75], simple mode [76], and weighted mode [77]) were employed to comprehensively assess the impact of meat consumers and different types of meat intake on aging-related phenotypes [75,78,79]. Causality was judged based on two criteria: (1) consistent direction of causal effect estimates (β values or B values) across various methods; (2) statistical significance primarily reliant on a P_{ivw} value of less than 0.05. Each direction of analysis was conducted independently, and no multiple comparison corrections were applied, with significance determined using the original p-values.

2.4. Sensitivity Analysis

Sensitivity analyses included heterogeneity analysis and pleiotropy testing, performed using Cochran's Q test and the MR-Egger method [80,81]. The MR-Egger intercept test was used to evaluate horizontal pleiotropy [82], with a significance level set at $p < 0.05$. For IVW analysis, a fixed-effect model was used in the absence of heterogeneity, whereas a random-effect model was applied when heterogeneity was present ($P_{Q\ test} < 0.05$) [83]. Additionally, to clarify the independent causal effects of each IV on the outcome variable in various analytical directions, we assessed the causal effect of individual SNP. We also conducted a leave-one-out analysis to observe the causal effect on the outcome variable after sequentially removing each SNP. Both sets of results are presented using forest plots. Funnel plots and scatter plots were utilized for visual assessments of heterogeneity and pleiotropy in the MR analyses.

3. Results

3.1. Causal Relationship between Meat Consumers and Aging-Related Phenotypes

According to MR guidelines, 27 SNPs significantly associated with meat consumers (ukb-b-12276) were identified, with an average F-statistic of 24.64 (Table S2). The MR analysis results (Table 1) revealed a significant positive causal effect between meat consumers and DNA methylation PhenoAge acceleration (all five MR methods yielded positive B values, and $P_{IVW} = 0.02$). However, no significant causal effects were observed between meat consumers and telomere length, mitochondrial DNA copy number, or DNA methylation Hannum age acceleration ($P_{IVW} > 0.05$). These results suggest that meat consumers may have a more direct impact on certain specific aging biomarkers, such as DNA methylation PhenoAge acceleration, while its effects on other biomarkers might be limited. Despite the lack of significant associations with telomere length and mitochondrial DNA copy number, the potential complex and indirect impacts of meat consumers on these indicators cannot be ruled out. These findings highlight the differential effects of meat consumers on various aging phenotypes, suggesting that dietary guidelines should carefully consider the health effects related to total meat intake to delay biological aging and reduce the risk of age-related diseases.

Table 1. The results of Mendelian randomization (MR) analysis between meat consumers and aging-related phenotypes.

Directions and Methods	N_{snp}	*B*	*SE*	*p*-Value
Meat consumers to telomere length				
MR-Egger	22	0.08	0.07	0.28
Weighted median	22	−0.02	0.05	0.67
Inverse-variance weighted	22	−0.03	0.04	0.36
Simple mode	22	−0.02	0.11	0.85
Weighted mode	22	−0.03	0.10	0.79
Meat consumers to mitochondrial DNA copy number				
MR-Egger	22	0.03	0.07	0.66
Weighted median	22	0.02	0.05	0.63
Inverse-variance weighted	22	−0.02	0.04	0.63
Simple mode	22	0.05	0.10	0.63
Weighted mode	22	0.05	0.10	0.61
Meat consumers to DNA methylation Hannum age acceleration				
MR-Egger	23	−0.33	1.33	0.81
Weighted median	23	0.24	0.82	0.77
Inverse-variance weighted	23	0.23	0.58	0.69
Simple mode	23	0.41	1.67	0.81
Weighted mode	23	0.41	1.62	0.80
Meat consumers to DNA methylation PhenoAge acceleration				
MR-Egger	23	1.10	1.79	0.55
Weighted median	23	2.20	1.07	0.04
Inverse-variance weighted	23	1.73	0.77	0.02 *
Simple mode	23	2.12	2.12	0.33
Weighted mode	23	2.25	1.93	0.26

* statistical significance; N_{snp}, number of single nucleotide polymorphisms (SNPs) in the MR analysis; *B*, the causal effects estimated by different MR methods; *SE*, standard error.

3.2. Sensitivity Analysis

To ensure the reliability and robustness of the causal relationship between meat consumers and aging-related phenotypes, sensitivity analyses were conducted. The results of pleiotropy and heterogeneity analyses (Table 2) indicated no significant heterogeneity or pleiotropy in the MR analyses of meat consumers and the four aging-related phenotypes ($p > 0.05$). Evaluating the causal relationships from the perspective of individual SNPs showed diverse causal effects on different aging-related phenotypes (Figure S1). This diversity further supports the complex role of meat consumers in biological aging. The leave-one-out analysis confirmed that, excluding any single SNP, the remaining SNPs maintained consistent causal effects on the outcome variables in the analyses of meat consumers with telomere length (Figure 2A), mitochondrial DNA copy number (Figure 2B), and DNA methylation PhenoAge acceleration (Figure 2C). However, in the analysis of meat consumers and DNA methylation Hannum age acceleration (Figure 2D), excluding rs11690184 reversed the causal effect of the remaining SNPs, suggesting potential heterogeneity in this direction. Combining individual SNP effect estimates and leave-one-out analysis results confirmed that some SNPs exhibited broad and differential effects on certain aging biomarkers, but the overall causal effect remained consistent.

Table 2. The results of pleiotropy and heterogeneity test in the Mendelian randomization (MR) analysis between meat consumers (exposure) and different aging-related phenotypes (outcomes).

Outcomes	Pleiotropy			Heterogeneity		
	Intercept *	SE	*p*-Value	Q Test	Q_df	*p*-Value
Telomere length	−0.003	0.001	0.083	22.953	21	0.347
mtDNA copy number	−0.001	0.001	0.431	14.716	21	0.837
DNAm Hannum age acceleration	0.011	0.022	0.644	17.088	22	0.759
DNAm PhenoAge acceleration	0.012	0.030	0.701	17.166	22	0.754

* egger_intercept, mtDNA, mitochondrial DNA, DNAm, DNA methylation.

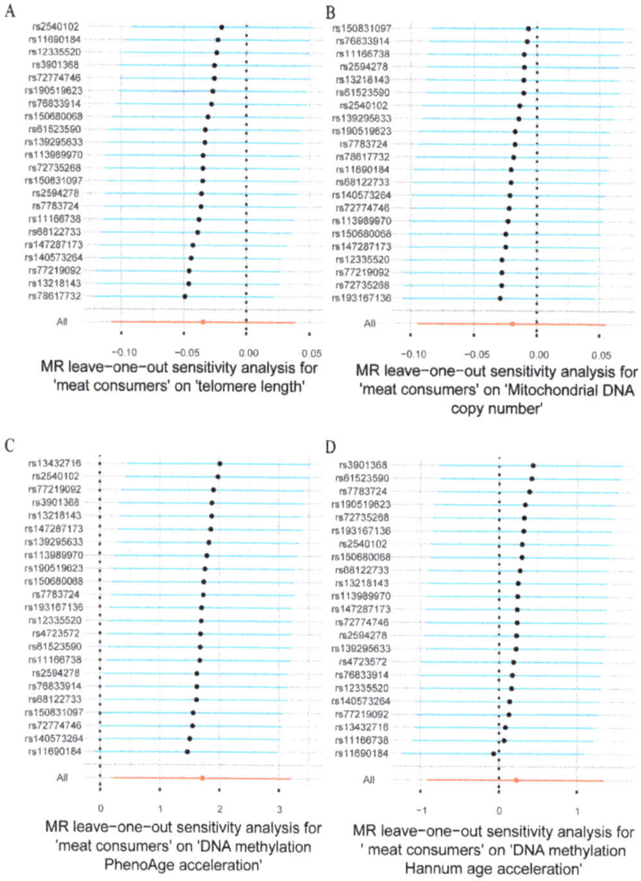

Figure 2. Leave-one-out plots of the causal association between meat consumers and aging-related phenotypes ((**A**) telomere length; (**B**) mitochondrial DNA copy number; (**C**) DNA methylation PhenoAge acceleration; (**D**) DNA methylation Hannum age acceleration). In the plot, each black point represents the recalculated pooled effect size when the corresponding single nucleotide polymorphism is excluded from the analysis. The blue horizontal line represents the maximum (right end) and minimum (left end) estimated values of causal effects. Red points represent cumulative estimates of causal effects, while red horizontal lines represent maximum (right end) and minimum (left end) levels of causal effects.

Scatter plots depicted the potential associations between the causal effects of the instrumental variables on the exposure and the outcome. The results (Figure 3) showed no significant slope changes in the relationships between meat consumers and telomere length (Figure 3A) or mitochondrial DNA copy number (Figure 3B), confirming the absence of significant causal relationships. For aging-related phenotypes, the SNP effect estimates showed no significant slope changes in the relationship between meat consumers and DNA methylation Hannum age acceleration (Figure 3C), confirming no significant causal relationship, whereas a significant causal relationship was observed with DNA methylation PhenoAge acceleration (Figure 3D). Funnel plots for each analysis direction (Figure S2) did not show obvious abnormal distributions of SNPs. In conclusion, the sensitivity analysis results confirmed that the causal relationships observed in the MR analysis remained consistent across different analytical methods, providing robust support for our conclusion that meat consumption influences the aging process, with multifactorial impacts and significant individual-level differences.

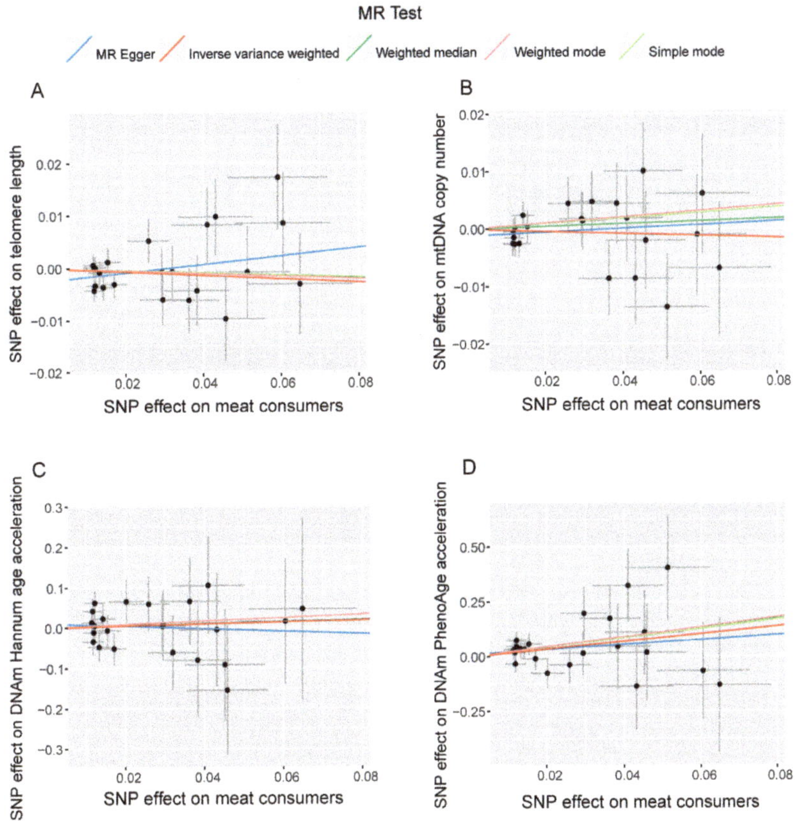

Figure 3. Scatter plots of SNP effects on the meat consumers and the various aging-related phenotypes ((**A**) telomere length; (**B**) mitochondrial DNA copy number; (**C**) DNA methylation PhenoAge acceleration; (**D**) DNA methylation Hannum age acceleration). The X-axis represents the causal effect of single nucleotide polymorphisms (SNPs) on exposure; the Y-axis represents the causal effect of SNPs on outcome; the dots represent SNP for causal estimation in the analysis direction; the different colored lines represent different MR analysis methods as shown in the legend.

3.3. Causal Relationship between Red Meat Intake and Aging-Related Phenotypes

Following MR guidelines, SNPs closely associated with red meat intake were identified, with 168 SNPs for pork intake (average F-statistic = 22.82), 209 SNPs for beef intake (average F-statistic = 23.04), and 281 SNPs for lamb intake (average F-statistic = 23.66) (Table S3). The results (Figure 4) of MR analysis demonstrated a significant positive causal relationship between lamb intake and mitochondrial DNA copy number (P_{IVW} = 0.02). However, no significant results were observed in other analysis directions. Detailed MR analysis results for each direction are provided in Table S4. This finding highlights a potential important link between lamb intake and mitochondrial function, suggesting dietary habits' potential impact on mitochondrial health.

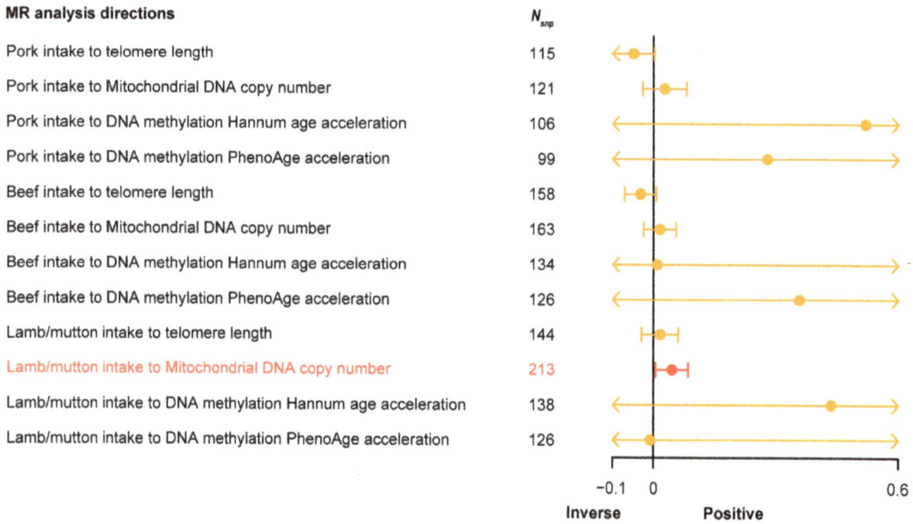

Figure 4. Forest plot shows the result of Mendelian randomization (MR) analysis between red meat (pork, beef, and lamb) intake and four types of aging-related phenotypes. The X-axis represents the direction of causal effect. The vertical line at the 0 scale represents an invalid line. The yellow point in each yellow line represents the causal effect value obtained by the inverse-variance weighted (IVW) method in the current analysis direction, and the yellow horizontal line represents the maximum (right endpoint) and minimum (left endpoint) causal effects estimated by the IVW method in the current analysis direction. If the maximum and minimum values are within the range marked on the X-axis, the end is represented by a small vertical line. When the maximum or minimum value exceeds the range marked on the X-axis, the end is indicated by an arrow. The red section emphasizes that the analysis direction is statistically significant. N_{snp}, number of single nucleotide polymorphisms (SNPs) in the analysis direction.

3.4. Causal Relationship between White Meat Intake and Aging-Related Phenotypes

SNPs closely associated with white meat intake were identified, with 145 SNPs for poultry intake (average F-statistic = 22.26), 268 SNPs for oily fish intake (average F-statistic = 27.42), and 124 SNPs for non-oily fish intake (average F-statistic = 23.89) (Table S5). The results (Figure 5) showed no significant causal effects between poultry, oily fish, and non-oily fish intake and the four different aging phenotypes (P_{IVW} > 0.05). Detailed MR analysis results for each direction are provided in Table S4. This finding suggests that although these foods are considered potentially beneficial for health, their impact on the aging process may not be as significant as expected, prompting further investigation into the specific effects and biological mechanisms of different food categories on aging.

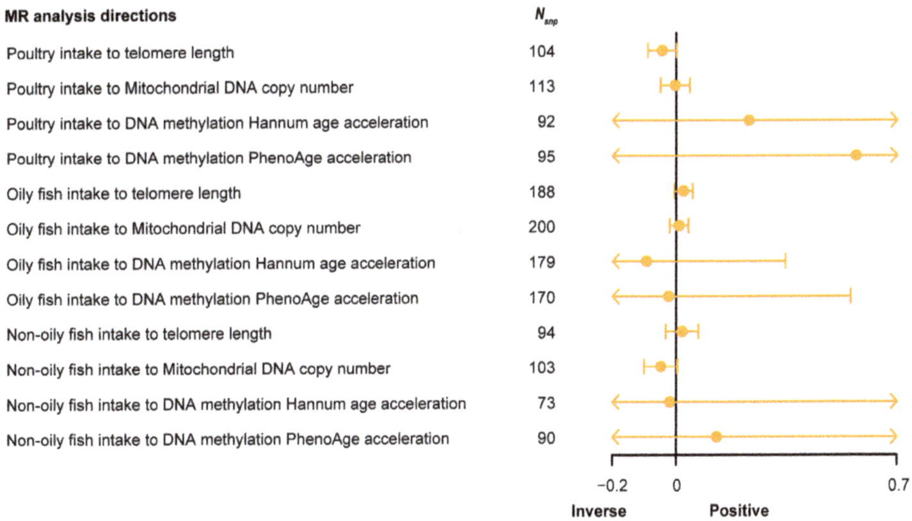

Figure 5. Forest plot shows the result of Mendelian randomization (MR) analysis between white meat (poultry, oily fish, and non-oily fish) intake and four types of aging-related phenotypes. The X-axis represents the direction of causal effect. The vertical line at the 0 scale represents an invalid line. The yellow point in each yellow line represents the causal effect value obtained by the inverse-variance weighted (IVW) method in the current analysis direction, and the yellow horizontal line represents the maximum (right endpoint) and minimum (left endpoint) causal effects estimated by the IVW method in the current analysis direction. If the maximum and minimum values are within the range marked on the X-axis, the end is represented by a small vertical line. When the maximum or minimum value exceeds the range marked on the X-axis, the end is indicated by an arrow. N_{snp}, number of single nucleotide polymorphisms (SNPs) in the analysis direction.

3.5. Causal Relationship between Processed Meat Intake and Aging-Related Phenotypes

SNPs closely associated with processed meat consumption were identified, with 85 SNPs for overall processed meat intake (average F-statistic = 23.10), 45 SNPs for ham intake (average F-statistic = 20.02), 44 SNPs for bacon intake (average F-statistic = 20.89), and 49 SNPs for sausage intake (average F-statistic = 20.19) (Table S6). The MR analysis results (Table 3) confirmed a significant inverse causal relationship between processed meat intake and telomere length (Table 3, $P_{IVW} < 0.05$), but no significant causal effects were observed for the other three aging-related phenotypes (Table 3, $P_{IVW} > 0.05$). In subgroup analyses of processed meats (ham, bacon, sausage intake), no significant causal effects on the four aging-related phenotypes were observed (Figure 6). Detailed MR analysis results for each direction are provided in Table S4. This evidence indicates that while single types of processed meat may not significantly impact the aging process, the cumulative effect of overall processed meat consumption should not be ignored. These findings emphasize the importance of controlling total processed meat intake and suggest dietary adjustments to reduce processed meat consumption, advocating for a more balanced and healthy diet to prevent health risks associated with processed meat intake.

Table 3. The results of Mendelian randomization (MR) analysis between processed meat consumers and aging-related phenotypes.

Directions and Methods	N_{snp}	*B*	*SE*	*p*-Value
Processed meat consumers to telomere length				
MR-Egger	71	−0.33	0.37	0.37
Weighted median	71	−0.19	0.12	0.12
Inverse-variance weighted	71	−0.19	0.09	0.03 *
Simple mode	71	−0.35	0.33	0.30
Weighted mode	71	−0.33	0.31	0.28
Processed meat consumers to mitochondrial DNA copy number				
MR-Egger	76	−0.11	0.40	0.78
Weighted median	76	−0.15	0.13	0.26
Inverse-variance weighted	76	0.01	0.09	0.89
Simple mode	76	−0.20	0.35	0.56
Weighted mode	76	−0.21	0.34	0.54
Processed meat consumers to DNA methylation Hannum age acceleration				
MR-Egger	67	−0.63	7.63	0.94
Weighted median	67	1.44	1.88	0.44
Inverse-variance weighted	67	1.41	1.38	0.31
Simple mode	67	−4.23	4.79	0.38
Weighted mode	67	−3.91	4.48	0.39
Processed meat consumers to DNA methylation PhenoAge acceleration				
MR-Egger	64	1.25	9.43	0.89
Weighted median	64	2.32	2.46	0.35
Inverse-variance weighted	64	0.97	1.83	0.59
Simple mode	64	4.78	5.99	0.43
Weighted mode	64	5.75	6.05	0.35

* statistical significance; N_{snp}, Number of Single nucleotide polymorphisms (SNPs) in the MR analysis; *B*, the causal effects estimated by different MR methods; *SE*, standard error.

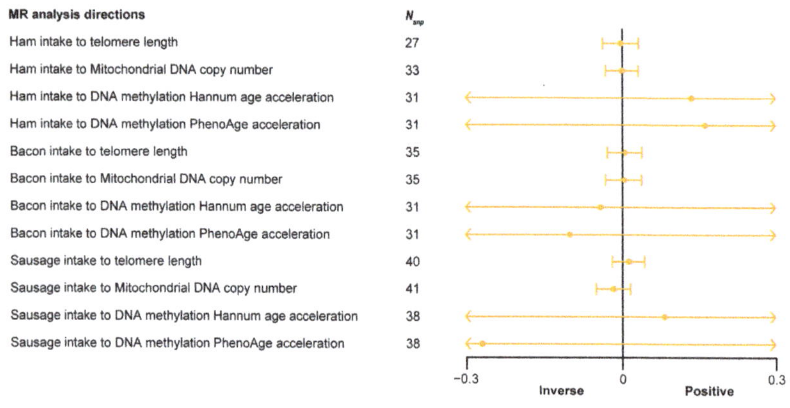

Figure 6. Forest plot shows the result of Mendelian randomization (MR) analysis between processed meat (ham, bacon, and sausage) intake and four types of aging-related phenotypes. The X-axis represents the direction of causal effect. The vertical line at the 0 scale represents an invalid line. The yellow point in each yellow line represents the causal effect value obtained by the inverse-variance weighted (IVW) method in the current analysis direction, and the yellow horizontal line represents the maximum (right endpoint) and minimum (left endpoint) causal effects estimated by the IVW method in the current analysis direction. If the maximum and minimum values are within the range marked on the X-axis, the end is represented by a small vertical line. When the maximum or minimum value exceeds the range marked on the X-axis, the end is indicated by an arrow. N_{snp}, number of single nucleotide polymorphisms (SNPs) in the analysis direction.

3.6. Heterogeneity and Pleiotropy in Subgroup Analyses

In the sensitivity analyses of subgroups, no significant pleiotropy (Table S7, $p > 0.05$) or heterogeneity (Table S7, $p > 0.05$) was observed between different types of meat intake and the four aging-related phenotypes. This stage focused on the distribution of instrumental variables that showed significance in the MR analysis to ensure the reliability and validity of the MR results. In the MR analysis of lamb intake and mitochondrial DNA copy number, although individual SNPs showed differential causal effects on mitochondrial DNA copy number (Figure S3), the remaining SNPs continued to show significant causal effects after excluding any single SNP, consistently positioned to the right of the null line (Figure S4). Furthermore, the scatter plot indicated a significant positive trend in the causal relationship between exposure-related SNPs and the outcome variable (IVW method fitting curve showing a distinct positive slope, Figure 7A), with no significant abnormal SNP distribution observed in the funnel plot (Figure 7B). Similarly, in the MR analysis of overall processed meat intake and telomere length, individual SNPs displayed differential causal effects on telomere length (Figure S5), but the remaining SNPs maintained an overall consistent causal effect after excluding any single SNP, positioned to the left of the null line (Figure S6). The scatter plot showed a significant negative trend in the causal relationship between exposure-related SNPs and the outcome variable (IVW method fitting curve showing a distinct negative slope, Figure 7C), with no significant abnormal SNP distribution observed in the funnel plot (Figure 7D). These results not only confirm the reliability and robustness of the MR analysis but also enhance the credibility of these findings, further supporting their value in guiding dietary recommendations.

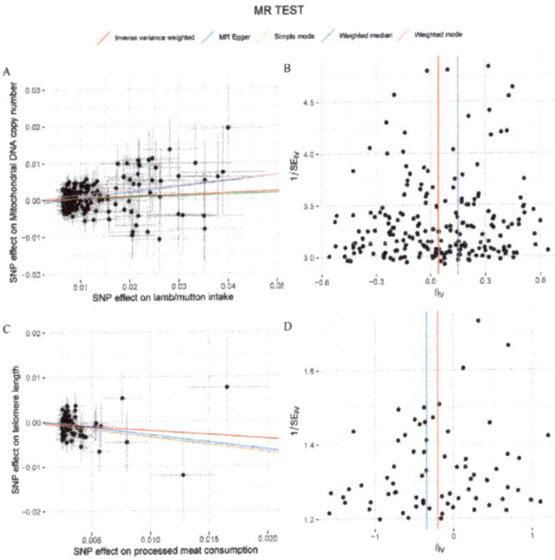

Figure 7. The scatter plots and funnel plots for the Mendelian randomization (MR) analysis between lamb intake and mitochondrial DNA copy number (**A**,**B**), between processed meat consumption and telomere length (**C**,**D**). For scatter plots (**A**,**C**), the X-axis represents the causal effect of single nucleotide polymorphisms (SNPs) on exposure; the Y-axis represents the causal effect of SNPs on outcome; the dots represent SNP for causal estimation in the analysis direction; the different colored lines represent different MR analysis methods as shown in the legend. For funnel plots (**B**,**D**), the X-axis represents the causal effect of single nucleotide polymorphisms (SNPs) on exposure; the Y-axis represents the ratio of 1/standard error (SE) of SNPs from the exposure variable; the dots represent SNP for causal estimation in the analysis direction; the different colored lines represent different MR analysis methods as shown in the legend.

4. Discussion

This study delves into the causal relationship between different types of meat intake and aging phenotypes (such as telomere length, mitochondrial DNA copy number, and DNA methylation levels related to human age). The results first revealed a positive causal relationship between meat consumption and accelerated DNA methylation aging in meat consumers, suggesting that increased meat intake may accelerate the biological aging process. This finding sets the premise for our subsequent analyses. To further evaluate the causal effects of different types of meat consumption on aging phenotypes, we conducted subgroup analyses based on three categories: red meat, white meat, and processed meat. For red meat, we selected the most commonly consumed types in daily life, namely beef, lamb, and pork. Mendelian randomization (MR) analysis confirmed a positive causal effect of lamb intake on mitochondrial DNA copy number. However, no significant effects were observed for beef and pork on the four aging-related phenotypes. For white meat, we chose the most common types of meat in daily diets: chicken and fish. MR analysis showed no significant association between the types of white meat included in the study (chicken, oily fish, and non-oily fish) and the four aging phenotypes. Processed meat, which includes commonly consumed types such as ham, sausage, and bacon, was also analyzed. MR analysis indicated a negative causal relationship between overall processed meat consumption and telomere length, suggesting that higher intake of processed meat may shorten telomere length. However, no significant associations were found between processed meat consumption and the other three aging phenotypes (mitochondrial DNA copy number and DNA methylation levels related to human age). Subgroup analyses for processed meat also did not reveal any significant causal links between consumption levels and the four aging-related phenotypes. These findings emphasize that the total consumption of processed meat may impact the aging process mediated by telomere length, while the cumulative effect of meat consumption in general may be more significant than the impact of individual types of meat. Therefore, although specific types of meat consumption have varying effects on aging markers, the overall quantity and frequency of meat intake may be more crucial in influencing biological aging. In summary, this study provides new insights into the relationship between different types of meat intake and aging phenotypes, offering scientific evidence for public health recommendations. It emphasizes the importance of choosing healthy meat products and highlights that the potential harm associated with meat consumption is closely linked to cooking methods and the degree of processing. We advise the public to pay closer attention to meat selection and consumption methods to promote healthy aging. This distinction is crucial in guiding more nuanced and effective public health strategies.

The primary advantage of the MR method is its ability to overcome confounding factors and reverse causation issues common in traditional observational studies [84]. Selecting appropriate IVs is crucial for ensuring the validity and reliability of MR analysis [85]. In this study, genetic variants strongly associated with meat intake were selected from large genomic databases, ensuring their independence and pleiotropy-free nature. The F-statistic was calculated to evaluate the strength of the IVs, ensuring sufficient statistical power for reliable causal inference. Sensitivity analyses were conducted to validate the robustness of the MR analysis, using various MR methods to assess the validity of causal estimates [81]. Pleiotropy bias was evaluated using the MR-Egger regression intercept test [86], and heterogeneity was assessed using Cochran's Q test to ensure consistency among the IVs [87]. Sensitivity analysis results showed no significant issues, confirming the robustness of the main analysis results and the reliability of the study conclusions. Overall, this study adheres to MR analysis standards, employing rigorous IV screening, quality control, and sensitivity analyses to investigate the impact of meat intake on aging phenotypes. These findings enhance our understanding of the causal relationship between meat intake and aging, providing a theoretical basis for strategies to delay aging.

Despite establishing causal relationships between certain meat intakes and the aging process through MR analysis, the specific biological mechanisms involved require further investigation. Recent studies suggest that meat consumption significantly impacts human metabolic processes, which may play a crucial role in aging. For instance, red and processed meats contain saturated fatty acids and cholesterol, which are linked to various metabolic diseases [88–90]. Meat consumption influences aging through two primary pathways: first, diets with high meat content affect gut microbiota composition and function, indirectly impacting metabolic status and aging [91–94]. Reduced gut microbiota diversity and increased pathogenic bacteria can induce low-grade chronic inflammation, accelerating aging through various mechanisms [95–99]. Short-chain fatty acids (SCFAs) produced by gut microbiota are vital for metabolic health, and high-meat diets significantly reduce SCFA production, affecting energy metabolism and immune function [100–102]. Second, the different nutritional contents of meats can influence aging-related phenotypes by regulating key metabolic pathways. Red meat intake increases *N*-nitroso compounds, associated with cancer and other age-related diseases [103,104]. Additionally, red meat's L-carnitine and its metabolite trimethylamine-*N*-oxide (TMAO) are linked to increased cardiovascular disease and mortality [105–107]. These metabolites impact cardiovascular health and may accelerate aging through oxidative stress and inflammation pathways [108]. Further research indicates that meat intake affects insulin signaling pathways, altering glucose and lipid metabolism, thus impacting the aging process [109,110]. High-meat diets are associated with insulin resistance and a higher incidence of type 2 diabetes, which are considered significant drivers of aging [111–117]. This study found a significant inverse causal relationship between overall processed meat consumption and telomere length, possibly due to high levels of additives and preservatives in processed meats that increase oxidative stress and cellular DNA damage, leading to accelerated telomere shortening [118–124]. High saturated fat and salt content in processed meats may also induce chronic inflammation [125–127], a critical factor in telomere shortening [21,128–130]. These findings highlight the potential negative health impacts of processed meat consumption, emphasizing the importance of reducing processed meat intake and increasing antioxidant and anti-inflammatory foods to delay aging.

In our analysis, a causal relationship was found between overall meat intake and health risks, but subgroup analyses did not show that specific types of meat significantly increased health risks, except for lamb intake, which increased mitochondrial DNA copy number. Mitochondrial DNA copy number is essential for maintaining mitochondrial function and is closely linked to mitochondrial energy production [131]. This result underscores lamb intake's potential positive impact on mitochondrial health. Several factors may explain this causal relationship: lamb is a rich source of various nutrients [132–135], including iron, protein, B vitamins (especially B12 and riboflavin), and minerals like selenium, which are vital for normal cell metabolism and biosynthesis. Iron is a necessary cofactor in the mitochondrial electron transport chain, while vitamin B12 is involved in mitochondrial DNA synthesis and maintenance [136]. Specific fatty acids and proteins in lamb might stimulate mitochondrial biogenesis, enhancing energy metabolism efficiency [137–139]. Additionally, the antioxidant components in lamb (e.g., selenium) may reduce oxidative stress on mitochondria, protecting mitochondrial DNA and increasing its stability and copy number. The anti-inflammatory properties of certain compounds in lamb might also reduce chronic inflammation, protecting mitochondria and maintaining higher DNA copy numbers [140]. Overall, dietary factors' regulatory relationships with health are complex, but our evidence supports lamb's potential benefits in maintaining mitochondrial function and metabolic health. In conclusion, meat intake has multiple effects on metabolism and aging. Future research should focus on uncovering the underlying mechanisms to provide scientific evidence for dietary guidelines and aging intervention strategies.

This study's findings, derived from MR analysis, reveal causal relationships between increased meat intake and individual aging risks, offering significant guidance for public health policy and clinical practice. First, the results emphasize the importance of dietary structure in healthy aging, suggesting that public health policies encourage reducing red and processed meat intake while increasing alternatives like fish and poultry. These alternatives are not only nutritious but also lower in saturated fat and cholesterol, helping to reduce the risk of chronic diseases associated with aging. Second, the findings indicate that clinicians should consider potential health impacts from the increase in meat intake when devising personalized dietary plans. For elderly and chronic disease patients, reducing red and processed meat intake and increasing dietary fiber and antioxidant intake can help slow the aging process and improve overall health. This dietary adjustment can improve metabolic status and reduce inflammation, aiding in delaying aging. Third, lamb's potential in increasing mitochondrial DNA copy number suggests it could be a valuable part of healthy dietary plans, particularly due to its unique nutritional components like conjugated linoleic acid (CLA) and omega-3 fatty acids, which enhance mitochondrial function and replication efficiency. Therefore, dietary recommendations for specific populations might include lamb to optimize energy production and cellular function.

In summary, this study offers a comprehensive perspective on the nutritional value and health impacts of different types of meat, supporting the formulation of more targeted and effective dietary recommendations. These findings can help optimize individual health and provide scientific evidence for public health policies, promoting healthy aging across society. However, there are limitations to this study. First, MR analysis relies on the reasonableness and independence of genetic variants; some variants may influence multiple phenotypes, causing bias. In addition, there may also be differences in the results produced by both MR analysis methods and statistical methods, such as whether or not to utilize *p*-value corrections. Therefore, the findings of this study still require further validation with real-world data. Second, the study samples mainly come from Western populations, which may limit the applicability of the results to other ethnic groups and regions. Future research should include more diverse samples. Third, due to the complexity of dietary habits and lifestyles, it is challenging to completely exclude confounding factors. Fourth, this study primarily focuses on the impact of different types of meat intake on aging-related phenotypes and does not encompass the diversity of other dietary habits, which may limit our understanding of the overall effects of diet. Future research should combine long-term longitudinal data and various statistical models to further verify the robustness of the causal relationships. Fifth, this study categorizes both fish and poultry under the same classification of "white meat". This approach does not account for the distinct lipid profiles associated with fish consumption, which differ significantly from those of poultry. Fish is rich in omega-3 fatty acids, which have different health implications compared to the predominantly protein-based profile of poultry. This distinction could affect the associations with aging phenotypes observed in our study. Hence, subsequent studies need to focus on multidisciplinary approaches, utilizing modern biotechnological methods to explore the complex relationship between meat consumption and aging, especially the specific impacts of different types of meat. This will provide theoretical support and practical strategies for achieving healthy aging.

5. Conclusions

MR analysis has demonstrated a positive causal effect between meat consumers and accelerated DNA methylation PhenoAge, suggesting that increased meat intake may expedite the biological aging process. Additionally, lamb intake appears to positively influence mtDNA copy number, potentially supporting mitochondrial health and energy metabolism, while processed meats negatively affect telomere length, underscoring the risks associated with their consumption. These variations highlight the significant impact that processing and cooking methods have on meat's role in health and aging. Consequently, we recommend reducing the intake of processed meats and, where possible, opting for minimally processed alternatives that are cooked in a manner preserving nutritional integrity and reducing harmful effects. This study enhances our understanding of how specific types of meat and their preparation affect the aging process, providing a theoretical basis for dietary strategies aimed at delaying aging and enhancing quality of life.

Supplementary Materials: The following supporting information can be downloaded at: https://www.mdpi.com/article/10.3390/nu16152433/s1, Figure S1: The causal effect for single SNP for the MR analysis between meat consumers and aging phenotypes; Figure S2: Funnel plots for the MR analysis between meat consumers and aging phenotypes; Figure S3: Single SNP forest plot of MR analysis between lamb intake and mtDNA copy number; Figure S4: Leave-one-out plot of MR analysis between lamb intake and mtDNA copy number; Figure S5: Single SNP forest plot of MR analysis between processed meat intake and telonere length; Figure S6: Leave-one-out plot of MR analysis between processed meat intake and telonere length; Table S1: The datasets used in the study; Table S2: IVs_for_ukb-b-12276_added_F_Value; Table S3: IVs2_for_Pork intake, Beef intake, Lamb intake, added F_Values; Table S4: The all results of MR analysis for the subgroups of meat consumers; Table S5: IVs2_for Poultry intake, Oily fish intake, Non-oily fish intake and added F_Values; Table S6: IVs2_for Processed meat consumption, Ham intake, Bacon intake, Sausage intake and added F_Values; Table S7: The results of sensitivity analysis for subgroup of meat consumers.

Author Contributions: Z.F. (Zhijun Feng) and M.Z. designed the study. S.L. and Y.D. analyzed the data and drafted the paper. S.L. and Y.D. critically revised the paper. H.L., Z.F. (Zhengzheng Fu) and Y.W. completed the visualization. All authors have read and agreed to the published version of the manuscript.

Funding: This research was funded by the National Natural Science Foundation of China (grant number: 82273582 and 82103785).

Institutional Review Board Statement: Not applicable.

Informed Consent Statement: Patient consent was not required as the research utilized publicly available GWAS summary statistics data, which had already obtained informed consent from all participating studies in accordance with approved protocols by their respective institutional review boards.

Data Availability Statement: All data used in this study are available in the public repository. The code involved in the data analysis process can be obtained by contacting the corresponding author.

Acknowledgments: We thank the team of OpenGWAS, and UK Biobank database for making the summary data publicly available, and we would like to acknowledge the principal investigators of the studies who made their data openly accessible for research.

Conflicts of Interest: The authors declare no conflicts of interest.

References

1. Talens, C.; Llorente, R.; Simó-Boyle, L.; Odriozola-Serrano, I.; Tueros, I.; Ibargüen, M. Hybrid Sausages: Modelling the Effect of Partial Meat Replacement with Broccoli, Upcycled Brewer's Spent Grain and Insect Flours. *Foods* **2022**, *11*, 3396. [CrossRef] [PubMed]
2. Yan, Y.; Ning, J.; Cheng, X.; Lv, Q.; Teng, S.; Wang, W. Rapid and High-Throughput Determination of Sixteen β-agonists in Livestock Meat Using One-Step Solid-Phase Extraction Coupled with UHPLC-MS/MS. *Foods* **2022**, *12*, 76. [CrossRef] [PubMed]
3. Spyrelli, E.D.; Doulgeraki, A.I.; Argyri, A.A.; Tassou, C.C.; Panagou, E.Z.; Nychas, G.E. Implementation of Multispectral Imaging (MSI) for Microbiological Quality Assessment of Poultry Products. *Microorganisms* **2020**, *8*, 552. [CrossRef] [PubMed]
4. Jeong, D.; Seo, J.W.; Lee, H.G.; Jung, W.K.; Park, Y.H.; Bae, H. Efficient Myogenic/Adipogenic Transdifferentiation of Bovine Fibroblasts in a 3D Bioprinting System for Steak-Type Cultured Meat Production. *Adv. Sci.* **2022**, *9*, e2202877. [CrossRef] [PubMed]
5. Ivancovsky-Wajcman, D.; Fliss-Isakov, N.; Grinshpan, L.S.; Salomone, F.; Lazarus, J.V.; Webb, M.; Shibolet, O.; Kariv, R.; Zelber-Sagi, S. High Meat Consumption Is Prospectively Associated with the Risk of Non-Alcoholic Fatty Liver Disease and Presumed Significant Fibrosis. *Nutrients* **2022**, *14*, 3533. [CrossRef] [PubMed]
6. Damigou, E.; Kosti, R.I.; Panagiotakos, D.B. White Meat Consumption and Cardiometabolic Risk Factors: A Review of Recent Prospective Cohort Studies. *Nutrients* **2022**, *14*, 5213. [CrossRef] [PubMed]
7. Pan, L.; Chen, L.; Lv, J.; Pang, Y.; Guo, Y.; Pei, P.; Du, H.; Yang, L.; Millwood, I.Y.; Walters, R.G.; et al. Association of Red Meat Consumption, Metabolic Markers, and Risk of Cardiovascular Diseases. *Front. Nutr.* **2022**, *9*, 833271. [CrossRef] [PubMed]
8. Pérez, L.; Pincay, R.; Salazar, D.; Flores, N.; Escolastico, C. Evaluation of the Quality and Lipid Content of Artisan Sausages Produced in Tungurahua, Ecuador. *Foods* **2023**, *12*, 4288. [CrossRef] [PubMed]
9. Alegria-Lertxundi, I.; Bujanda, L.; Arroyo-Izaga, M. Role of Dairy Foods, Fish, White Meat, and Eggs in the Prevention of Colorectal Cancer: A Systematic Review of Observational Studies in 2018-2022. *Nutrients* **2022**, *14*, 3430. [CrossRef] [PubMed]
10. Chao, A.; Thun, M.J.; Connell, C.J.; McCullough, M.L.; Jacobs, E.J.; Flanders, W.D.; Rodriguez, C.; Sinha, R.; Calle, E.E. Meat consumption and risk of colorectal cancer. *JAMA* **2005**, *293*, 172–182. [CrossRef]

11. Ma, H.; Qi, X. Red Meat Consumption and Cancer Risk: A Systematic Analysis of Global Data. *Foods* **2023**, *12*, 4164. [CrossRef] [PubMed]
12. Van Blarigan, E.L.; Ou, F.S.; Bainter, T.M.; Fuchs, C.S.; Niedzwiecki, D.; Zhang, S.; Saltz, L.B.; Mayer, R.J.; Hantel, A.; Benson, A.B., 3rd; et al. Associations Between Unprocessed Red Meat and Processed Meat With Risk of Recurrence and Mortality in Patients With Stage III Colon Cancer. *JAMA Netw. Open* **2022**, *5*, e220145. [CrossRef] [PubMed]
13. Bučan Nenadić, D.; Radić, J.; Kolak, E.; Vučković, M.; Novak, I.; Selak, M.; Radić, M. Phase Angle Association with Dietary Habits and Metabolic Syndrome in Diabetic Hypertensive Patients: A Cross-Sectional Study. *Nutrients* **2022**, *14*, 5058. [CrossRef] [PubMed]
14. Cifuentes, M.; Vahid, F.; Devaux, Y.; Bohn, T. Biomarkers of food intake and their relevance to metabolic syndrome. *Food Funct.* **2024**, *15*, 7271–7304. [CrossRef]
15. Kim, Y.; Je, Y. Meat Consumption and Risk of Metabolic Syndrome: Results from the Korean Population and a Meta-Analysis of Observational Studies. *Nutrients* **2018**, *10*, 390. [CrossRef] [PubMed]
16. Chan, D.S.; Lau, R.; Aune, D.; Vieira, R.; Greenwood, D.C.; Kampman, E.; Norat, T. Red and processed meat and colorectal cancer incidence: Meta-analysis of prospective studies. *PLoS ONE* **2011**, *6*, e20456. [CrossRef] [PubMed]
17. Bajracharya, R.; Kaaks, R.; Katzke, V. Food Sources of Animal Protein in Relation to Overall and Cause-Specific Mortality-Causal Associations or Confounding? An Analysis of the EPIC-Heidelberg Cohort. *Nutrients* **2023**, *15*, 3322. [CrossRef] [PubMed]
18. Etemadi, A.; Sinha, R.; Ward, M.H.; Graubard, B.I.; Inoue-Choi, M.; Dawsey, S.M.; Abnet, C.C. Mortality from different causes associated with meat, heme iron, nitrates, and nitrites in the NIH-AARP Diet and Health Study: Population based cohort study. *BMJ* **2017**, *357*, j1957. [CrossRef] [PubMed]
19. Kim, K.; Hyeon, J.; Lee, S.A.; Kwon, S.O.; Lee, H.; Keum, N.; Lee, J.K.; Park, S.M. Role of Total, Red, Processed, and White Meat Consumption in Stroke Incidence and Mortality: A Systematic Review and Meta-Analysis of Prospective Cohort Studies. *J. Am. Heart Assoc.* **2017**, *6*, e005983. [CrossRef] [PubMed]
20. Lupoli, R.; Vitale, M.; Calabrese, I.; Giosuè, A.; Riccardi, G.; Vaccaro, O. White Meat Consumption, All-Cause Mortality, and Cardiovascular Events: A Meta-Analysis of Prospective Cohort Studies. *Nutrients* **2021**, *13*, 676. [CrossRef] [PubMed]
21. López-Otín, C.; Blasco, M.A.; Partridge, L.; Serrano, M.; Kroemer, G. Hallmarks of aging: An expanding universe. *Cell* **2023**, *186*, 243–278. [CrossRef] [PubMed]
22. Regulski, M.J. Cellular Senescence: What, Why, and How. *Wounds* **2017**, *29*, 168–174. [PubMed]
23. Bao, H.; Cao, J.; Chen, M.; Chen, M.; Chen, W.; Chen, X.; Chen, Y.; Chen, Y.; Chen, Y.; Chen, Z.; et al. Biomarkers of aging. *Sci. China Life Sci.* **2023**, *66*, 893–1066. [CrossRef] [PubMed]
24. Moqri, M.; Herzog, C.; Poganik, J.R.; Justice, J.; Belsky, D.W.; Higgins-Chen, A.; Moskalev, A.; Fuellen, G.; Cohen, A.A.; Bautmans, I.; et al. Biomarkers of aging for the identification and evaluation of longevity interventions. *Cell* **2023**, *186*, 3758–3775. [CrossRef]
25. Carey, A.; Niedernhofer, L.; Camell, C. Telomeres are a life-extending gift. *Nat. Cell Biol.* **2022**, *24*, 1449–1450. [CrossRef] [PubMed]
26. Chakravarti, D.; LaBella, K.A.; DePinho, R.A. Telomeres: History, health, and hallmarks of aging. *Cell* **2021**, *184*, 306–322. [CrossRef]
27. Borghini, A.; Ndreu, R.; Canale, P.; Campolo, J.; Marinaro, I.; Mercuri, A.; Turchi, S.; Andreassi, M.G. Telomere Length, Mitochondrial DNA, and Micronucleus Yield in Response to Oxidative Stress in Peripheral Blood Mononuclear Cells. *Int. J. Mol. Sci.* **2024**, *25*, 1428. [CrossRef] [PubMed]
28. Fu, M.; Wang, C.; Hong, S.; Guan, X.; Meng, H.; Feng, Y.; Xiao, Y.; Zhou, Y.; Liu, C.; Zhong, G.; et al. Multiple metals exposure and blood mitochondrial DNA copy number: A cross-sectional study from the Dongfeng-Tongji cohort. *Environ. Res.* **2023**, *216*, 114509. [CrossRef]
29. Li, R.; Chen, G.; Liu, X.; Pan, M.; Kang, N.; Hou, X.; Liao, W.; Dong, X.; Yuchi, Y.; Mao, Z.; et al. Aging biomarkers: Potential mediators of association between long-term ozone exposure and risk of atherosclerosis. *J. Intern. Med.* **2022**, *292*, 512–522. [CrossRef] [PubMed]
30. Field, A.E.; Robertson, N.A.; Wang, T.; Havas, A.; Ideker, T.; Adams, P.D. DNA Methylation Clocks in Aging: Categories, Causes, and Consequences. *Mol. Cell* **2018**, *71*, 882–895. [CrossRef] [PubMed]
31. Ming, X.; Zhang, Z.; Zou, Z.; Lv, C.; Dong, Q.; He, Q.; Yi, Y.; Li, Y.; Wang, H.; Zhu, B. Kinetics and mechanisms of mitotic inheritance of DNA methylation and their roles in aging-associated methylome deterioration. *Cell Res.* **2020**, *30*, 980–996. [CrossRef] [PubMed]
32. Xu, H.; Li, S.; Liu, Y.S. Roles and Mechanisms of DNA Methylation in Vascular Aging and Related Diseases. *Front. Cell Dev. Biol.* **2021**, *9*, 699374. [CrossRef] [PubMed]
33. Srinivas, N.; Rachakonda, S.; Kumar, R. Telomeres and Telomere Length: A General Overview. *Cancers* **2020**, *12*, 558. [CrossRef] [PubMed]
34. Prasad, K.N.; Wu, M.; Bondy, S.C. Telomere shortening during aging: Attenuation by antioxidants and anti-inflammatory agents. *Mech. Ageing Dev.* **2017**, *164*, 61–66. [CrossRef] [PubMed]
35. Codd, V.; Wang, Q.; Allara, E.; Musicha, C.; Kaptoge, S.; Stoma, S.; Jiang, T.; Hamby, S.E.; Braund, P.S.; Bountziouka, V.; et al. Polygenic basis and biomedical consequences of telomere length variation. *Nat. Genet.* **2021**, *53*, 1425–1433. [CrossRef] [PubMed]
36. Ashar, F.N.; Zhang, Y.; Longchamps, R.J.; Lane, J.; Moes, A.; Grove, M.L.; Mychaleckyj, J.C.; Taylor, K.D.; Coresh, J.; Rotter, J.I.; et al. Association of Mitochondrial DNA Copy Number With Cardiovascular Disease. *JAMA Cardiol.* **2017**, *2*, 1247–1255. [CrossRef] [PubMed]

37. Park, S.J.; Gavrilova, O.; Brown, A.L.; Soto, J.E.; Bremner, S.; Kim, J.; Xu, X.; Yang, S.; Um, J.H.; Koch, L.G.; et al. DNA-PK Promotes the Mitochondrial, Metabolic, and Physical Decline that Occurs During Aging. *Cell Metab.* **2017**, *25*, 1135–1146.e1137. [CrossRef]
38. Yan, X.; Yang, P.; Li, Y.; Liu, T.; Zha, Y.; Wang, T.; Zhang, J.; Feng, Z.; Li, M. New insights from bidirectional Mendelian randomization: Causal relationships between telomere length and mitochondrial DNA copy number in aging biomarkers. *Aging* **2024**, *16*, 7387–7404. [CrossRef] [PubMed]
39. Cancello, R.; Rey, F.; Carelli, S.; Cattaldo, S.; Fontana, J.M.; Goitre, I.; Ponzo, V.; Merlo, F.D.; Zuccotti, G.; Bertoli, S.; et al. Telomere Length and Mitochondrial DNA Copy Number Variations in Patients with Obesity: Effect of Diet-Induced Weight Loss-A Pilot Study. *Nutrients* **2022**, *14*, 4293. [CrossRef] [PubMed]
40. Humphreys, K.L.; Sisk, L.M.; Manczak, E.M.; Lin, J.; Gotlib, I.H. Depressive Symptoms Predict Change in Telomere Length and Mitochondrial DNA Copy Number Across Adolescence. *J. Am. Acad. Child Adolesc. Psychiatry* **2020**, *59*, 1364–1370.e1362. [CrossRef] [PubMed]
41. Liu, C.; Chen, Y.J.; Sun, B.; Chen, H.G.; Mustieles, V.; Messerlian, C.; Sun, Y.; Meng, T.Q.; Lu, W.Q.; Pan, X.F.; et al. Blood trihalomethane concentrations in relation to sperm mitochondrial DNA copy number and telomere length among 958 healthy men. *Environ. Res* **2023**, *216*, 114737. [CrossRef]
42. Vostatek, R.; Hohensinner, P.; Nopp, S.; Haider, P.; Englisch, C.; Pointner, J.; Pabinger, I.; Ay, C. Association of telomere length and mitochondrial DNA copy number, two biomarkers of biological aging, with the risk of venous thromboembolism. *Thromb. Res.* **2023**, *223*, 168–173. [CrossRef] [PubMed]
43. Becerra, C.Y.; Wells, R.K.; Kunihiro, B.P.; Lee, R.H.; Umeda, L.; Allan, N.P.; Rubas, N.C.; McCracken, T.A.; Nunokawa, C.K.L.; Lee, M.H.; et al. Examining the immunoepigenetic-gut microbiome axis in the context of self-esteem among Native Hawaiians and other Pacific Islanders. *Front. Genet.* **2023**, *14*, 1125217. [CrossRef] [PubMed]
44. Chiavellini, P.; Canatelli-Mallat, M.; Lehmann, M.; Gallardo, M.D.; Herenu, C.B.; Cordeiro, J.L.; Clement, J.; Goya, R.G. Aging and rejuvenation—A modular epigenome model. *Aging* **2021**, *13*, 4734–4746. [CrossRef] [PubMed]
45. Hernández Cordero, A.I.; Yang, C.X.; Yang, J.; Horvath, S.; Shaipanich, T.; MacIsaac, J.; Lin, D.T.S.; Kobor, M.S.; Guillemi, S.; Harris, M.; et al. Airway Aging and Methylation Disruptions in HIV-associated Chronic Obstructive Pulmonary Disease. *Am. J. Respir. Crit. Care Med.* **2022**, *206*, 150–160. [CrossRef] [PubMed]
46. McCartney, D.L.; Min, J.L.; Richmond, R.C.; Lu, A.T.; Sobczyk, M.K.; Davies, G.; Broer, L.; Guo, X.; Jeong, A.; Jung, J.; et al. Genome-wide association studies identify 137 genetic loci for DNA methylation biomarkers of aging. *Genome Biol.* **2021**, *22*, 194. [CrossRef] [PubMed]
47. Murach, K.A.; Dimet-Wiley, A.L.; Wen, Y.; Brightwell, C.R.; Latham, C.M.; Dungan, C.M.; Fry, C.S.; Watowich, S.J. Late-life exercise mitigates skeletal muscle epigenetic aging. *Aging Cell* **2022**, *21*, e13527. [CrossRef] [PubMed]
48. Shi, W.; Gao, X.; Cao, Y.; Chen, Y.; Cui, Q.; Deng, F.; Yang, B.; Lin, E.Z.; Fang, J.; Li, T.; et al. Personal airborne chemical exposure and epigenetic ageing biomarkers in healthy Chinese elderly individuals: Evidence from mixture approaches. *Environ. Int.* **2022**, *170*, 107614. [CrossRef] [PubMed]
49. Andreo-López, M.C.; Contreras-Bolívar, V.; Muñoz-Torres, M.; García-Fontana, B.; García-Fontana, C. Influence of the Mediterranean Diet on Healthy Aging. *Int. J. Mol. Sci.* **2023**, *24*, 4491. [CrossRef] [PubMed]
50. Díaz-Hernández, V.; Montaño, L.M.; Caldelas, I.; Marmolejo-Valencia, A. A High-Fat and High-Carbohydrate Diet Promotes Reminiscent Hallmarks of an Aging Ovary in the Rabbit Model. *Biomedicines* **2022**, *10*, 3068. [CrossRef] [PubMed]
51. Grande de França, N.A.; Rolland, Y.; Guyonnet, S.; de Souto Barreto, P. The role of dietary strategies in the modulation of hallmarks of aging. *Ageing Res. Rev.* **2023**, *87*, 101908. [CrossRef]
52. Longo, V.D.; Anderson, R.M. Nutrition, longevity and disease: From molecular mechanisms to interventions. *Cell* **2022**, *185*, 1455–1470. [CrossRef] [PubMed]
53. Thomas, D.C.; Conti, D.V. Commentary: The concept of 'Mendelian Randomization'. *Int. J. Epidemiol.* **2004**, *33*, 21–25. [CrossRef]
54. Gurung, R.L.; Dorajoo, R.; Liu, S.; M, Y.; Liu, J.J.; Wang, L.; Guo, L.; Yu, X.; Liu, J.J.; Lim, S.C. Genetic markers for urine haptoglobin is associated with decline in renal function in type 2 diabetes in East Asians. *Sci. Rep.* **2018**, *8*, 5109. [CrossRef] [PubMed]
55. Emdin, C.A.; Khera, A.V.; Kathiresan, S. Mendelian Randomization. *JAMA* **2017**, *318*, 1925–1926. [CrossRef] [PubMed]
56. Huang, P.; Gu, Q.; Hu, Y.; Li, H.; Wu, Z.; Liu, W.; Zhu, Z.; Yuan, P.; Duan, L.; Zhou, Y.; et al. Genetic Analysis of a Collection of Rice Germplasm (*Oryza sativa* L.) through High-Density SNP Array Provides Useful Information for Further Breeding Practices. *Genes* **2022**, *13*, 830. [CrossRef] [PubMed]
57. Burgess, S.; Small, D.S.; Thompson, S.G. A review of instrumental variable estimators for Mendelian randomization. *Stat. Methods Med. Res.* **2017**, *26*, 2333–2355. [CrossRef] [PubMed]
58. Skrivankova, V.W.; Richmond, R.C.; Woolf, B.A.R.; Yarmolinsky, J.; Davies, N.M.; Swanson, S.A.; VanderWeele, T.J.; Higgins, J.P.T.; Timpson, N.J.; Dimou, N.; et al. Strengthening the Reporting of Observational Studies in Epidemiology Using Mendelian Randomization: The STROBE-MR Statement. *JAMA* **2021**, *326*, 1614–1621. [CrossRef] [PubMed]
59. Pirastu, N.; McDonnell, C.; Grzeszkowiak, E.J.; Mounier, N.; Imamura, F.; Merino, J.; Day, F.R.; Zheng, J.; Taba, N.; Concas, M.P.; et al. Using genetic variation to disentangle the complex relationship between food intake and health outcomes. *PLoS Genet.* **2022**, *18*, e1010162. [CrossRef] [PubMed]
60. Yun, Z.; Nan, M.; Li, X.; Liu, Z.; Xu, J.; Du, X.; Dong, Q.; Hou, L. Processed meat, red meat, white meat, and digestive tract cancers: A two-sample Mendelian randomization study. *Front. Nutr.* **2023**, *10*, 1078963. [CrossRef] [PubMed]

61. Allaire, P.; He, J.; Mayer, J.; Moat, L.; Gerstenberger, P.; Wilhorn, R.; Strutz, S.; Kim, D.S.L.; Zeng, C.; Cox, N.; et al. Genetic and clinical determinants of telomere length. *HGG Adv.* **2023**, *4*, 100201. [CrossRef] [PubMed]
62. Chong, M.; Mohammadi-Shemirani, P.; Perrot, N.; Nelson, W.; Morton, R.; Narula, S.; Lali, R.; Khan, I.; Khan, M.; Judge, C.; et al. GWAS and ExWAS of blood mitochondrial DNA copy number identifies 71 loci and highlights a potential causal role in dementia. *eLife* **2022**, *11*, e70382. [CrossRef] [PubMed]
63. Glymour, M.M.; Tchetgen Tchetgen, E.J.; Robins, J.M. Credible Mendelian randomization studies: Approaches for evaluating the instrumental variable assumptions. *Am. J. Epidemiol.* **2012**, *175*, 332–339. [CrossRef] [PubMed]
64. Chen, D.; Wu, H.; Wang, X.; Huang, T.; Jia, J. Shared Genetic Basis and Causal Relationship Between Television Watching, Breakfast Skipping and Type 2 Diabetes: Evidence From a Comprehensive Genetic Analysis. *Front. Endocrinol.* **2022**, *13*, 836023. [CrossRef]
65. Hemani, G.; Zheng, J.; Elsworth, B.; Wade, K.H.; Haberland, V.; Baird, D.; Laurin, C.; Burgess, S.; Bowden, J.; Langdon, R.; et al. The MR-Base platform supports systematic causal inference across the human phenome. *eLife* **2018**, *7*, e34408. [CrossRef] [PubMed]
66. Byrska-Bishop, M.; Evani, U.S.; Zhao, X.; Basile, A.O.; Abel, H.J.; Regier, A.A.; Corvelo, A.; Clarke, W.E.; Musunuri, R.; Nagulapalli, K.; et al. High-coverage whole-genome sequencing of the expanded 1000 Genomes Project cohort including 602 trios. *Cell* **2022**, *185*, 3426–3440.e3419. [CrossRef] [PubMed]
67. Burgess, S.; Thompson, S.G. Avoiding bias from weak instruments in Mendelian randomization studies. *Int. J. Epidemiol.* **2011**, *40*, 755–764. [CrossRef] [PubMed]
68. Liu, D.; Gao, X.; Pan, X.F.; Zhou, T.; Zhu, C.; Li, F.; Fan, J.G.; Targher, G.; Zhao, J. The hepato-ovarian axis: Genetic evidence for a causal association between non-alcoholic fatty liver disease and polycystic ovary syndrome. *BMC Med.* **2023**, *21*, 62. [CrossRef] [PubMed]
69. Lin, S.-H.; Brown, D.W.; Machiela, M.J. LDtrait: An Online Tool for Identifying Published Phenotype Associations in Linkage Disequilibrium. *Cancer Res.* **2020**, *80*, 3443–3446. [CrossRef] [PubMed]
70. Bowden, J.; Spiller, W.; Del Greco M, F.; Sheehan, N.; Thompson, J.; Minelli, C.; Davey Smith, G. Improving the visualization, interpretation and analysis of two-sample summary data Mendelian randomization via the Radial plot and Radial regression. *Int. J. Epidemiol.* **2018**, *47*, 1264–1278. [CrossRef]
71. Verbanck, M.; Chen, C.Y.; Neale, B.; Do, R. Detection of widespread horizontal pleiotropy in causal relationships inferred from Mendelian randomization between complex traits and diseases. *Nat. Genet.* **2018**, *50*, 693–698. [CrossRef]
72. Wu, P.F.; Lu, H.; Zhou, X.; Liang, X.; Li, R.; Zhang, W.; Li, D.; Xia, K. Assessment of causal effects of physical activity on neurodegenerative diseases: A Mendelian randomization study. *J. Sport Health Sci.* **2021**, *10*, 454–461. [CrossRef] [PubMed]
73. Burgess, S.; Thompson, S.G. Interpreting findings from Mendelian randomization using the MR-Egger method. *Eur. J. Epidemiol.* **2017**, *32*, 377–389. [CrossRef] [PubMed]
74. Bowden, J.; Davey Smith, G.; Haycock, P.C.; Burgess, S. Consistent Estimation in Mendelian Randomization with Some Invalid Instruments Using a Weighted Median Estimator. *Genet. Epidemiol.* **2016**, *40*, 304–314. [CrossRef] [PubMed]
75. Burgess, S.; Butterworth, A.; Thompson, S.G. Mendelian randomization analysis with multiple genetic variants using summarized data. *Genet. Epidemiol.* **2013**, *37*, 658–665. [CrossRef] [PubMed]
76. Morrison, J.; Knoblauch, N.; Marcus, J.H.; Stephens, M.; He, X. Mendelian randomization accounting for correlated and uncorrelated pleiotropic effects using genome-wide summary statistics. *Nat. Genet.* **2020**, *52*, 740–747. [CrossRef] [PubMed]
77. Burgess, S.; Foley, C.N.; Allara, E.; Staley, J.R.; Howson, J.M.M. A robust and efficient method for Mendelian randomization with hundreds of genetic variants. *Nat. Commun.* **2020**, *11*, 376. [CrossRef] [PubMed]
78. Burgess, S.; Davey Smith, G.; Davies, N.M.; Dudbridge, F.; Gill, D.; Glymour, M.M.; Hartwig, F.P.; Kutalik, Z.; Holmes, M.V.; Minelli, C.; et al. Guidelines for performing Mendelian randomization investigations: Update for summer 2023. *Wellcome Open Res.* **2019**, *4*, 186. [CrossRef] [PubMed]
79. Burgess, S.; Dudbridge, F.; Thompson, S.G. Combining information on multiple instrumental variables in Mendelian randomization: Comparison of allele score and summarized data methods. *Stat. Med.* **2016**, *35*, 1880–1906. [CrossRef] [PubMed]
80. VanderWeele, T.J. Mediation Analysis: A Practitioner's Guide. *Annu. Rev. Public Health* **2016**, *37*, 17–32. [CrossRef] [PubMed]
81. Burgess, S.; Bowden, J.; Fall, T.; Ingelsson, E.; Thompson, S.G. Sensitivity Analyses for Robust Causal Inference from Mendelian Randomization Analyses with Multiple Genetic Variants. *Epidemiology* **2017**, *28*, 30–42. [PubMed]
82. Bowden, J.; Davey Smith, G.; Burgess, S. Mendelian randomization with invalid instruments: Effect estimation and bias detection through Egger regression. *Int. J. Epidemiol.* **2015**, *44*, 512–525. [CrossRef] [PubMed]
83. Bowden, J.; Del Greco, M.F.; Minelli, C.; Davey Smith, G.; Sheehan, N.; Thompson, J. A framework for the investigation of pleiotropy in two-sample summary data Mendelian randomization. *Stat. Med.* **2017**, *36*, 1783–1802. [CrossRef] [PubMed]
84. Von Hinke Kessler Scholder, S.; Smith, G.D.; Lawlor, D.A.; Propper, C.; Windmeijer, F. Mendelian randomization: The use of genes in instrumental variable analyses. *Health Econ.* **2011**, *20*, 893–896. [CrossRef] [PubMed]
85. Pingault, J.B.; O'Reilly, P.F.; Schoeler, T.; Ploubidis, G.B.; Rijsdijk, F.; Dudbridge, F. Using genetic data to strengthen causal inference in observational research. *Nat. Rev. Genet.* **2018**, *19*, 566–580. [CrossRef]
86. Van Kippersluis, H.; Rietveld, C.A. Pleiotropy-robust Mendelian randomization. *Int. J. Epidemiol.* **2018**, *47*, 1279–1288. [CrossRef]
87. Rees, J.M.B.; Wood, A.M.; Dudbridge, F.; Burgess, S. Robust methods in Mendelian randomization via penalization of heterogeneous causal estimates. *PLoS ONE* **2019**, *14*, e0222362. [CrossRef] [PubMed]

88. Feskens, E.J.; Sluik, D.; van Woudenbergh, G.J. Meat consumption, diabetes, and its complications. *Curr. Diabetes Rep.* **2013**, *13*, 298–306. [CrossRef]
89. Laskowski, W.; Górska-Warsewicz, H.; Kulykovets, O. Meat, Meat Products and Seafood as Sources of Energy and Nutrients in the Average Polish Diet. *Nutrients* **2018**, *10*, 1412. [CrossRef] [PubMed]
90. Paik, J.M.; Mir, S.; Alqahtani, S.A.; Younossi, Y.; Ong, J.P.; Younossi, Z.M. Dietary Risks for Liver Mortality in NAFLD: Global Burden of Disease Data. *Hepatol. Commun.* **2022**, *6*, 90–100. [CrossRef] [PubMed]
91. Ali, Q.; Ma, S.; Farooq, U.; Niu, J.; Li, F.; Li, D.; Wang, Z.; Sun, H.; Cui, Y.; Shi, Y. Pasture intake protects against commercial diet-induced lipopolysaccharide production facilitated by gut microbiota through activating intestinal alkaline phosphatase enzyme in meat geese. *Front. Immunol.* **2022**, *13*, 1041070. [CrossRef] [PubMed]
92. Neuffer, J.; González-Domínguez, R.; Lefèvre-Arbogast, S.; Low, D.Y.; Driollet, B.; Helmer, C.; Du Preez, A.; de Lucia, C.; Ruigrok, S.R.; Altendorfer, B.; et al. Exploration of the Gut-Brain Axis through Metabolomics Identifies Serum Propionic Acid Associated with Higher Cognitive Decline in Older Persons. *Nutrients* **2022**, *14*, 4688. [CrossRef] [PubMed]
93. Ong, J.S.; Lew, L.C.; Hor, Y.Y.; Liong, M.T. Probiotics: The Next Dietary Strategy against Brain Aging. *Prev. Nutr. Food Sci.* **2022**, *27*, 1–13. [CrossRef] [PubMed]
94. Yaskolka Meir, A.; Keller, M.; Hoffmann, A.; Rinott, E.; Tsaban, G.; Kaplan, A.; Zelicha, H.; Hagemann, T.; Ceglarek, U.; Isermann, B.; et al. The effect of polyphenols on DNA methylation-assessed biological age attenuation: The DIRECT PLUS randomized controlled trial. *BMC Med.* **2023**, *21*, 364. [CrossRef] [PubMed]
95. Calder, P.C.; Ortega, E.F.; Meydani, S.N.; Adkins, Y.; Stephensen, C.B.; Thompson, B.; Zwickey, H. Nutrition, Immunosenescence, and Infectious Disease: An Overview of the Scientific Evidence on Micronutrients and on Modulation of the Gut Microbiota. *Adv. Nutr.* **2022**, *13*, S1–S26. [CrossRef] [PubMed]
96. Wang, Y.; Du, W.; Hu, X.; Yu, X.; Guo, C.; Jin, X.; Wang, W. Targeting the blood-brain barrier to delay aging-accompanied neurological diseases by modulating gut microbiota, circadian rhythms, and their interplays. *Acta Pharm. Sin. B* **2023**, *13*, 4667–4687. [CrossRef] [PubMed]
97. Warman, D.J.; Jia, H.; Kato, H. The Potential Roles of Probiotics, Resistant Starch, and Resistant Proteins in Ameliorating Inflammation during Aging (Inflammaging). *Nutrients* **2022**, *14*, 747. [CrossRef] [PubMed]
98. Yu, L.; Pan, J.; Guo, M.; Duan, H.; Zhang, H.; Narbad, A.; Zhai, Q.; Tian, F.; Chen, W. Gut microbiota and anti-aging: Focusing on spermidine. *Crit. Rev. Food Sci. Nutr.* **2023**, 1–19, *Online ahead of print*. [CrossRef]
99. Zeng, X.; Li, X.; Li, X.; Wei, C.; Shi, C.; Hu, K.; Kong, D.; Luo, Q.; Xu, Y.; Shan, W.; et al. Fecal microbiota transplantation from young mice rejuvenates aged hematopoietic stem cells by suppressing inflammation. *Blood* **2023**, *141*, 1691–1707. [CrossRef] [PubMed]
100. Lee, S.Y.; Lee, D.Y.; Kang, J.H.; Kim, J.H.; Jeong, J.W.; Kim, H.W.; Oh, D.H.; Yoon, S.H.; Hur, S.J. Relationship between gut microbiota and colorectal cancer: Probiotics as a potential strategy for prevention. *Food Res. Int.* **2022**, *156*, 111327. [CrossRef] [PubMed]
101. Liang, Y.; Zhang, H.; Tian, L.; Shi, C.; Zheng, Y.; Wang, J.; Tan, Y.; Luo, Y.; Hong, H. Gut microbiota and metabolic profile as affected by Maillard reaction products derived from bighead carp meat hydrolysates with galactose and galacto-oligosaccharides during in vitro pig fecal fermentation. *Food Chem.* **2023**, *398*, 133905. [CrossRef] [PubMed]
102. Lv, G.; Wang, H.; Wei, X.; Lu, M.; Yang, W.; Aalim, H.; Capanoglu, E.; Zou, X.; Battino, M.; Zhang, D. Cooking-Induced Oxidation and Structural Changes in Chicken Protein: Their Impact on In Vitro Gastrointestinal Digestion and Intestinal Flora Fermentation Characteristics. *Foods* **2023**, *12*, 4322. [CrossRef] [PubMed]
103. De La Pomélie, D.; Santé-Lhoutellier, V.; Gatellier, P. Mechanisms and kinetics of heme iron nitrosylation in an in vitro gastro-intestinal model. *Food Chem.* **2018**, *239*, 86–93. [CrossRef] [PubMed]
104. Zhang, Y.; Zhang, Y.; Jia, J.; Peng, H.; Qian, Q.; Pan, Z.; Liu, D. Nitrite and nitrate in meat processing: Functions and alternatives. *Curr. Res. Food Sci.* **2023**, *6*, 100470. [CrossRef] [PubMed]
105. James, K.L.; Gertz, E.R.; Cervantes, E.; Bonnel, E.L.; Stephensen, C.B.; Kable, M.E.; Bennett, B.J. Diet, Fecal Microbiome, and Trimethylamine N-Oxide in a Cohort of Metabolically Healthy United States Adults. *Nutrients* **2022**, *14*, 1376. [CrossRef] [PubMed]
106. Fu, B.C.; Hullar, M.A.J.; Randolph, T.W.; Franke, A.A.; Monroe, K.R.; Cheng, I.; Wilkens, L.R.; Shepherd, J.A.; Madeleine, M.M.; Le Marchand, L.; et al. Associations of plasma trimethylamine N-oxide, choline, carnitine, and betaine with inflammatory and cardiometabolic risk biomarkers and the fecal microbiome in the Multiethnic Cohort Adiposity Phenotype Study. *Am. J. Clin. Nutr.* **2020**, *111*, 1226–1234. [CrossRef] [PubMed]
107. Stubbs, J.R.; House, J.A.; Ocque, A.J.; Zhang, S.; Johnson, C.; Kimber, C.; Schmidt, K.; Gupta, A.; Wetmore, J.B.; Nolin, T.D.; et al. Serum Trimethylamine-N-Oxide is Elevated in CKD and Correlates with Coronary Atherosclerosis Burden. *J. Am. Soc. Nephrol. JASN* **2016**, *27*, 305–313. [CrossRef] [PubMed]
108. Kronenberg, F. Telomere length and chronic kidney disease: Cause or consequence? *Kidney Int.* **2021**, *100*, 980–983. [CrossRef]
109. Hernández-Martín, M.; Macho-González, A.; Garcimartín, A.; López-Oliva, M.E.; Bocanegra, A.; Redondo-Castillejo, R.; Bastida, S.; Benedí, J.; Sánchez-Muniz, F.J. Silicon-enriched meat positively improves plasma lipidaemia and lipoproteinaemia, LDLr, and insulin capability and the signalling pathway induced by an atherogenic diet in late-stage type 2 diabetes mellitus rats. *Food Funct.* **2024**, *15*, 1513–1526. [CrossRef] [PubMed]

110. Reynolds, C.M.; Toomey, S.; McBride, R.; McMonagle, J.; Morine, M.J.; Belton, O.; Moloney, A.P.; Roche, H.M. Divergent effects of a CLA-enriched beef diet on metabolic health in ApoE-/- and ob/ob mice. *J. Nutr. Biochem.* **2013**, *24*, 401–411. [CrossRef] [PubMed]
111. Feinman, R.D. Red meat and type 2 diabetes mellitus. *JAMA Intern. Med.* **2014**, *174*, 646. [CrossRef] [PubMed]
112. Kremer, D. Red meat and type 2 diabetes: Do methodological advancements sufficiently increase the certainty of evidence? *Am. J. Clin. Nutr.* **2023**, *118*, 1075–1076. [CrossRef]
113. Lescinsky, H.; Afshin, A.; Ashbaugh, C.; Bisignano, C.; Brauer, M.; Ferrara, G.; Hay, S.I.; He, J.; Iannucci, V.; Marczak, L.B.; et al. Health effects associated with consumption of unprocessed red meat: A Burden of Proof study. *Nat. Med.* **2022**, *28*, 2075–2082. [CrossRef] [PubMed]
114. Al-Daghri, N.M.; Abdi, S.; Sabico, S.; Alnaami, A.M.; Wani, K.A.; Ansari, M.G.A.; Khattak, M.N.K.; Khan, N.; Tripathi, G.; Chrousos, G.P.; et al. Gut-Derived Endotoxin and Telomere Length Attrition in Adults with and without Type 2 Diabetes. *Biomolecules* **2021**, *11*, 1693. [CrossRef] [PubMed]
115. Demircan, K.; Hybsier, S.; Chillon, T.S.; Vetter, V.M.; Rijntjes, E.; Demuth, I.; Schomburg, L. Sex-specific associations of serum selenium and selenoprotein P with type 2 diabetes mellitus and hypertension in the Berlin Aging Study II. *Redox Biol.* **2023**, *65*, 102823. [CrossRef] [PubMed]
116. Induri, S.N.R.; Kansara, P.; Thomas, S.C.; Xu, F.; Saxena, D.; Li, X. The Gut Microbiome, Metformin, and Aging. *Annu. Rev. Pharmacol. Toxicol.* **2022**, *62*, 85–108. [CrossRef] [PubMed]
117. Zhao, Y.; Yue, R. Aging adipose tissue, insulin resistance, and type 2 diabetes. *Biogerontology* **2024**, *25*, 53–69. [CrossRef] [PubMed]
118. Ruggeri, R.M.; Giovinazzo, S.; Barbalace, M.C.; Cristani, M.; Alibrandi, A.; Vicchio, T.M.; Giuffrida, G.; Aguennouz, M.H.; Malaguti, M.; Angeloni, C.; et al. Influence of Dietary Habits on Oxidative Stress Markers in Hashimoto's Thyroiditis. *Thyroid Off. J. Am. Thyroid Assoc.* **2021**, *31*, 96–105. [CrossRef] [PubMed]
119. Singh, R.B.; Fedacko, J.; Pella, D.; Fatima, G.; Elkilany, G.; Moshiri, M.; Hristova, K.; Jakabcin, P.; Vaňova, N. High Exogenous Antioxidant, Restorative Treatment (Heart) for Prevention of the Six Stages of Heart Failure: The Heart Diet. *Antioxidants* **2022**, *11*, 1464. [CrossRef]
120. Zelber-Sagi, S.; Ivancovsky-Wajcman, D.; Fliss Isakov, N.; Webb, M.; Orenstein, D.; Shibolet, O.; Kariv, R. High red and processed meat consumption is associated with non-alcoholic fatty liver disease and insulin resistance. *J. Hepatol.* **2018**, *68*, 1239–1246. [CrossRef] [PubMed]
121. Armstrong, E.; Boonekamp, J. Does oxidative stress shorten telomeres in vivo? A meta-analysis. *Ageing Res. Rev.* **2023**, *85*, 101854. [CrossRef] [PubMed]
122. De Vos-Houben, J.M.; Ottenheim, N.R.; Kafatos, A.; Buijsse, B.; Hageman, G.J.; Kromhout, D.; Giltay, E.J. Telomere length, oxidative stress, and antioxidant status in elderly men in Zutphen and Crete. *Mech. Ageing Dev.* **2012**, *133*, 373–377. [CrossRef] [PubMed]
123. Jitjumnong, M.; Chalermkitpanit, P.; Suantawee, T.; Dechsupa, S.; Vajarintarangoon, L.; Honsawek, S. Telomere Shortening and Increased Oxidative Stress in Lumbar Disc Degeneration. *Int. J. Mol. Sci.* **2022**, *23*, 10125. [CrossRef] [PubMed]
124. Zhou, M.; Zhu, L.; Cui, X.; Feng, L.; Zhao, X.; He, S.; Ping, F.; Li, W.; Li, Y. Influence of diet on leukocyte telomere length, markers of inflammation and oxidative stress in individuals with varied glucose tolerance: A Chinese population study. *Nutr. J.* **2016**, *15*, 39. [CrossRef] [PubMed]
125. Gupta, K.; Testa, H.; Greenwood, T.; Kostek, M.; Haushalter, K.; Kris-Etherton, P.M.; Petersen, K.S. The effect of herbs and spices on risk factors for cardiometabolic diseases: A review of human clinical trials. *Nutr. Rev.* **2022**, *80*, 400–427. [CrossRef] [PubMed]
126. Hong, S.; Nagayach, A.; Lu, Y.; Peng, H.; Duong, Q.A.; Pham, N.B.; Vuong, C.A.; Bazan, N.G. A high fat, sugar, and salt Western diet induces motor-muscular and sensory dysfunctions and neurodegeneration in mice during aging: Ameliorative action of metformin. *CNS Neurosci. Ther.* **2021**, *27*, 1458–1471. [CrossRef] [PubMed]
127. Lastuvkova, H.; Faradonbeh, F.A.; Schreiberova, J.; Hroch, M.; Mokry, J.; Faistova, H.; Nova, Z.; Hyspler, R.; Igreja Sa, I.C.; Nachtigal, P.; et al. Atorvastatin Modulates Bile Acid Homeostasis in Mice with Diet-Induced Nonalcoholic Steatohepatitis. *Int. J. Mol. Sci.* **2021**, *22*, 6468. [CrossRef] [PubMed]
128. Jurk, D.; Wilson, C.; Passos, J.F.; Oakley, F.; Correia-Melo, C.; Greaves, L.; Saretzki, G.; Fox, C.; Lawless, C.; Anderson, R.; et al. Chronic inflammation induces telomere dysfunction and accelerates ageing in mice. *Nat Commun* **2014**, *2*, 4172. [CrossRef] [PubMed]
129. Reljic, D.; Koller, A.; Herrmann, H.J.; Ekici, A.B.; Neurath, M.F.; Zopf, Y. Differential Effects of Very-Low-Volume Exercise Modalities on Telomere Length, Inflammation, and Cardiometabolic Health in Obese Metabolic Syndrome Patients: A Subanalysis from Two Randomized Controlled Trials. *Antioxidants* **2023**, *12*, 1847. [CrossRef] [PubMed]
130. Zhang, J.; Rane, G.; Dai, X.; Shanmugam, M.K.; Arfuso, F.; Samy, R.P.; Lai, M.K.; Kappei, D.; Kumar, A.P.; Sethi, G. Ageing and the telomere connection: An intimate relationship with inflammation. *Ageing Res. Rev.* **2016**, *25*, 55–69. [CrossRef] [PubMed]
131. Pfanner, N.; Warscheid, B.; Wiedemann, N. Mitochondrial proteins: From biogenesis to functional networks. *Nat. Rev. Mol. Cell Biol.* **2019**, *20*, 267–284. [CrossRef] [PubMed]
132. Charlebois, E.; Pantopoulos, K. Nutritional Aspects of Iron in Health and Disease. *Nutrients* **2023**, *15*, 2441. [CrossRef] [PubMed]
133. Cabrera, M.C.; Saadoun, A. An overview of the nutritional value of beef and lamb meat from South America. *Meat Sci.* **2014**, *98*, 435–444. [CrossRef] [PubMed]

134. Hassan, A.A.; Sandanger, T.M.; Brustad, M. Selected vitamins and essential elements in meat from semi-domesticated reindeer (*Rangifer tarandus tarandus* L.) in mid- and northern Norway: Geographical variations and effect of animal population density. *Nutrients* **2012**, *4*, 724–739. [CrossRef] [PubMed]
135. Vahid, F.; Hatami, M.; Sadeghi, M.; Ameri, F.; Faghfoori, Z.; Davoodi, S.H. The association between the Index of Nutritional Quality (INQ) and breast cancer and the evaluation of nutrient intake of breast cancer patients: A case-control study. *Nutrition* **2018**, *45*, 11–16. [CrossRef] [PubMed]
136. Feng, Z.; Wang, Y.; Fu, Z.; Liao, J.; Liu, H.; Zhou, M. Exploring the Causal Effects of Mineral Metabolism Disorders on Telomere and Mitochondrial DNA: A Bidirectional Two-Sample Mendelian Randomization Analysis. *Nutrients* **2024**, *16*, 1417. [CrossRef] [PubMed]
137. Gong, S.; Yin, Y.; Han, M.; Guo, L.; Duan, Y.; Guo, Q.; Yin, J.; Li, F. Dietary leucine and fish oil cooperatively regulate skeletal myofiber type transformation via the CaMKII signaling pathway of pigs. *Food Funct.* **2023**, *14*, 133–147. [CrossRef] [PubMed]
138. Hu, J.; Kyrou, I.; Tan, B.K.; Dimitriadis, G.K.; Ramanjaneya, M.; Tripathi, G.; Patel, V.; James, S.; Kawan, M.; Chen, J.; et al. Short-Chain Fatty Acid Acetate Stimulates Adipogenesis and Mitochondrial Biogenesis via GPR43 in Brown Adipocytes. *Endocrinology* **2016**, *157*, 1881–1894. [CrossRef] [PubMed]
139. Mancusso, R.; Gregorio, G.G.; Liu, Q.; Wang, D.N. Structure and mechanism of a bacterial sodium-dependent dicarboxylate transporter. *Nature* **2012**, *491*, 622–626. [CrossRef]
140. Polak-Szczybyło, E.; Tabarkiewicz, J. IL-17A, IL-17E and IL-17F as Potential Biomarkers for the Intensity of Low-Grade Inflammation and the Risk of Cardiovascular Diseases in Obese People. *Nutrients* **2022**, *14*, 643. [CrossRef]

nutrients

MDPI

Article

Association of Methyl Donor Nutrients' Dietary Intake and Cognitive Impairment in the Elderly Based on the Intestinal Microbiome

Qianqian Chen [1,2,†], Rui Fan [1,2,†], Lixia Song [1,2], Shuyue Wang [1,2], Mei You [1,2], Meng Cai [1,2], Yuxiao Wu [1,2], Yong Li [1,2] and Meihong Xu [1,2,*]

1 Department of Nutrition and Food Hygiene, School of Public Health, Peking University, Beijing 100191, China; 2211210120@stu.pku.edu.cn (Q.C.); fanruirf@bjmu.edu.cn (R.F.)
2 Beijing Key Laboratory of Toxicological Research and Risk Assessment for Food Safety, Peking University, Beijing 100191, China
* Correspondence: xumeihong@bjmu.edu.cn; Tel.: +86-010-8280-5243
† These authors contributed equally to this work.

Abstract: Globally, cognitive impairment (CI) is the leading cause of disability and dependency among the elderly, presenting a significant public health concern. However, there is currently a deficiency in pharmacological interventions that can effectively cure or significantly reverse the progression of cognitive impairment. Methyl donor nutrients (MDNs), including folic acid, choline, and vitamin B12, have been identified as potential enhancers of cognitive function. Nevertheless, there remains a dearth of comprehensive research investigating the connection between the dietary intake of MDNs and CI. In our study, we comprehensively assessed the relationship between MDNs' dietary intake and CI in older adults, utilizing 16S rRNA gene sequencing to investigate the potential underlying mechanisms. The results showed an obvious difference in the methyl-donor nutritional quality index (MNQI) between the dementia (D) group and the dementia-free (DF) group. Specifically, there was a lower MNQI in the D group than that in the DF group. For the gut microbiome, the beta diversity of gut flora exhibited higher levels in the high methyl-donor nutritional quality (HQ) group as opposed to the low methyl-donor nutritional quality (LQ) group, and lower levels in the D group in comparison to the DF group. Subsequently, we performed a correlation analysis to examine the relationship between the relative abundance of microbiota, the intake of MDNs, and Montreal Cognitive Assessment (MoCA) scores, ultimately identifying ten genera with potential regulatory functions. Additionally, KEGG pathway analyses suggested that the one-carbon metabolism, chronic inflammation, and DNA synthesis potentially serve as pathways through which MDNs may be promising for influencing cognitive function. These results implied that MDNs might have the potential to enhance cognitive function through the regulation of microbiota homeostasis. This study offers dietary recommendations for the prevention and management of CI in the elderly.

Keywords: cognitive impairment; gut microbiome; one-carbon metabolism; nutrition; aging

1. Introduction

It is expected that dementia prevalence will rise as the global population ages, with projections indicating a tripling in the number of individuals living with this condition by the year 2050 from the estimated 47 million in 2015 [1]. Cognitive impairment (CI) includes both mild cognitive impairment (MCI) and dementia, the latter characterized by a more pronounced deterioration in cognitive abilities that hinders daily functioning and social engagement [2]. Long-term caring for individuals with CI is a significant burden on patients, their families, and society. The estimated global cost of dementia in 2015 was USD 818 billion, a figure that is expected to rise as the prevalence of dementia increases [3]. This underscores the growing urgency of dementia as a significant global public health concern.

Historically, dementia has been regarded as an incurable disease. However, emerging evidence suggests that a significant proportion of dementia cases may be preventable [4]. MCI represents a transitional stage between normal cognitive function and dementia, often serving as an early indicator of the latter [5]. Consequently, early intervention has emerged as a promising approach for the prevention and management of dementia. Several studies have provided evidence suggesting that dietary patterns may confer protective benefits against cognitive decline, particularly in individuals with MCI who undergo intervention [6]. Furthermore, lower serum folate levels [7,8] and higher homocysteine levels [9,10] have been linked to an increased risk of conversion from any type of MCI to all-cause dementia, a relationship that is associated with one-carbon metabolism.

One-carbon metabolism (OCM) has been identified as playing a significant role in the initiation and progression of CI [11,12]. OCM is a complex system of biochemical reactions that supply methyl donors with diverse biosynthetic pathways, supporting various physiological functions in the brain such as DNA synthesis, neurotransmitter synthesis, epigenetic regulation, and antioxidant defense mechanisms. Additionally, OCM serves as a pivotal point connecting multiple pathways [13]. The consumption of methyl donor nutrients (MDNs) such as protein, folate, choline, betaine, vitamin B2 (VB_2), VB_6, VB_{12}, and zinc plays a crucial role in OCM [14,15]. Inadequacies in any of these nutrients can interfere with the intricate regulatory system responsible for maintaining the OCM, resulting in impairments in brain function [16]. Recent studies have consistently reported the potential therapeutic effects of MDNs in CI. B vitamins, particularly folate, VB_6, and VB_{12}, play a role in homocysteine metabolism and can decrease the risk of CI [17,18]. Additionally, choline supplementation has been shown to elevate brain acetylcholine levels and suppress neuroinflammation, leading to enhancements in learning and memory [19]. Betaine has been found to inhibit the overactivation of hippocampal microglia and decrease oxidative stress, thereby serving as a preventive measure against CI [20]. A deep understanding of the correlation between MDNs and CI may provide valuable insights into effective dietary strategies for preventing and managing dementia. Nevertheless, there is a lack of research that has thoroughly investigated the relationship between MDNs and CI.

The gut microbiota, a crucial element of the gut–brain axis, play a significant role in modulating cognitive functions through various mechanisms such as modulating neurotransmitter synthesis and metabolism, as well as regulating inflammatory and immune responses [21]. Recent research has increasingly demonstrated a correlation between gut microbial imbalance and the onset of CI [22]. While a definitive standard for gut microbiota homeostasis remains elusive, individuals with chronic illnesses often exhibit an elevated presence of pro-inflammatory bacteria and a diminished presence of beneficial bacteria. Patients with Alzheimer's disease (AD), for example, commonly display dysbiosis in their gut microbiota, characterized by the reduced abundance and diversity of flora, heightened levels of pro-inflammatory bacteria, and an elevated Firmicutes/Bacteroidetes ratio [21,23,24]. Prior research has indicated that diets deficient in folate and VB_{12} may lead to a reduction in B vitamin-producing bacteria and an increase in inflammation-associated bacteria [25]. Thus, we posited that integrated MDNs may exert an influence on the gut microbiota, thereby potentially impacting CI.

To illustrate the relationship between dietary MDNs and CI in the elderly, and if dietary MDNs may affect CI, we systematically assessed the association between the dietary intake of MDNs and CI in the elderly, and explored the potential mechanisms based on the intestinal microbiota. This study has the potential to propose intervention strategies aimed at preventing CI in the elderly through a nutritional lens, as well as offering novel insights for the targeted manipulation of the intestinal microbiota.

2. Methods

2.1. Study Population

A cohort of 301 older adults aged 60–70 years was enrolled in the TALENTs trial (Targeting Aging and Longevity with Exogenous Nucleotides), a pragmatic four-month prospective single-center randomized controlled trial. Research protocols have been reported [26]. Participants in the study were administered the Montreal Cognitive Assessment (MoCA) to assess cognitive health, with a MoCA score of 18 or lower indicating membership in the dementia (D) group, while those scoring above 18 were categorized as belonging to the dementia-free (DF) group [27]. Demographic characteristics and medical history pertaining to chronic diseases were obtained through the administration of a questionnaire. Subsequently, fecal samples from all participants were collected, processed, and preserved at a temperature of −80 °C for future analysis. A total of 290 valid fecal samples were acquired, corresponding to the inclusion of 290 individuals in the ultimate analysis. All participants provided written consent by signing an informed consent form. The TALENTs trial has been approved by Peking University's Biomedical Ethics Committee (IRB00001052-21114) and registered by Clincal Trials.gov (NCT05243018).

2.2. *Assessment of Methyl Donor Nutrients Dietary Intake*

The Photo-Assisted Dietary Intake Assessment (PAD) method was utilized to gather dietary intake data for MDNs. Previous research has demonstrated the accuracy and feasibility of the PAD method, particularly in its application to elderly populations [28]. Participants underwent standardized training and received real-time professional guidance. Subsequently, photographs of participants' meals were taken before and after consumption over a period of three consecutive days to assess intake of both regular and additional foods. Subsequently, we determined the mean daily consumption levels of protein, folate, choline, riboflavin, VB_6, VB_{12}, betaine, and zinc. Subsequently, we calculated the methyl-donor nutritional quality index (MNQI) by considering the intake of these eight MDNs in accordance with the Chinese Dietary Reference Intakes (version 2023), in order to comprehensively evaluate the dietary intake of MDNs [29]. The methodology utilized in this study was derived from a prior research investigation that demonstrated a notable association between the MNQI and OCM, thus validating the MNQI as a reliable instrument for assessing the nutritional quality of dietary MDNs sources in a comprehensive manner. As per the findings of the study, an MNQI score equal to or exceeding 6 is classified as high methyl-donor nutritional quality (HQ), while a score below 6 indicates low methyl-donor nutritional quality (LQ) [29].

2.3. *Analysis of Intestinal Flora Test*

2.3.1. 16S rRNA Gene Sequencing Analysis

We extracted fecal microbial DNA using the MGIEasy fecal genomic DNA (meta) extraction kit (BGI, Shenzhen, China). Subsequently, we used the primers PF5′-ACTCCTACG GGGAGGCAGCAG-3′ and PR5′-GGACTACNNGGGTATCTAAT-3′ on the polymerase chain reaction (PCR) amplification of the bacterial V3–V4 highly variable region of the 16SrRNA gene. The PCR amplification products were purified and dissolved in Elution Buffer using Agencourt AMPure XP magnetic beads (Beckman Coulter, UK) and labeled to complete the library construction. The fragment range and concentration of the libraries were detected using an Agilent 2100 Bioanalyzer (Agilent, CA, USA). The qualified libraries were sequenced on the MGISEQ-2000 platform (BGI, Shenzhen, China) according to the size of the inserted fragments and the paired ends were sequenced. The raw data underwent filtering and splicing through the utilization of FLASH (Fast Length Adjustment of Short reads, v1.2.11), which facilitated the assembly of paired reads derived from double-end sequencing into a unified sequence based on overlap relationships to acquire Tags within the high-variance region. Subsequently, USEARCH (v7.0.1090) was employed to amalgamate the spliced Tags, which were then clustered at a 97% sequence similarity threshold to produce Operational Taxonomic Units (OTUs), with chimeric sequences being eliminated through the application of UCHIME (v4.2.40). After obtaining the OTU representative sequences, the OTU representative sequences were compared with the Ribosomal Database Project (RDP, http://rdp.cme.msu.edu/, accessed on 20 April 2024) for species annotation by RDP classifier (v2.2) software, with the confidence threshold set to 0.6.

2.3.2. Bioinformatics Analysis

In order to study species diversity within one sample, Alpha Diversity analysis was performed, and The Alpha Diversity Index at the OTU level was calculated using mothur software (v1.31.2), including the Chao1 index, ACE index, and Shannon index. An analysis of the beta diversity was performed to compare the diversity of species between the samples, and the 16S rRNA gene beta diversity was analyzed using QIIME (v1.80) software. A predictive functional analysis of microbiota was performed using PICRUSt2 (v2.3.0-b) software to relate species to their functions.

Intergroup Venn diagrams were plotted to illustrate the overlap of OTUs between groups using the VennDiagram package in R (v3.4.1). The beta diversity was expressed as unweighted and weighted UniFrac distances and tested for intergroup differences using ANOSIM with the QIIME (v1.80) software. In addition, Partial Least Squares Discriminant Analysis (PLSDA) was performed using the mixOmics package to assess differences in the beta diversity. The pheatmap package facilitated the correlation of heatmap clusters while the Wilcoxon rank sum test was also utilized to determine differences in microbial abundance among taxa. Differences in microbial abundance were used to identify important species or functions.

2.3.3. Statistical Analysis

The values of the numerical variables are given as the mean + standard deviation, and categorical variables are given as percentages. Nonparametric tests (including Wilcoxon rank-sum tests and Mann–Whitney U tests), independent sample *t*-tests, and Pearson's chi-square tests were conducted to assess the differences between groups. Correlations between the numerical variables were analyzed using Spearman's correlation analysis. We performed all statistical analyses using R (v3.4.1). Statistical significance was defined as p less than 0.05.

3. Results

3.1. *Demographic Data for D Group and DF Group*

The distribution of the general demographic characteristics for the D and DF groups was examined and no significant differences were found in terms of gender, age, education level, living alone or not, or body mass index (BMI), and the history of chronic diseases between the two groups is shown in Table 1.

Table 1. Demographic characteristics between dementia (D) and dementia-free (DF) groups.

Characteristics	D (n = 45)	DF (n = 245)	p-Value
Gender, male (%)	13 (28.9)	74 (30.2)	0.860
Age (years) x ± s	65.44 ± 2.72	65.67 ± 2.66	0.604
Highest educational level (n) %			
Primary school or below	4 (8.9)	18 (7.3)	0.958
Junior high or above	41 (91.1)	227 (92.7)	
Living alone or not (n) %			
Yes	6 (13.3)	19 (7.8)	0.245
No	39 (86.7)	226 (92.2)	
BMI (n) %			
Normal	16 (35.6)	119 (48.6)	0.108
Overweight or obese	29 (64.4)	126 (51.4)	
Chronic diseases history (n)%			
Detected	9 (20)	68 (27.8)	0.279
Not detected	36 (80)	177 (72.2)	

The differences in gender, highest educational level, living alone or not, BMI, and chronic disease history between groups were analyzed by Pearson's chi-square test. The differences in age were examined by an independent samples *t*-test.

3.2. *Correlation between Dietary Intake of MDNs and Cognitive Function*

The distribution of the MNQI was examined in both the D and DF groups. Table 2 illustrated the disparities in the median dietary intake of MNQI between the D and DF groups. Compared to the DF group, the D group scored significantly lower on the MNQI ($p < 0.05$), with notable differences observed in the protein, choline, riboflavin, and zinc intake compared to the DF group ($p < 0.05$). Furthermore, there was a noticeable decrease in the intake of folate, VB_6, VB_{12}, and betaine among participants in the D group. This suggests a potential correlation between the intake of methyl-donor nutrients (MDNs) and cognitive function, with a higher MDN intake potentially benefitting cognitive function.

Table 2. Differences in MDNs intake between dementia (D) and dementia-free (DF) groups.

	D (n = 45)	DF (n = 245)	p-Value
MNQI	5.84 ± 3.13	6.86 ± 3.21	0.030 *
Protein	59.38 ± 25.40	71.64 ± 36.97	0.033 *
Folate	319.53 ± 194.05	375.93 ± 246.79	0.093
Choline	287.36 ± 123.06	327.51 ± 141.19	0.016 *
Riboflavin	0.88 ± 0.47	1.01 ± 0.52	0.017 *
VB_6	2.02 ± 0.86	2.17 ± 1.08	0.342
VB_{12}	1.80 ± 1.21	2.05 ± 2.28	0.595
Betaine	73.86 ± 63.31	75.44 ± 68.96	0.922
Zinc	7.31 ± 3.99	8.71 ± 4.94	0.028 *

* $p < 0.05$, according to the Mann–Whitney U test.

3.3. *Potential Role of Intestinal Flora in the Association between MDNs and Cognitive Function*

3.3.1. Differences in Intestinal Flora between Groups D and DF

To determine whether dietary MDNs affect cognitive function by the intestinal flora, a comparison of the microbial composition of groups D and DF was conducted. The findings revealed a total of 1652 operational taxonomic units (OTUs) in the D group and 1968 OTUs in the DF group, with 1596 OTUs shared between both groups (Figure 1A). For the phylum level, the composition of the intestinal flora in both groups included 28 phyla, with Firmicutes, Bacteroidetes, Proteobacteria, Actinobacteria, Fusobacteria, and Verrucomicrobia being the dominant phyla. However, there were variations in the distribution of these phyla between the two groups. In comparison with the DF group, the D group showed a higher ratio of Firmicutes to Bacteroidetes (Figure 1B).

Furthermore, an evaluation was conducted to compare the gut flora diversity between the two groups. The alpha diversity of the flora was assessed utilizing Chao1, ACE, and Shannon indices. D and DF groups showed no statistically significant variances in alpha diversity ($p > 0.05$, Figure 2A–C). The beta diversity index was utilized to assess the variation in species abundance distribution between the groups. The findings indicated a notable divergence in the composition of intestinal flora between the D and DF groups (Figure 2E,F), with the D group exhibiting lower beta diversity compared to the DF group ($p < 0.05$, Figure 2D,E). Therefore, there is a notable discrepancy in the abundance distribution of intestinal flora between the two groups. Specifically, patients with dementia exhibit decreased diversity in their intestinal flora, meaning specific alterations in the distribution of abundance of certain flora.

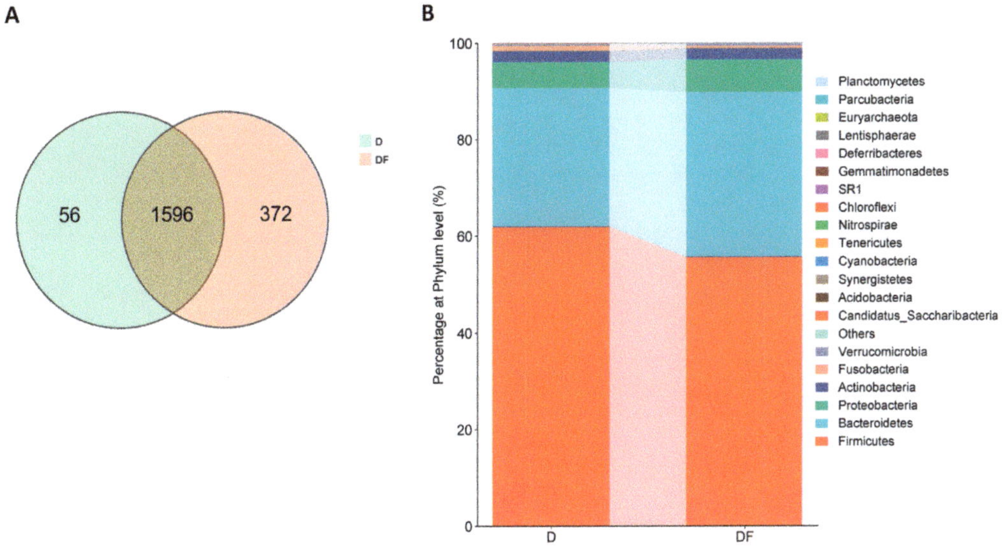

Figure 1. Distribution of identified intestinal flora in dementia (D) and dementia-free (DF) groups. (**A**) Venn diagram showing the distribution of OTUs between D and DF group. (**B**) Intestinal flora composition of the two groups at the phylum level.

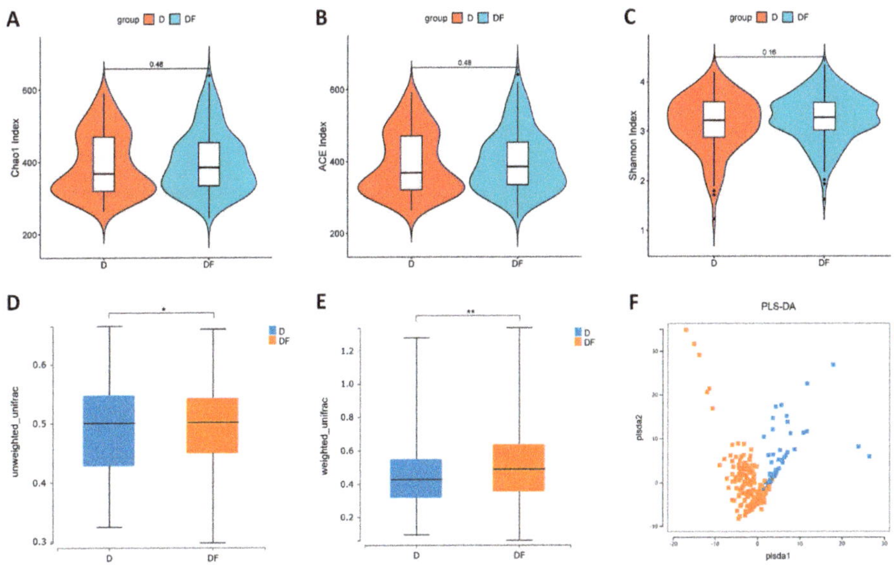

Figure 2. Comparisons of intestinal flora diversity between dementia (D) and dementia−free (DF) groups. (**A**) Chao1 index, (**B**) ACE index, and (**C**) Shannon index for assessing alpha diversity. *p*−value indicated differential clustering assessed by Wilcoxon rank−sum test. The beta diversity based on (**D**) unweighted and (**E**) weighted UniFrac distances analysis between the two groups. (**F**) PLS−DA analysis of intestinal flora between the two groups. *p*-value indicated differential clustering assessed by ADONIS test. * $p < 0.05$, ** $p < 0.001$.

3.3.2. Relationship between MDNs' Intake and Intestinal Flora

In this study, we employed beta diversity indices to assess the impact of dietary MDNs on the intestinal flora composition. The results demonstrated a significant difference in the distribution of intestinal flora abundance between the LQ and HQ groups, with the LQ group exhibiting lower beta diversity compared to the HQ group ($p < 0.001$, Figure 3A). Furthermore, the findings indicate that altering the abundance distribution of gut flora may have a close relationship to cognitive function, particularly given the lower beta diversity observed in the D group.

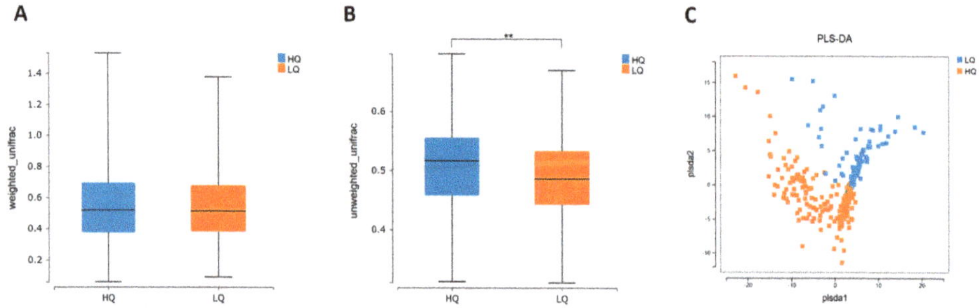

Figure 3. Comparisons of intestinal flora diversity between the LQ and HQ groups. The beta diversity based on (**A**) unweighted and (**B**) weighted UniFrac distances analysis between HQ and LQ group. (**C**) PLS−DA analysis of intestinal flora between HQ and LQ group. ** $p < 0.001$.

Subsequently, in order to investigate the relationship between the specific gut microbiota and MDNs, the relative abundance of gut microbiota was analyzed at the genus level. We then conducted a preliminary screening to identify gut microbiota influenced by micronutrient-dense nutrients, based on the presence of statistically significant differences in the correlation between the genera and MNQI or MDNs' intake. Given the necessity of incorporating a greater number of potential flora in the screening phase to facilitate a broader selection for subsequent comprehensive investigations, it was determined that a statistical divergence in flora screening occurred at a significance level of $p < 0.1$. Figure 4 illustrates the initial screening of 115 candidate genera, which were subsequently categorized into two clusters based on their correlation with MDNs' intake. Specifically, 85 genera exhibited a positive correlation with MDNs' intake, while 30 genera showed a negative correlation. Among these genera, Gp16, Sphingomonas, Gp6, and Allobaculum exhibited significant positive correlations with the intake of MDNs, particularly demonstrating strong associations with the MNQI and folate intake. Conversely, genera such as Clostridium_XI and Maribacter displayed notable negative correlations with the MNQI. Furthermore, a consistent trend was observed where the MNQI and MDNs' intake influenced the relative abundance of these genera, either up-regulating or down-regulating their presence.

3.3.3. Screening for Specific Genera That May Mediate the Association of MDNs with Cognitive Function

In order to investigate potential genera that may play a role in the relationship between MDNs and dementia, we conducted an analysis of the correlation between MoCA scores and the relative abundance of the 115 genera initially identified. It was determined that a statistical divergence at a significance level of p less than 0.1 would determine statistical differences during the screening phase for the aforementioned reasons. Table 3 displayed the screening results of 10 potential genera, of which 7 were identified as having cognitive benefits. These genera, including Victivallis, Turicibacter, Phascolarctobacterium, Snodgrassella, Terrimonas, Planomicrobium, and Centipeda, exhibited a positive correlation in relative abundance with both MoCA scores and the intake of MDNs. Additionally, MoCA scores and the intake of MDNs were negatively correlated with the relative abundance of three cognitively harmful genera (Porphyromonas, Peptoniphilus, and Howardella). These findings imply that MDNs may have a significant impact on cognition through these genera. Thus, the study implied that dietary MDNs may improve cognitive function to some extent.

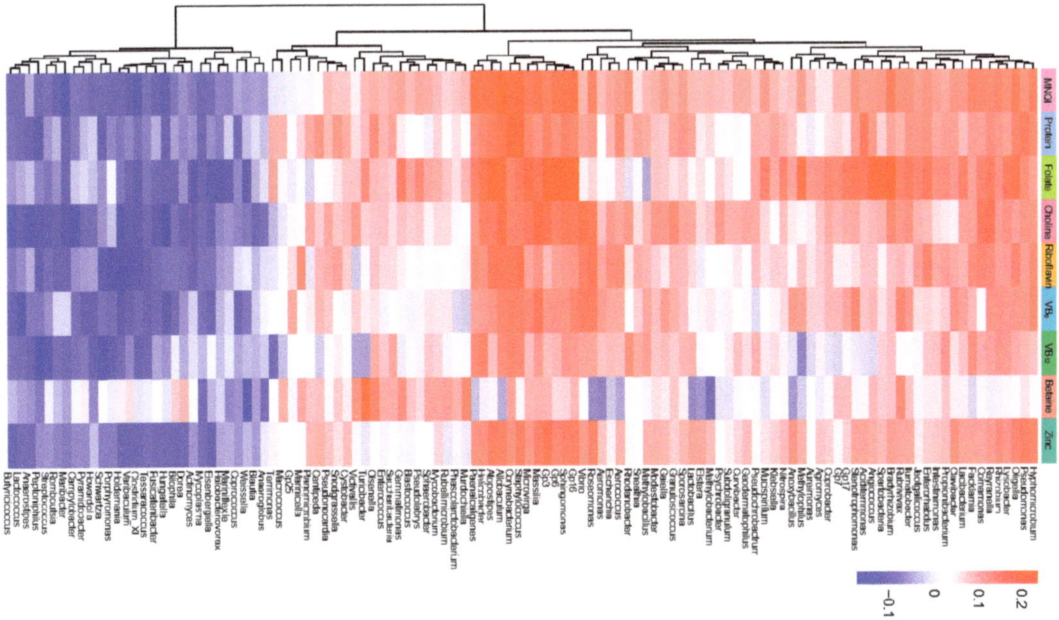

Figure 4. Heatmap clustering plot of correlations between the relative abundance of the initially screened genera and MDNs' intake. Red: correlation is positive, blue: correlation is negative.

Table 3. Correlations between the relative abundance of screened genera and MoCA, MNQI, and MDNs' intake.

CLR-Transformed Abundance		vs. Quality Score or Nutrient Intake									
		MoCA	MNQI	Protein	Folate	Choline	Riboflavin	VB_6	VB_{12}	Betaine	Zinc
Victivallis	r	0.127	0.048	0.060	0.074	0.116	−0.023	0.070	−0.047	0.130	0.018
	p value	0.030 *	0.410	0.312	0.208	0.048 *	0.696	0.236	0.422	0.028 *	0.766
Turicibacter	r	0.100	0.086	0.084	0.132	0.068	0.019	0.049	−0.047	0.237	−0.001
	p value	0.088 *	0.144	0.152	0.024 *	0.248	0.744	0.404	0.420	<0.001 *	0.982
Porphyromonas	r	−0.172	−0.092	−0.061	0.025	−0.019	−0.123	−0.079	−0.114	0.011	−0.071
	p value	0.004 *	0.118	0.304	0.668	0.744	0.036 *	0.182	0.052 *	0.848	0.228
Peptoniphilus	r	−0.103	−0.094	−0.122	−0.066	−0.116	−0.110	−0.074	−0.173	−0.073	−0.060
	p value	0.082 *	0.108	0.038 *	0.266	0.048 *	0.060 *	0.212	0.004 *	0.214	0.308
Howardella	r	−0.103	−0.094	−0.017	−0.025	−0.085	−0.043	−0.101	−0.108	−0.066	−0.027
	p value	0.080 *	0.110	0.778	0.666	0.150	0.464	0.086 *	0.068 *	0.264	0.646
Phascolarctobacterium	r	0.102	0.075	0.060	0.060	0.019	0.037	−0.015	0.038	0.097	0.013
	p value	0.082 *	0.200	0.306	0.308	0.744	0.526	0.794	0.524	0.098 *	0.820
Snodgrassella	r	0.135	0.076	0.120	0.055	0.077	0.106	0.031	0.049	0.075	0.058
	p value	0.022 *	0.200	0.040 *	0.348	0.190	0.072 *	0.596	0.404	0.204	0.324
Terrimonas	r	0.126	−0.001	0.094	0.106	−0.035	−0.003	0.031	−0.102	0.036	0.046
	p value	0.032 *	0.986	0.110	0.070 *	0.550	0.954	0.600	0.084 *	0.544	0.432
Planomicrobium	r	0.109	0.014	0.105	−0.019	0.104	−0.003	0.051	0.039	0.080	0.091
	p value	0.064 *	0.810	0.076 *	0.746	0.078 *	0.960	0.390	0.512	0.174	0.122

Table 3. *Cont.*

CLR-Transformed Abundance		vs. Quality Score or Nutrient Intake									
		MoCA	MNQI	Protein	Folate	Choline	Riboflavin	VB_6	VB_{12}	Betaine	Zinc
Centipeda	r	0.098	0.035	0.124	0.039	0.070	0.088	0.090	−0.024	0.082	0.097
	p value	0.098 *	0.548	0.034 *	0.512	0.234	0.132	0.126	0.680	0.166	0.100 *

* $p < 0.1$, according to the Spearman rank correlation analysis test.

3.3.4. Predicting Possible Mechanisms by which the Screened Genera Play a Role in Mediating the Association of MDNs with Cognitive Function

Finally, we performed PICRUSt functional prediction analysis on the obtained 16S rRNA sequences to determine the enrichment of functional genes in the KEGG pathway, in order to speculate the possible mechanisms by which the screened intestinal flora play a role in mediating the effects of MDNs on dementia. Due to the limited sample size, $p < 0.1$ was considered statistically different. The findings indicated a significantly higher relative abundance of genes associated with 17 pathways, such as Sulfur metabolism, Phenylalanine metabolism, Biotin metabolism, beta-Alanine metabolism, Glyoxylate and dicarboxylate metabolism, and Glutathione metabolism, in the D group compared to the DF group. The pathways that were predominantly enriched among the up-regulated genes included Biotin metabolism, Folate biosynthesis, Selenocompound metabolism, Lipopolysaccharide biosynthesis, and Sulfur metabolism. In comparison to the DF group, the D group exhibited a significantly lower relative abundance of genes involved in 21 pathways, such as Peptidoglycan biosynthesis, RNA polymerase, Lysine biosynthesis, Ribosome, Aminoacyl-tRNA biosynthesis, D-Glutamine and D-glutamate metabolism, Fatty acid biosynthesis, and Nucleotide excision repair. Among the down-regulated pathways, Valine, leucine and isoleucine biosynthesis, D-Glutamine and D-glutamate metabolism, Peptidoglycan biosynthesis, Mismatch repair, and Lysine biosynthesis were mainly enriched (Figure 5).

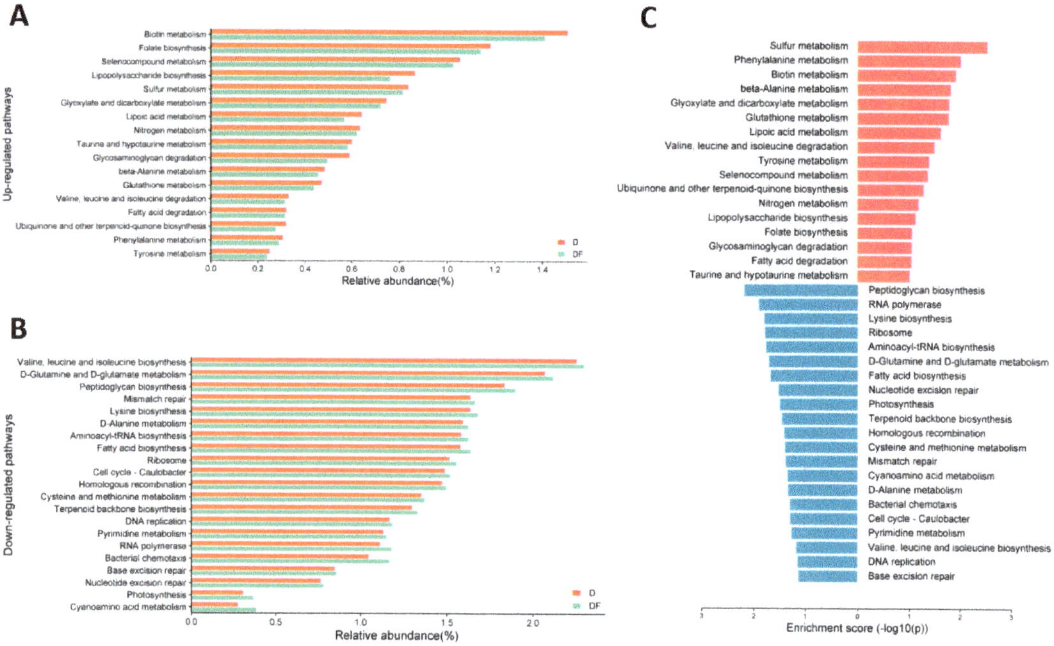

Figure 5. Significant differences between the D and DF groups at the third level of the KEGG pathway, $p < 0.1$. Compared with the DF group, the relative abundance of (**A**) up−regulated pathways and

(**B**) down−regulated pathways of the KEGG pathway's functional genes in D group. (**C**) Differences in the enrichment level of the KEGG pathway's functional genes. Up−regulated functional genes in the D group compared to the DF group are labeled by red and down−regulated by blue.

4. Discussion

OCM plays a crucial role in preserving brain health. The inadequate consumption of dietary folate and VB_{12} has been linked to disturbances in OCM, leading to the dysregulation of the methylation pathway, increased oxidative stress, and enhanced protein deposition, ultimately hastening the development and advancement of CI [11]. Through interactions within the OCM pathway, methyl donors regulate metabolism, immune responses, and epigenetic events in animals [30]. Consequently, it is imperative to thoroughly investigate the association between the overall intake of MDNs and health outcomes. This study represents the first comprehensive assessment of the relationship between the MDNs' intake and cognitive function. The findings revealed a significant correlation between cognitive function and dietary MDNs in older adults, suggesting that the intake of these nutrients may have a positive impact on the cognitive health status.

There is growing evidence linking imbalances in gut microbiota to the onset of neurodegenerative diseases [22], including AD [31], Parkinson's disease [32], and multiple sclerosis [33]. Individuals with dementia have been observed to display dysbiosis in their gut microbiota, characterized by diminished abundance and diversity of flora [34]. Compared with those without dementia, individuals with dementia had significantly less gut flora beta diversity, with distinct differences in the flora composition and abundance between the two groups, consistent with previous research [35]. Additionally, dietary factors can influence the composition and function of gut microbiota, impacting the metabolic, immune, and neurological functions of the host through metabolically active substances that may contribute to the etiology and pathogenesis of CI [36–38]. Moreover, it suggested that diets enriched with MDNs may alter the intestinal environment and change the composition and distribution of the intestinal flora. Our research revealed a notable disparity in the beta diversity of intestinal flora between individuals with a low versus high MDN intake. Consequently, our findings propose that individuals with CI may exhibit distinct changes in their gut microbiota, and that dietary MDNs may influence cognitive function through the modulation of host gut–brain axis signaling [39].

Subsequently, utilizing correlation analysis, we identified particular genera that exhibited a significant association with both the intake of MDNs and the MoCA scores. Our findings suggest that heightened levels of Victivallis, Turicibacter, Phascolarctobacterium, Snodgrassella, Terrimonas, Planomicrobium, and Centipeda may potentially enhance cognitive function, and that the consumption of substantial quantities of MDNs could potentially elevate their prevalence. On the other hand, the elevated presence of Porphyromonas, Peptoniphilus, and Howardella has been linked to impaired cognitive function, whereas the consumption of substantial quantities of MDNs may reduce their prevalence, consequently enhancing cognitive well-being. Research on the impact of microbes on host health is providing insight into the roles of specific bacterial genera. For instance, a study examining Victivallis, a bacterium believed to have cognitive benefits, discovered that older adults with preclinical AD had lower levels of Victivallis in their gut microbiota compared to those without the disease, particularly in individuals testing positive for amyloid-β (Aβ), a pathological marker of AD [40]. A subsequent cross-sectional study involving individuals aged 70 and older revealed a decrease in Victivallis relative abundance among those with MCI [41]. The findings of this study align with previous research indicating a potential association between Turicibacter, known for its role in heightened bile acid production, and the disruption of bile acid homeostasis in the pathogenesis of AD, suggesting a possible neuroprotective effect of Turicibacter [42,43]. Studies have indicated that the screen-detected pathogenic bacterium Porphyromonas has the ability to colonize the brain and induce neuroinflammation, resulting in elevated levels of Aβ1-42, a protein associated with dementia, particularly in cases of chronic periodontitis [44,45].

Furthermore, Peptoniphilus, an opportunistic pathogen, is frequently identified as the principal causative agent of inflammation resulting from diverse infections, notably chronic wounds like diabetic foot ulcers and chronic compression [46]. This pathogen is also linked to the energy metabolism, leading to the production of the short-chain fatty acid butyrate [47,48]. Also, colonic microbes have the capability to synthesize B vitamins, supply nutrients to both the host and microbiota, and influence immune cell activity, as well as impact the nervous system through the production of neurotransmitters [16]. Thus, based on the flora screened, MDNs may affect cognitive function through a variety of pathways such as the modulation of neuroinflammation and immunity, the maintenance of bile acid homeostasis, the provision of OCM nutrients, energy metabolism, and other effects through the gut microbiota. The full extent of the functions of these flora remains unknown, given the evolving understanding of the gut microbiome in medical contexts. Consequently, the intricate pathways through which gut flora influence brain health have yet to be fully elucidated, necessitating the further exploration of the specific regulatory mechanisms involved.

Hence, in order to seek the potential pathways through which dietary MDNs may impact cognitive function via the gut microbiota, we utilized gut microbiomics for gene function annotation. In our study, it was observed that metabolic pathways associated with OCM, including Folate biosynthesis, Sulfur metabolism, Taurine and hypotaurine metabolism, and Glutathione metabolism, among others, were found to be significantly enriched in the D group. Conversely, the Cysteine and methionine metabolism pathway exhibited significantly lower enrichment in this group. Interestingly, these findings indicate that a deficiency in MDNs may lead to inadequate substrates for crucial reactions in the OCM pathway, subsequently increasing the body's requirement for MDNs and prompting compensatory alterations in OCM-related pathways [49]. Additionally, the heightened oxidative stress and buildup of harmful substances associated with neurodegenerative conditions could amplify the body's response to stress [50,51]. These indicate that the disruption of the complex regulatory network responsible for maintaining the OCM due to a deficiency in any individual MDN contributes to the accelerated progression of CI [16,52].

Additionally, some studies have shown that there are interactions between MDNs (folate and VB_{12}) and the gut microbiota. Specific flora within the gut have been found to contribute significantly to the daily intake of folic acid, as well as various B vitamins including biotin, cobalamin, niacin, pantothenate, pyridoxine, riboflavin, and thiamine [53]. Conversely, diets lacking in folate and VB_{12} over an extended period can result in the reduced production of B vitamins by gut bacteria [25]. Based on the previous finds [25,53], the research indicates that a deficiency in MDNs may lead to a reduction in B vitamin-producing bacteria, which might impaire OCM in brain tissue and subsequent impacts on the nervous system. Notably, the significantly higher gene enrichment of the Lipopolysaccharide biosynthesis pathway was observed in the D group compared to the DF group. Porphyromonas was more abundant in HQ individuals compared to LQ individuals. It has been reported that folic acid supplementation reduces gut inflammation and mitigates oxidative stress and neurotoxicity, ultimately protecting neurons from damage [54]. Thus, it is suggested that high intake levels of MDNs may potentially help to slow down the progression of CI through reducing inflammatory flora, thereby attenuating chronic neuroinflammation, which was accordant with the previous report [54]. Furthermore, our study revealed a decrease in the gene enrichment of six pathways associated with DNA synthesis and repair, specifically Mismatch repair, Homologous recombination, DNA replication, Pyrimidine metabolism, Base excision repair, and Nucleotide excision repair, in the D group as compared to the DF group. Additionally, it is important to note that DNA synthesis and repair processes are intricately linked to OCM [55]. In a state of folate deficiency, the aberrant activation of serine hydroxymethyltransferase 1 (SHMT1) disrupts thymidylate biosynthesis, potentially resulting in cognitive dysfunction [56]. Taken together, the study may provide some clue that the modulation of intestinal flora homeostasis, the maintenance of OCM, the reduction in inflammation and oxidative stress, and the regulation of DNA synthesis and repair by MDNs may have a positive effect on cognitive function (see Figure 6 for illustration). Nonetheless, the intricate interplay among these pathways remains incompletely understood, necessitating additional mechanistic investigations.

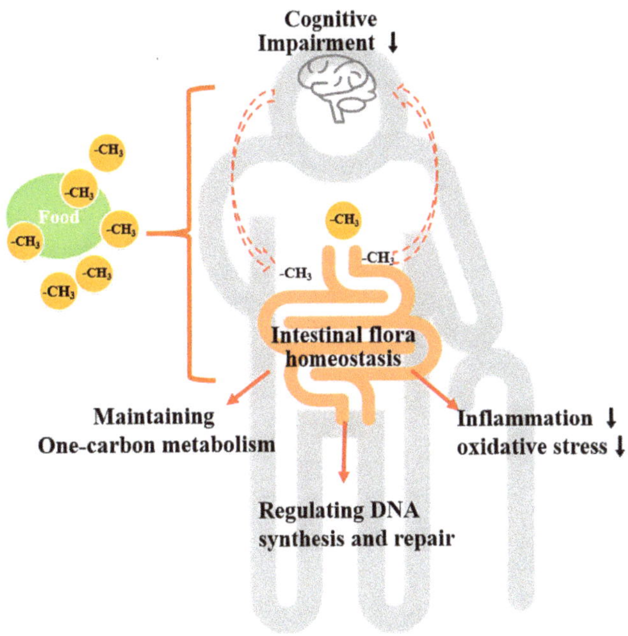

Figure 6. The potential mechanisms of dietary MDNs for cognitive function improvement through gut microbiota.

Furthermore, in order to illustrate the significance of early intervention, an analysis was conducted on the distribution of the MNQI among D, MCI, and normal cognition (NC) populations. The results from the Supplementary Material revealed that over half of the individuals assessed were at risk of developing MCI, indicating a substantial portion of the population at heightened risk for dementia [57]. Moreover, there was no statistically significant difference between the MCI and NC groups in the MNQI, which was reflected as a mildly lower MNQI in the MCI group than the NC group. In contrast, the MNQI of the D group was significantly lower than both the MCI and NC groups. This, in conjunction with the reversible nature of MCI, underscores the significance of early intervention with MDNs during the MCI stage to mitigate the risk of progression to dementia [2]. In addition, our study revealed that the consumption of MDNs among the elderly fell significantly below the recommended levels (see Supplementary Tables S2 and S3). This deficiency of MDNs may be particularly pronounced in the neurological system as digestive and absorptive capabilities weaken with age. Consequently, it is advised that elderly individuals consider taking appropriate preventive measures, such as supplementation with MDNs, to potentially mitigate the decline in cognitive function.

We applied the MNQI for the first time to assess the relationship between the MDNs' intake and cognitive function in older adults and explored the potential mechanisms of the gut microbiome, and screened for specific genera of bacteria, which has rarely been reported in prior studies. The present study initiatively utilized the MNQI for the assessment of the association between the intake of MDNs and cognitive function in elderly individuals. Additionally, the investigation delved into the potential mechanisms involving the gut microbiome and conducted a screening for specific bacterial genera, a topic that has received limited attention in prior research. Nevertheless, this study is constrained by certain limitations. Specifically, the sample size and study design are inadequate, leading to limited statistical validity, particularly in the realms of bacterial screening and functional prediction. Moreover, for community-based studies, this study is unable to obtain conclusive causation and segmented clinical diagnoses or treatment recommendations, as in clinical trial studies. Consequently, the present findings primarily serve as a foundation for future research. To more precisely elucidate the dynamic interplay between the nutrient intake, gut microbiota, and cognitive function, it is advisable to conduct future longitudinal studies with larger sample sizes. Furthermore, the inability to obtain the serum exposure levels of MDNs in the current study due to methodological constraints necessitates careful consideration in future research endeavors. Additionally, it is recommended that upcoming studies integrate multi-omics data to elucidate the effect of the gut microbiota on cognitive function, including its role in processes such as one-carbon metabolism, nutrient metabolism, immune modulation, and DNA synthesis. Ultimately, the gut flora identified through the screening conducted in this study offers preliminary insights for directing gut flora interventions aimed at enhancing cognitive function. It is advisable that future research endeavors encompass gut flora-focused animal or population intervention trials to substantiate their efficacy.

5. Conclusions

In conclusion, our research offers initial findings on the possible relationship between MDNs' consumption and CI in the elderly, highlighting the significant influence of the gut microbiota composition and diversity in this correlation. The study underscores the importance of the early adoption of MDNs to preserve cognitive function in older individuals, and identifies specific gut flora strains that may provide positive help to the cognitive performance. These findings will help us obtain a deep understanding of the intricate relationship between the gut microbiota balance and neurological well-being in the aging population.

Supplementary Materials: The following supporting information can be downloaded at: https://www.mdpi.com/article/10.3390/nu16132061/s1, Figure S1: Differences in MNQI among the D, MCI, and NC groups; Table S1: Demographic characteristics among the D, MCI and NC groups; Table S2: Evaluation of MDNs intake aged 60–64 years; Table S3: Evaluation of MDNs intake aged 65–70 years.

Author Contributions: Q.C. and R.F.: Investigation, Methodology, Writing, Visualization, Validation. L.S.: Investigation, Validation. S.W.: Investigation, Validation. M.Y.: Investigation, Validation. M.C.: Investigation, Validation. Y.W.: Validation. Y.L.: Conceptualization. M.X.: Conceptualization, Funding acquisition, Resources, Supervision, Writing—review and editing. All authors have read and agreed to the published version of the manuscript.

Funding: This study is commissioned and funded by Zhen-Ao Biotechnology Co., Ltd. (Dalian, China). The sponsor provided assistance with subject recruitment and provided medical professionals to participate in the entire health consultation process, but did not participate in the design, methodology, analysis, or paper writing. And this research is partially supported by Beijing Natural Science Foundation (Project Number: 7242186).

Institutional Review Board Statement: The study was conducted in accordance with the Declaration of Helsinki, and approved by Peking University's Biomedical Ethics Committee (approved No. IRB00001052-21114) and date of approval (17 November 2021).

Informed Consent Statement: Informed consent was obtained from all subjects involved in the study.

Data Availability Statement: The original contributions presented in the study are included in the article/supplementary material. Due to specific legal restrictions, the data are not publicly available, and further inquiries can be directed to the corresponding author.

Acknowledgments: The authors are grateful to Mengqi Li, Jichen Zou, Maochong Lin, Dan Zhang, and Lili Chang for their assistance during the investigation.

Conflicts of Interest: The authors declare that this study received funding from Zhen-Ao Biotechnology Co., Ltd. The funder was not involved in the study design, collection, analysis, interpretation of data, the writing of this article or the decision to submit it for publication.

Abbreviations

OCM	one-carbon metabolism
D	dementia
DF	dementia-free
MCI	mild cognitive impairment
AD	Alzheimer's disease
MNQI	methyl-donor nutritional quality index
MDNs	methyl donor nutrients
PAD	photo-assisted dietary intake assessment
HQ	high methyl-donor nutritional quality
LQ	low methyl-donor nutritional quality
MoCA	Montreal Cognitive Assessment
OTUs	operational taxonomic units
BMI	body mass index
KEGG	Kyoto Encyclopedia of Genes and Genomes
rRNA	ribosomal ribonucleic acid
VB	vitamin B
PCR	polymerase chain reaction
PLS-DA	partial least squares discrimination analysis
Aβ	amyloid-β
SHMT1	hydroxymethyltransferase 1

References

1. Prince, M.J.; Guerchet, M.; Prina, M.A. The Global Impact of Dementia 2013–2050. Available online: https://www.alzint.org/resource/policy-brief-the-global-impact-of-dementia-2013-2050/ (accessed on 6 May 2024).
2. Jongsiriyanyong, S.; Limpawattana, P. Mild Cognitive Impairment in Clinical Practice: A Review Article. *Am. J. Alzheimers Dis. Other Dement.* **2018**, *33*, 500–507. [CrossRef] [PubMed]
3. Winblad, B.; Amouyel, P.; Andrieu, S.; Ballard, C.; Brayne, C.; Brodaty, H.; Cedazo-Minguez, A.; Dubois, B.; Edvardsson, D.; Feldman, H.; et al. Defeating Alzheimer's disease and other dementias: A priority for European science and society. *Lancet Neurol.* **2016**, *15*, 455–532. [CrossRef]
4. Livingston, G.; Sommerlad, A.; Orgeta, V.; Costafreda, S.G.; Huntley, J.; Ames, D.; Ballard, C.; Banerjee, S.; Burns, A.; Cohen-Mansfield, J.; et al. Dementia prevention, intervention, and care. *Lancet* **2017**, *390*, 2673–2734. [CrossRef]
5. WHO. International Statistical Classification of Diseases and Related Health Problems (11th Revision). Available online: https://www.who.int/classifications/icd/en/ (accessed on 27 May 2024).
6. Cooper, C.; Sommerlad, A.; Lyketsos, C.G.; Livingston, G. Modifiable predictors of dementia in mild cognitive impairment: A systematic review and meta-analysis. *Am. J. Psychiatry* **2015**, *172*, 323–334. [CrossRef]
7. Ravaglia, G.; Forti, P.; Maioli, F.; Martelli, M.; Servadei, L.; Brunetti, N.; Pantieri, G.; Mariani, E. Conversion of mild cognitive impairment to dementia: Predictive role of mild cognitive impairment subtypes and vascular risk factors. *Dement. Geriatr. Cogn. Disord.* **2006**, *21*, 51–58. [CrossRef]
8. Blasko, I.; Hinterberger, M.; Kemmler, G.; Jungwirth, S.; Krampla, W.; Leitha, T.; Heinz Tragl, K.; Fischer, P. Conversion from mild cognitive impairment to dementia: Influence of folic acid and vitamin B12 use in the VITA cohort. *J. Nutr. Health Aging* **2012**, *16*, 687–694. [CrossRef]
9. Hansson, O.; Zetterberg, H.; Buchhave, P.; Londos, E.; Blennow, K.; Minthon, L. Association between CSF biomarkers and incipient Alzheimer's disease in patients with mild cognitive impairment: A follow-up study. *Lancet Neurol.* **2006**, *5*, 228–234. [CrossRef]
10. Gavrilova, S.I.; Fedorova, Y.B.; Roshchina, I.F.; Korovaitseva, G.I. Prognosis of mild cognitive impairment syndrome: Data from a two-year clinical follow-up study. *Neurosci. Behav. Physiol.* **2008**, *38*, 129–134. [CrossRef]
11. Lionaki, E.; Ploumi, C.; Tavernarakis, N. One-Carbon Metabolism: Pulling the Strings behind Aging and Neurodegeneration. *Cells* **2022**, *11*, 214. [CrossRef]
12. Bou Ghanem, A.; Hussayni, Y.; Kadbey, R.; Ratel, Y.; Yehya, S.; Khouzami, L.; Ghadieh, H.E.; Kanaan, A.; Azar, S.; Harb, F. Exploring the complexities of 1C metabolism: Implications in aging and neurodegenerative diseases. *Front. Aging Neurosci.* **2023**, *15*, 1322419. [CrossRef]
13. Ducker, G.S.; Rabinowitz, J.D. One-Carbon Metabolism in Health and Disease. *Cell Metab.* **2017**, *25*, 27–42. [CrossRef]

14. McKee, S.E.; Reyes, T.M. Effect of supplementation with methyl-donor nutrients on neurodevelopment and cognition: Considerations for future research. *Nutr. Rev.* **2018**, *76*, 497–511. [CrossRef]
15. Wu, M.M.; Yang, F. Research Advances in the Association between Maternal Intake of Methyl Donor Nutrients during Pregnancy and DNA Methylation in Offspring. *Zhongguo Dang Dai Er Ke Za Zhi* **2017**, *19*, 601–606. [CrossRef]
16. Brachet, P.; Chanson, A.; Demigné, C.; Batifoulier, F.; Alexandre-Gouabau, M.C.; Tyssandier, V.; Rock, E. Age-associated B vitamin deficiency as a determinant of chronic diseases. *Nutr. Res. Rev.* **2004**, *17*, 55–68. [CrossRef]
17. Chen, H.; Liu, S.; Ge, B.; Zhou, D.; Li, M.; Li, W.; Ma, F.; Liu, Z.; Ji, Y.; Huang, G. Effects of Folic Acid and Vitamin B12 Supplementation on Cognitive Impairment and Inflammation in Patients with Alzheimer's Disease: A Randomized, Single-Blinded, Placebo-Controlled Trial. *J. Prev. Alzheimers Dis.* **2021**, *8*, 249–256. [CrossRef]
18. Han, H.; Wang, F.; Chen, J.; Li, X.; Fu, G.; Zhou, J.; Zhou, D.; Wu, W.; Chen, H. Changes in Biothiol Levels Are Closely Associated with Alzheimer's Disease. *J. Alzheimers Dis.* **2021**, *82*, 527–540. [CrossRef]
19. Blusztajn, J.K.; Slack, B.E.; Mellott, T.J. Neuroprotective Actions of Dietary Choline. *Nutrients* **2017**, *9*, 815. [CrossRef]
20. Zhang, Y.; Jia, J. Betaine Mitigates Amyloid-β-Associated Neuroinflammation by Suppressing the NLRP3 and NF-κB Signaling Pathways in Microglial Cells. *J. Alzheimers Dis.* **2023**, *94*, S9–S19. [CrossRef]
21. Jiang, C.; Li, G.; Huang, P.; Liu, Z.; Zhao, B. The Gut Microbiota and Alzheimer's Disease. *J. Alzheimers Dis.* **2017**, *58*, 1–15. [CrossRef]
22. Frausto, D.M.; Forsyth, C.B.; Keshavarzian, A.; Voigt, R.M. Dietary Regulation of Gut-Brain Axis in Alzheimer's Disease: Importance of Microbiota Metabolites. *Front. Neurosci.* **2021**, *15*, 736814. [CrossRef]
23. Goyal, D.; Ali, S.A.; Singh, R.K. Emerging Role of Gut Microbiota in Modulation of Neuroinflammation and Neurodegeneration with Emphasis on Alzheimer's disease. *Prog. Neuropsychopharmacol. Biol. Psychiatry* **2021**, *106*, 110112. [CrossRef] [PubMed]
24. Vogt, N.M.; Kerby, R.L.; Dill-McFarland, K.A.; Harding, S.J.; Merluzzi, A.P.; Johnson, S.C.; Carlsson, C.M.; Asthana, S.; Zetterberg, H.; Blennow, K.; et al. Gut microbiome alterations in Alzheimer's disease. *Sci. Rep.* **2017**, *7*, 13537. [CrossRef] [PubMed]
25. Rubini, E.; Schenkelaars, N.; Rousian, M.; Sinclair, K.D.; Wekema, L.; Faas, M.M.; Steegers-Theunissen, R.P.M.; Schoenmakers, S. Maternal obesity during pregnancy leads to derangements in one-carbon metabolism and the gut microbiota: Implications for fetal development and offspring wellbeing. *Am. J. Obstet. Gynecol.* **2022**, *227*, 392–400. [CrossRef] [PubMed]
26. Wang, S.; Song, L.; Fan, R.; Chen, Q.; You, M.; Cai, M.; Wu, Y.; Li, Y.; Xu, M. Targeting Aging and Longevity with Exogenous Nucleotides (TALENTs): Rationale, Design, and Baseline Characteristics from a Randomized Controlled Trial in Older Adults. *Nutrients* **2024**, *16*, 1343. [CrossRef] [PubMed]
27. Tian, J.; Xie, H.; Wang, L.; Wang, Y.; Wang, H.; Shi, J.; Qin, B.; Fan, D.; Ni, J.; Sun, Y.; et al. Chinese guideline for the diagnosis and treatment of Alzheimer's disease dementia (2020). *Chin. J. Geriatr.* **2021**, *40*, 269–283. [CrossRef]
28. Fan, R.; Chen, Q.; Song, L.; Wang, S.; You, M.; Cai, M.; Wang, X.; Li, Y.; Xu, M. The Validity and Feasibility of Utilizing the Photo-Assisted Dietary Intake Assessment among College Students and Elderly Individuals in China. *Nutrients* **2024**, *16*, 211. [CrossRef] [PubMed]
29. Li, N.; Teng, Y.; Dong, S.; Lu, L.; Hu, R.; Sun, J.; Zhu, W.; Pan, L.; Jiang, W.; Zeng, X.; et al. Development and assessment of methyl-donor nutritional quality index in pregnant women. *Zhongguo Sheng Yu Jian Kang Za Zhi* **2022**, *33*, 407–414.
30. Coleman, D.N.; Alharthi, A.S.; Liang, Y.; Lopes, M.G.; Lopreiato, V.; Vailati-Riboni, M.; Loor, J.J. Multifaceted role of one-carbon metabolism on immunometabolic control and growth during pregnancy, lactation and the neonatal period in dairy cattle. *J. Anim. Sci. Biotechnol.* **2021**, *12*, 27. [CrossRef]
31. Lin, L.; Zheng, L.J.; Zhang, L.J. Neuroinflammation, Gut Microbiome, and Alzheimer's Disease. *Mol. Neurobiol.* **2018**, *55*, 8243–8250. [CrossRef]
32. Scheperjans, F.; Derkinderen, P.; Borghammer, P. The Gut and Parkinson's Disease: Hype or Hope? *J. Park. Dis.* **2018**, *8*, S31–S39. [CrossRef]
33. Castillo-Álvarez, F.; Marzo-Sola, M.E. Role of intestinal microbiota in the development of multiple sclerosis. *Neurologia* **2017**, *32*, 175–184. [CrossRef] [PubMed]
34. Hu, X.; Wang, T.; Jin, F. Alzheimer's disease and gut microbiota. *Sci. China Life Sci.* **2016**, *59*, 1006–1023. [CrossRef] [PubMed]
35. Li, B.; He, Y.; Ma, J.; Huang, P.; Du, J.; Cao, L.; Wang, Y.; Xiao, Q.; Tang, H.; Chen, S. Mild cognitive impairment has similar alterations as Alzheimer's disease in gut microbiota. *Alzheimers Dement.* **2019**, *15*, 1357–1366. [CrossRef] [PubMed]
36. Sun, X.; Wen, J.; Guan, B.; Li, J.; Luo, J.; Li, J.; Wei, M.; Qiu, H. Folic acid and zinc improve hyperuricemia by altering the gut microbiota of rats with high-purine diet-induced hyperuricemia. *Front. Microbiol.* **2022**, *13*, 907952. [CrossRef] [PubMed]
37. Wan, Z.; Zheng, J.; Zhu, Z.; Sang, L.; Zhu, J.; Luo, S.; Zhao, Y.; Wang, R.; Zhang, Y.; Hao, K.; et al. Intermediate role of gut microbiota in vitamin B nutrition and its influences on human health. *Front. Nutr.* **2022**, *9*, 1031502. [CrossRef] [PubMed]
38. Barrea, L.; Muscogiuri, G.; Frias-Toral, E.; Laudisio, D.; Pugliese, G.; Castellucci, B.; Garcia-Velasquez, E.; Savastano, S.; Colao, A. Nutrition and immune system: From the Mediterranean diet to dietary supplementary through the microbiota. *Crit. Rev. Food Sci. Nutr.* **2021**, *61*, 3066–3090. [CrossRef] [PubMed]
39. Prajjwal, P.; Inban, P.; Sai, V.P.; Shiny, K.S.; Lam, J.R.; John, J.; Sulaimanov, M.; Tekuru, Y.; Wasi Ul Haq, M.; Marsool, M.D.M.; et al. The effects of the interplay between vitamins, antibiotics, and gut microbiota on the pathogenesis and progression of dementia: A systematic review and meta-analysis. *Health Sci. Rep.* **2024**, *7*, e1808. [CrossRef] [PubMed]
40. Jung, J.H.; Kim, G.; Byun, M.S.; Lee, J.H.; Yi, D.; Park, H.; Lee, D.Y. Gut microbiome alterations in preclinical Alzheimer's disease. *PLoS ONE* **2022**, *17*, e0278276. [CrossRef] [PubMed]

41. Hatayama, K.; Ebara, A.; Okuma, K.; Tokuno, H.; Hasuko, K.; Masuyama, H.; Ashikari, I.; Shirasawa, T. Characteristics of Intestinal Microbiota in Japanese Patients with Mild Cognitive Impairment and a Risk-Estimating Method for the Disorder. *Biomedicines* **2023**, *11*, 1789. [CrossRef]
42. MahmoudianDehkordi, S.; Arnold, M.; Nho, K.; Ahmad, S.; Jia, W.; Xie, G.; Louie, G.; Kueider-Paisley, A.; Moseley, M.A.; Thompson, J.W.; et al. Altered bile acid profile associates with cognitive impairment in Alzheimer's disease-An emerging role for gut microbiome. *Alzheimers Dement.* **2019**, *15*, 76–92. [CrossRef]
43. Lynch, J.B.; Gonzalez, E.L.; Choy, K.; Faull, K.F.; Jewell, T.; Arellano, A.; Liang, J.; Yu, K.B.; Paramo, J.; Hsiao, E.Y. Gut microbiota Turicibacter strains differentially modify bile acids and host lipids. *Nat. Commun.* **2023**, *14*, 3669. [CrossRef] [PubMed]
44. Dominy, S.S.; Lynch, C.; Ermini, F.; Benedyk, M.; Marczyk, A.; Konradi, A.; Nguyen, M.; Haditsch, U.; Raha, D.; Griffin, C.; et al. Porphyromonas gingivalis in Alzheimer's disease brains: Evidence for disease causation and treatment with small-molecule inhibitors. *Sci. Adv.* **2019**, *5*, eaau3333. [CrossRef] [PubMed]
45. Haditsch, U.; Roth, T.; Rodriguez, L.; Hancock, S.; Cecere, T.; Nguyen, M.; Arastu-Kapur, S.; Broce, S.; Raha, D.; Lynch, C.C.; et al. Alzheimer's Disease-Like Neurodegeneration in Porphyromonas gingivalis Infected Neurons with Persistent Expression of Active Gingipains. *J. Alzheimers Dis.* **2020**, *75*, 1361–1376. [CrossRef] [PubMed]
46. Dowd, S.E.; Sun, Y.; Secor, P.R.; Rhoads, D.D.; Wolcott, B.M.; James, G.A.; Wolcott, R.D. Survey of bacterial diversity in chronic wounds using pyrosequencing, DGGE, and full ribosome shotgun sequencing. *BMC Microbiol.* **2008**, *8*, 43. [CrossRef] [PubMed]
47. Han, Z.L.; Yang, M.; Fu, X.D.; Chen, M.; Su, Q.; Zhao, Y.H.; Mou, H.J. Evaluation of Prebiotic Potential of Three Marine Algae Oligosaccharides from Enzymatic Hydrolysis. *Mar. Drugs* **2019**, *17*, 173. [CrossRef] [PubMed]
48. Cho, E.; Park, S.N.; Kim, H.K.; Kim, D.S.; Jung, J.; Baek, J.H.; Lim, Y.K.; Jo, E.; Choi, M.H.; Chang, Y.H.; et al. Draft Genome Sequence of the Novel Peptoniphilus sp. Strain ChDC B134, Isolated from a Human Periapical Abscess Lesion. *Genome. Announc.* **2013**, *1*, e00822-13. [CrossRef] [PubMed]
49. Onisiforou, A.; Christodoulou, C.C.; Zamba-Papanicolaou, E.; Zanos, P.; Georgiou, P. Transcriptomic analysis reveals sex-specific patterns in the hippocampus in Alzheimer's disease. *Front. Endocrinol.* **2024**, *15*, 1345498. [CrossRef] [PubMed]
50. Houldsworth, A. Role of oxidative stress in neurodegenerative disorders: A review of reactive oxygen species and prevention by antioxidants. *Brain Commun.* **2024**, *6*, fcad356. [CrossRef] [PubMed]
51. Gao, Y.; Li, J.; Wu, Q.; Wang, S.; Yang, S.; Li, X.; Chen, N.; Li, L.; Zhang, L. Tetrahydroxy stilbene glycoside ameliorates Alzheimer's disease in APP/PS1 mice via glutathione peroxidase related ferroptosis. *Int. Immunopharmacol.* **2021**, *99*, 108002. [CrossRef]
52. Araújo, J.R.; Martel, F.; Borges, N.; Araújo, J.M.; Keating, E. Folates and aging: Role in mild cognitive impairment, dementia and depression. *Ageing Res. Rev.* **2015**, *22*, 9–19. [CrossRef]
53. Magnúsdóttir, S.; Ravcheev, D.; de Crécy-Lagard, V.; Thiele, I. Systematic genome assessment of B-vitamin biosynthesis suggests co-operation among gut microbes. *Front. Genet.* **2015**, *6*, 148. [CrossRef] [PubMed]
54. Tuska, R.M.; Helm, S.M.; Graf, C.F.; James, C.; Kong, G.; Stiemsma, L.T.; Green, D.B.; Helm, S.E. Surfeit folic acid, protein, and exercise modify oncogenic inflammatory biomarkers and fecal microbiota. *Front. Nutr.* **2022**, *9*, 1060212. [CrossRef] [PubMed]
55. Petrova, B.; Maynard, A.G.; Wang, P.; Kanarek, N. Regulatory mechanisms of one-carbon metabolism enzymes. *J. Biol. Chem.* **2023**, *299*, 105457. [CrossRef] [PubMed]
56. Abarinov, E.V.; Beaudin, A.E.; Field, M.S.; Perry, C.A.; Allen, R.H.; Stabler, S.P.; Stover, P.J. Disruption of shmt1 impairs hippocampal neurogenesis and mnemonic function in mice. *J. Nutr.* **2013**, *143*, 1028–1035. [CrossRef]
57. Pérez Palmer, N.; Trejo Ortega, B.; Joshi, P. Cognitive Impairment in Older Adults: Epidemiology, Diagnosis, and Treatment. *Psychiatr. Clin. N. Am.* **2022**, *45*, 639–661. [CrossRef]

nutrients

MDPI

Article

Associations of Dietary Diversity Trajectories with Frailty among Chinese Older Adults: A Latent Class Trajectory Analysis Based on a CLHLS Cohort

Chenyu Zhao, Yuping Wang, Xiaocan Jia, Jingwen Fan, Nana Wang, Yongli Yang *,† and Xuezhong Shi *,†

Department of Epidemiology and Biostatistics, College of Public Health, Zhengzhou University, Zhengzhou 450001, China; zcy8715@gs.zzu.edu.cn (C.Z.); 202011271030324@gs.zzu.edu.cn (Y.W.); jxc@zzu.edu.cn (X.J.); fanjingwen126@126.com (J.F.); wnn0924@126.com (N.W.)
* Correspondence: ylyang377@zzu.edu.cn (Y.Y.); xzshi@zzu.edu.cn (X.S.)
† These authors contributed equally to this work.

Abstract: Background: High dietary diversity has been found to be associated with frailty. However, the trajectory of dietary diversity intake in relation to frailty is unclear. Methods: Using the latent class trajectory modeling approach, we identified distinctive dietary variety trajectory groups among 2017 participants based on the Chinese Longitudinal Healthy Longevity Survey acquired at four time points within a 10-year period. Frailty status was assessed using a frailty index comprising 37 health deficits. Dietary diversity was quantified using the dietary variety score (DVS), based on food category consumption frequency. Logistic regression analyses were employed to explore the association between DVS change trajectories and frailty. Results: This study identified two distinct DVS trajectories: "Moderate-Slow decline-Slow growth", encompassing 810 (40.16%) individuals, and "Moderate-Slow growth-Accelerated decline", including 1207 (59.84%) individuals. After adjusting for covariates, the odds ratio for DVS in the "Moderate-Slow decline-Slow growth" group was 1.326 (95% confidence interval: 1.075–1.636) compared to the "Moderate-Slow growth-Accelerated decline" group. The "Moderate-Slow decline-Slow growth" trajectory continued to decrease and was maintained at a low level in the early stages of aging. Conclusion: Sustaining a high dietary diversity trajectory over time, particularly in the early stages of aging, could potentially decrease the risk of frailty among older Chinese adults.

Keywords: Chinese older adults; frailty; dietary variety score; latent class trajectory modeling

1. Introduction

Frailty is a condition characterized by a decline in the functioning of multiple body systems, rendering the body more susceptible to the negative effects of stressors [1]. The prevalence of frailty varies with age. Globally, among individuals aged 60 to 69, the prevalence is 23%. In the 70 to 79 age group, it is 25%, while among those aged 80 to 89, it reaches 32%. For individuals aged 90 and above, the prevalence of frailty is notably higher, at 61% [2]. Meanwhile, frailty is strongly linked to unfavorable health outcomes and a notable rise in healthcare expenditures [3,4]. The population with frailty has been associated with increased risks of aging-related diseases, such as sarcopenia [5], polypharmacy [6], and ischemic heart disease [7], which seriously affect the health of older people. The global impact of frailty is expected to increase due to an increasingly aging population. Consequently, the identification of risk factors and the proactive addressing of frailty represent pressing public health imperatives [8].

Unlike the natural aging process, frailty is a condition that can be mitigated or even reversed through suitable interventions and preventive measures. In addition to the demographic characteristics that are associated with frailty, addressing poor lifestyle choices and mitigating the effects of childhood adversity may help reduce the risk of frailty [9–11]. In recent years, many researchers have focused on the effects of dietary diversity on frailty. Previous studies have indicated that dietary diversity has the potential to improve the frailty status of older people. Three cross-sectional studies suggest that low dietary variety is associated with frailty in older people and contributes to a higher prevalence of frailty [12–14]. Particularly among older persons who live alone, there is a notable trend toward higher frailty scores and lower dietary diversity [15]. Findings from two cohort studies indicate a heightened need for dietary support to reduce the onset of frailty in older adults, with a particular emphasis on older women [16,17]. At the same time, studies focusing on older adults in China suggest that a high food variety in one's diet may contribute to a lower incidence of frailty [18,19]. Since dietary diversity is a dynamic and heterogeneous process, it may be better captured by repeatedly measuring the intake frequency of multiple foods over time. However, most studies have only collected information on dietary diversity at a single time, and few studies have examined the longitudinal pattern of dietary diversity as well as its associations with frailty. Latent class trajectory modeling (LCTM) offers a valuable approach to categorizing heterogeneous groups into more homogeneous ones while still allowing for discerning the distinctions between individuals and groups [20]. This modeling technique is commonly used to depict the trajectory of a quantitative variable over time [21]. Remarkably, there is limited research exploring the association between dietary diversity trajectories and frailty in older Chinese adults.

To enhance the quality of life during later years, older individuals should prioritize reducing the risk of frailty. This study conducted latent class trajectory analyses using longitudinal data from the Chinese Longitudinal Healthy Longevity Survey (CLHLS). The objective was to investigate the relationship between long-term dietary diversity development trajectories and frailty to identify potential optimal times for intervention and strategies for mitigating frailty in older adults.

2. Methods

2.1. Study Design and Participants

A prospective cohort study was utilized to construct a trajectory of dietary diversity and explore its association with frailty using data from CLHLS. The CLHLS is a comprehensive nationwide study conducted with randomly selected samples from half of the counties and cities distributed across 22 out of the 31 provinces in China. This extensive survey encompassed approximately 85% of the entire Chinese population. The questionnaires employed in this study were categorized into two distinct types: one designed for surviving respondents and the other for the families of deceased elderly individuals. The data quality of the CLHLS has been systematically evaluated [22]. Prior to their participation in the study, all subjects provided informed consent for inclusion. Additionally, the study received approval from the Biomedical Ethics Committee of Peking University (IRB00001052-13074).

In this study, individuals who were followed in all four survey waves (2008, 2011, 2014, and 2018) within the CLHLS were included in the cohort. After exclusion, a total of 2017 subjects were ultimately enrolled in this study (Figure 1).

2.2. Frailty Index

The frailty index (FI) was constructed following established protocols [23]. Inclusion of deficits associated with health status was contingent upon meeting specific criteria, which included the following: the deficit affecting multiple body systems and spanning various physiological areas; a tendency for the deficit's prevalence to rise with age; and the deficit not being nearly universal in middle-aged individuals.

The FI was defined as an unweighted count, representing the number of deficits divided by the total possible deficits for an individual. After reviewing self-reported or measured data, a total of 37 indicators encompassing various aspects of self-reported health status, cognitive function, depression, and various chronic diseases were identified in the 2008 and 2018 CLHLSs (Table S1). The components of FI closely resembled those employed in previous research [4,24–26]. Each individual item was dichotomized, with a code of one indicating the presence of a deficit. Furthermore, in line with prior studies [27], a score of two was assigned if the respondent had a serious illness that led to hospitalization or confinement to bed on two or more occasions. The FI was subsequently computed by summing all deficits and dividing the sum by the total number of possible deficits, resulting in a range of scores from zero to one. Then, participant classification into non-frail (FI $\leq$ 0.25) and frail (FI > 0.25) categories was determined based on previously established cutoff points [28].

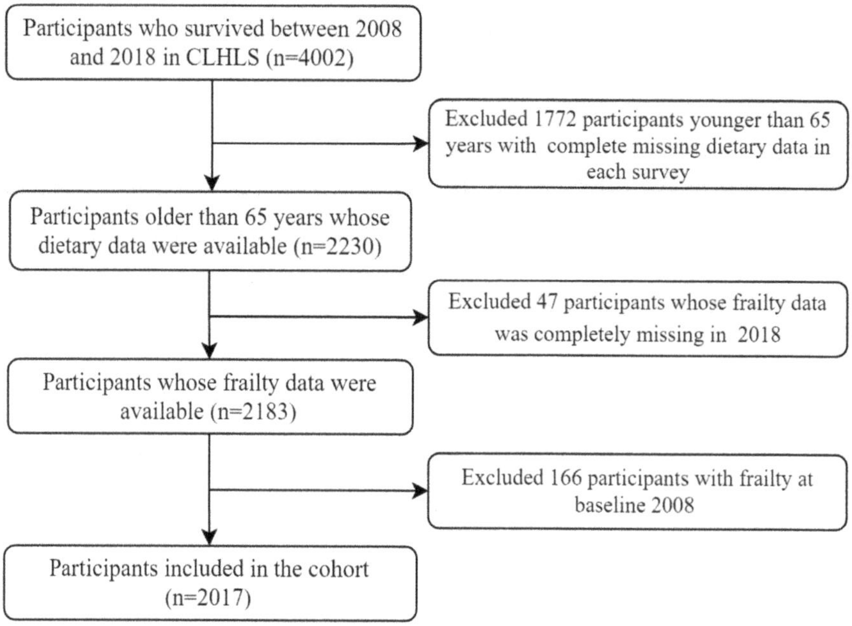

Figure 1. Flow chart for inclusion and exclusion of research subjects.

2.3. Dietary Variety Assessment

This study applied the dietary variety score (DVS) to evaluate the initial dietary diversity in the 2008, 2011, 2014, and 2018 CLHLS surveys [18,29]. The intake frequency was used to indicate the intake status of food groups, including fresh fruit, vegetables, meat, fish, eggs, bean products, salted vegetables, sugar, garlic, milk products, nut products, mushrooms or algae, and tea. The DVS was calculated according to the intake frequency of 13 food groups. The specific intake frequency and scoring criteria are shown in Table S2. The total DVS was the sum of the scores of the 13 food groups, with the lowest score being zero and the highest score being 13. The higher the score, the better the dietary diversity.

2.4. Covariates

Data for sociodemographic characteristics (e.g., age, sex, ethnicity, illiteracy, economic status, marital status, and co-residence), childhood life status (e.g., place of birth, only child, and hungry in childhood), and lifestyle factors (e.g., smoking, drinking, and physical activity) were collected by trained staff using a questionnaire. More detailed information on the data and measurements collected is available in previous research. The covariates were obtained from the baseline questionnaire [22,30].

2.5. Missing Data

To address partial missing data in frailty, diet, and covariates, a multiple imputation approach utilizing chained equations was employed. This approach resulted in the generation of five imputed datasets. The model coefficients were independently estimated within each of these imputed datasets. Subsequently, these estimated coefficients were combined across the imputed datasets using Rubin's rules [31].

2.6. Statistical Analysis

The examination of DVS was conducted at four distinct time points (2008, 2011, 2014, and 2018) using LCTM. The objective was to identify distinct subgroups of individuals whose DVS measurements exhibited similar patterns of change over time. Briefly, this method was designed to identify clusters of individuals following a similar developmental trajectory based on a semiparametric group-based approach. In this study, age, ranging from over 65 to 115, was employed as the timescale for the trajectories. Models based on two to five trajectories were examined, and the optimal model was determined using the Bayesian Information Criterion (BIC). A good model was defined as one with an average posterior probability (APP) exceeding 70% and with at least 5% of individuals belonging to each trajectory group. Three logistic regression models were used to investigate the association between the trajectory group and frailty, and the odds ratio (OR) and 95% confidence interval (CI) were computed. Model 1 was adjusted for sociodemographic characteristics including age, sex, ethnicity, illiteracy, economic status, marital status, and co-residence. Model 2 was additionally adjusted for lifestyle factors, including smoking, drinking, and physical activity. Model 3 was additionally adjusted for childhood life status, including place of birth, only child, and hungry in childhood.

The logistic regression models mentioned above were conducted within various strata defined by age groups (under 75 or 75 years and older), gender, ethnicity, literacy status, economic status, marital status, co-residence, smoking habits, alcohol consumption, exercise habits, place of birth, only-child status, and experiences of childhood hunger. Secondly, likelihood ratio tests were performed to investigate potential statistical interactions between DVS trajectories and place of birth, DVS trajectories and experiences of childhood hunger, as well as DVS trajectories and only-child status. Lastly, for the sensitivity analysis, after eliminating the DVS of zero and one in the baseline survey, latent class trajectory analyses and a logistic regression analysis were performed. Then, people over 100 years old in the 2018 survey were excluded. Moreover, the E-value was estimated to examine the magnitude of an unmeasured confounding factor that could affect the association between dietary diversity development trajectories and frailty by random chance [32].

The LCTM was constructed using the "lcmm" package in the R (version 4.1.0) software. Statistical analyses were performed using SAS (version 9.4), and all tests were two-sided with $p < 0.05$ indicating statistical significance.

3. Result

3.1. Estimated DVS Trajectory Modeling

Four distinct types of DVS trajectories were explored using LCTM to account for the heterogeneity in participants' DVS, as detailed in Table S3. The LCTM model with the lowest BIC value was observed to have two trajectories (BIC = 37,725.6), which was considered the most optimal choice. Figure 2 visually depicts these trajectories, illustrating two distinct DVS classes: Class 1 was labeled "Moderate-Slow decline-Slow growth", comprising 810 (40.16%) individuals; Class 2 was identified as "Moderate-Slow growth-Accelerated decline", including 1207 (59.84%) individuals.

3.2. Baseline Characteristics of Trajectory Subpopulation

The baseline characteristics of participants according to DVS trajectories are shown in Table 1. Notably, individuals belonging to the "Moderate-Slow decline-Slow growth" trajectory group exhibited a higher likelihood of being female, born in rural areas, experiencing childhood hunger, not possessing literacy skills, having a lower economic status, being widowed and never married, residing alone, and not engaging in smoking, alcohol consumption, or regular exercise, compared to those in the "Moderate-Slow growth-Accelerated decline" trajectory group.

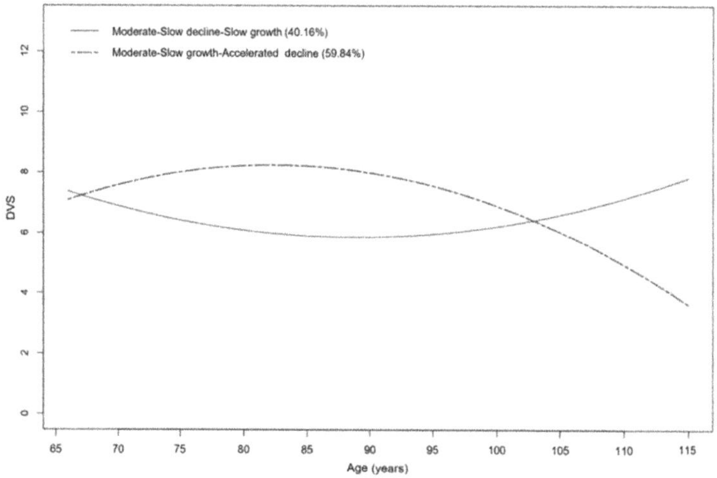

Figure 2. Trajectories of DVS.

Table 1. Baseline characteristics of participants based on the trajectories of DVS.

Variables	Overall n = 2017	Moderate-Slow Decline-Slow Growth n = 810	Moderate-Slow Growth-Accelerated Decline n = 1207	p-Value [a]
Age, M (P25, P75)	74 (69, 80)	75 (70, 80)	73 (69, 80)	<0.001
Sex, n (%)				
Male	969 (48.04)	317 (39.14)	652 (54.02)	<0.001
Female	1048 (51.96)	493 (60.86)	555 (45.98)	
Ethnicity, n (%)				
Han	1894 (93.9)	740 (91.36)	1154 (95.61)	<0.001
Hui	5 (0.25)	1 (0.12)	4 (0.33)	
Zhuang	74 (3.67)	47 (5.8)	27 (2.24)	
Yao	11 (0.55)	5 (0.62)	6 (0.5)	
Man	6 (0.3)	2 (0.25)	4 (0.33)	
Others	27 (1.34)	15 (1.85)	12 (0.99)	
Place of birth, n (%)				
Urban	155 (7.68)	42 (5.19)	113 (9.36)	<0.001
Rural	1860 (92.22)	767 (94.69)	1093 (90.56)	
Missing	2 (0.1)	1 (0.12)	1 (0.08)	
Only child, n (%)				
Yes	150 (7.44)	61 (7.53)	89 (7.37)	0.895
No	1867 (92.56)	749 (92.47)	1118 (92.63)	
Hungry in Childhood, n (%)				
Yes	1454 (72.09)	614 (75.8)	840 (69.59)	<0.001
No	515 (25.53)	171 (21.11)	344 (28.5)	
Missing	48 (2.38)	25 (3.09)	23 (1.91)	
Illiteracy, n (%)				
Yes	991 (49.13)	478 (59.01)	513 (42.5)	<0.001
No	1026 (50.87)	332 (40.99)	694 (57.5)	

Table 1. *Cont.*

Variables	Overall *n* = 2017	Moderate-Slow Decline-Slow Growth *n* = 810	Moderate-Slow Growth-Accelerated Decline *n* = 1207	*p*-Value [a]
Economic status, *n* (%)				
Very rich	22 (1.09)	5 (0.62)	17 (1.41)	<0.001
Rich	217 (10.76)	54 (6.67)	163 (13.5)	
Moderate	1469 (72.83)	581 (71.73)	888 (73.57)	
Poor	270 (13.39)	142 (17.53)	128 (10.6)	
Very poor	37 (1.83)	27 (3.33)	10 (0.83)	
Missing	2 (0.1)	1 (0.12)	1 (0.08)	
Current marital status, *n* (%)				
Married and living with spouse	1172 (58.11)	411 (50.74)	761 (63.05)	<0.001
Separated	61 (3.02)	26 (3.21)	35 (2.9)	
Divorced	5 (0.25)	2 (0.25)	3 (0.25)	
Widowed	758 (37.58)	361 (44.57)	397 (32.89)	
Never married	21 (1.04)	10 (1.23)	11 (0.91)	
Co-residence, *n* (%)				
With household member	1673 (82.94)	638 (78.77)	1035 (85.75)	<0.001
Alone	330 (16.36)	169 (20.86)	161 (13.34)	
In an institution	14 (0.69)	3 (0.37)	11 (0.91)	
Smoke at present, *n* (%)				
Yes	470 (23.3)	174 (21.48)	296 (24.52)	0.113
No	1547 (76.7)	636 (78.52)	911 (75.48)	
Smoke in the past, *n* (%)				
Yes	699 (34.66)	227 (28.02)	472 (39.11)	<0.001
No	1314 (65.15)	581 (71.73)	733 (60.73)	
Missing	4 (0.2)	2 (0.25)	2 (0.17)	
Drink at present, *n* (%)				
Yes	455 (22.56)	155 (19.14)	300 (24.86)	0.003
No	1562 (77.44)	655 (80.86)	907 (75.14)	
Drink in the past, *n* (%)				
Yes	621 (30.79)	203 (25.06)	418 (34.63)	<0.001
No	1392 (69.01)	605 (74.69)	787 (65.2)	
Missing	4 (0.2)	2 (0.25)	2 (0.17)	
Exercise at present, *n* (%)				
Yes	732 (36.29)	237 (29.26)	495 (41.01)	<0.001
No	1285 (63.71)	573 (70.74)	712 (58.99)	
Exercise in the past, *n* (%)				
Yes	556 (27.57)	183 (22.59)	373 (30.9)	<0.001
No	1455 (72.14)	624 (77.04)	831 (68.85)	
Missing	6 (0.3)	3 (0.37)	3 (0.25)	

[a] Based on a Wilcoxon signed-rank test or Chi-square statistics as appropriate.

3.3. Association of DVS Trajectories with Frailty

In this study, the overall prevalence of frailty was 36% (Table S4). Table 2 presents the OR for DVS in the "Moderate-Slow decline-Slow growth" group compared to the "Moderate-Slow growth-Accelerated decline" group. After adjusting for sociodemographic characteristics (Model 1), the OR for DVS in the "Moderate-Slow decline-Slow growth" group was 1.296 (95% CI: 1.054–1.594). Upon further adjustment for lifestyle factors (Model 2), the association between DVS and frailty remained statistically significant; the OR for DVS in the "Moderate-Slow decline-Slow growth" group was 1.323 (95% CI: 1.073–1.632). Furthermore, even after additional adjustment for childhood life status (Model 3), DVS continued to exhibit a significant association with frailty; the OR for DVS in the "Moderate-Slow decline-Slow growth" group was 1.326 (95% CI: 1.075–1.636). Additionally, in the "Moderate-Slow decline-Slow growth" group, it was worth noting that the trajectory of DVS continued to decrease from the early stages of aging, and at 85 to 95 years, it was maintained at a low level.

Table 2. Logistic regression models for frailty and DVS trajectories.

	Moderate-Slow Decline-Slow Growth	Moderate-Slow Growth-Accelerated Decline
Subjects, *n*	810	1207
Frailty cases, *n*	334	393
OR (95% CI)		
Model 1	1.296 (1.054–1.594)	Reference
Model 2	1.323 (1.073–1.632)	Reference
Model 3	1.326 (1.075–1.636)	Reference

Model 1 was adjusted for age, sex, ethnicity, illiteracy, economic status, marital status, and co-residence; Model 2 was adjusted for Model 1 + smoking, drinking, and physical activity; Model 3 was adjusted for Model 2 + place of birth, only child, and hungry in childhood; OR, odds ratio; CI, confidence interval.

3.4. Subgroup Analysis

Figure 3 displays the outcomes of subgroup analyses. These findings indicated a statistically significant relationship between the "Moderate-Slow decline-Slow growth" trajectories and frailty among specific subgroups including individuals who were male, of Han ethnicity, born in rural areas, experienced childhood hunger, possessed literacy skills, had a moderate economic status, lived with others, currently smoked, had a history of smoking, did not currently consume alcohol, had a history of not drinking alcohol, and did not engage in exercise in the past. Furthermore, this relationship remained unchanged regardless of whether one was an only child or their current marital status.

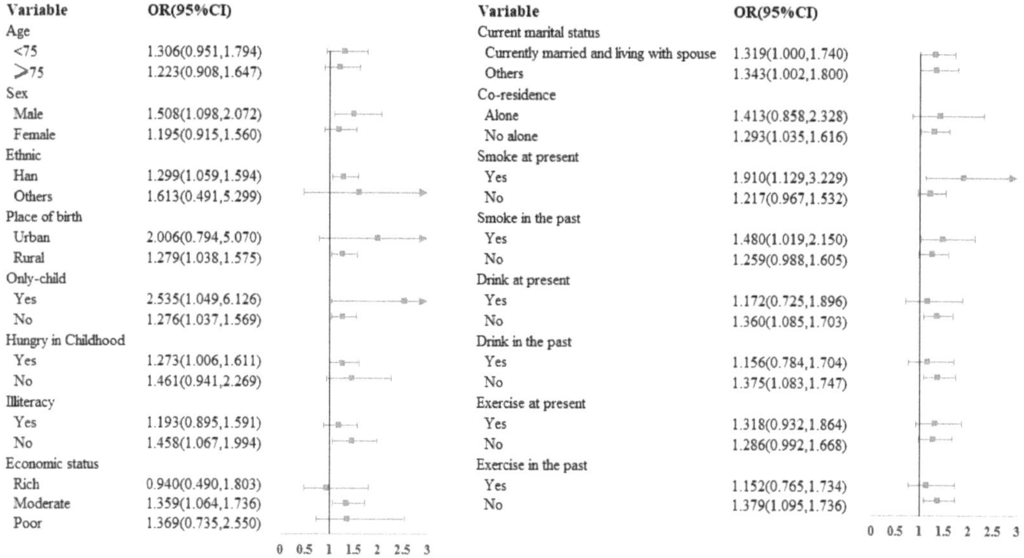

Figure 3. Subgroup analysis. Arrows indicate that the upper or lower limit of the 95% CI of the OR exceeds the horizontal boundary.

3.5. Interaction Analysis

Moreover, interaction analysis indicated that there was no interaction between the DVS trajectory and any of the three childhood life status variables (place of birth: *P*-interaction = 0.389, only child: *P*-interaction = 0.338, hungry in childhood: *P*-interaction = 0.650).

3.6. Sensitivity Analysis

After eliminating the DVS of zero and one in the baseline survey and excluding persons older than 100 years in the 2018 survey, the change trend of the two groups of trajectories (Figures S1 and S2) and the correlation analysis results (Tables S5 and S6) were still consistent with the original analysis results. The E-values of Model 1, Model 2, and Model 3 were 1.54 (confidence interval, 1.19), 1.57 (confidence interval, 1.23), and 1.57 (confidence interval, 1.23), respectively.

4. Discussion

In this study, data from the CLHLS cohort spanning from 2008 to 2018 were utilized to evaluate the influence of long-term dietary diversity trajectories on frailty among older Chinese adults. The analysis identified two distinct trajectories of dietary diversity, and a heightened risk of frailty was observed within the group labeled "Moderate-Slow decline-Slow growth" compared to the group labeled "Moderate-Slow growth-Accelerated decline". Furthermore, to prevent frailty in older adults, it is advisable to implement interventions before 65 years old, with a focus on increasing dietary diversity. These new findings hold significant clinical and public health implications for the prevention and intervention of frailty.

In this study, results showed that the prevalence of frailty was as high as 36%. The prevalence of frailty among Chinese older community dwellers was 10.1% in a previous study [33]. Even after standardizing the two datasets for age and sex, the prevalence of frailty in the study (age: 28.87%, sex: 36.62%) remained higher than what has been reported in previous studies (age: 9.12%, sex: 9.98%). The notable difference was attributed to two primary factors. Firstly, the utilization of the FI encompassed an extensive list of clinical conditions and diseases rather than relying on a few specific signs or symptoms. It drew from a wealth of information, potentially rendering it more sensitive and effective in identifying frailty [34]. Previous studies had suggested that frailty prevalence, as assessed using the FI, tended to be higher than when using the frailty phenotype [2]. Secondly, this study encompassed a larger proportion of older individuals from rural areas. In China, research in underdeveloped or rural regions is relatively limited. Given that frailty tended to be more prevalent in these less developed areas, it was possible that the overall frailty prevalence in China was underestimated in previous studies [33]. Hence, this study provided a more comprehensive representation of frailty prevalence among older people in China. Additionally, the findings of this study indicated that frailty was indeed a more severe and prevalent concern among older individuals in China. This underscored the importance of addressing and managing frailty as a significant public health issue in China.

Indeed, previous studies have investigated the relationship between dietary diversity and frailty. A study of Chinese older adults, also based on CLHLS, found that a more diverse diet was associated with a lower risk of frailty [19], which was consistent with the findings of the present study. However, it measured diet at a single time, ignoring the long-term effects of dietary diversity patterns. This oversight may result in biased estimates when assessing exposure–outcome relationships. For example, based on the identification of two dietary diversity trajectories in this study, the dynamic change trends were different before and after the intersection point. Better than measuring dietary profiles at a single time point, establishing dietary diversity trajectories based on dietary profiles measured over a longer time period may capture the process of dietary dynamics and heterogeneity and, expectedly, identify precisely different populations with varying risks of frailty. Compared to previous studies, the current study demonstrated the feasibility of estimating trends in dietary diversity over a long period of time in the Chinese population. Moreover, the timing of early intervention can be more accurately pinpointed. In the "Moderate-Slow decline-Slow growth" group, although people's DVS was moderate at age over 65, it has already entered a phase of decline, and the resulting risk of frailty began to increase. Although DVS began to rise after age 85, people were already in the later stages of aging, and the benefits of rising DVS were very limited. Therefore, it was suggested that promoting increased dietary diversity intake in the population in the early stages of aging may be a more effective strategy for preventing the onset of frailty [35].

Since diet represents a long-term cumulative process, it was essential to explore the impact of sustained long-term dietary changes or habits on frailty. The findings suggested that maintaining a high level of dietary diversity was advantageous for preventing frailty, especially in the older population. It may have the following explanations: Firstly, a wider variety of food consumption was associated with increased diversity in the gut microbiota among older individuals [36]. Gut dysbiosis, or an imbalance in the gut microbiota, can trigger an innate immune response and persistent low-grade inflammation [37]. This inflammation was linked to a range of age-related degenerative diseases, including frailty. Secondly, a diverse diet provided a broader spectrum of essential nutrients, ensuring more comprehensive nutritional intake for the body [38]. This enhanced nutritional adequacy was beneficial for preventing frailty in the elderly. Additionally, many nutrients in the body function in coordination with others [39]. The presence and balance of various nutrients influenced the roles of other nutrients. Only when dietary diversity is rich, can multiple nutrients achieve a balanced state, allowing for more effective interactions between nutrients. For instance, the intake of vitamin D, vitamin C, and vitamin K2 can enhance the body's absorption of calcium, contributing to the prevention of frailty [40–42].

In this study, a series of analyses were conducted to assess the reliability of the findings thoroughly. Firstly, consistent conclusions with the main analysis were observed when examining subgroups based on multiple characteristics. Secondly, previous research had consistently demonstrated that variables related to one's childhood were significantly associated with both phenotypic and functional aging [43]. However, in a previous study, no interaction was found between childhood status and DVS trajectory. The potential impact of extreme values of DVS in the baseline survey and people of older age on the results was also taken into account. Lastly, an innovative sensitivity analysis was introduced using the E-value [32]. When the association between confounders and exposure and outcome reached 1.57 (confidence interval, 1.23), the association between DVS trajectory and frailty in this study appeared to be invalid. Given that 1.57 represents a relatively high association strength, the results can be considered robust.

This study had several potential limitations. First, dietary information was self-reported, which may lead to recall bias. Second, CLHLS did not directly measure dietary intake, and dietary frequency was not exactly equal to dietary intake. Measurement errors in dietary intake assessments were a challenge in epidemiological research, despite ongoing efforts to improve methods. These errors were unlikely to be entirely eliminated due to factors like day-to-day variation and self-reporting limitations [44,45]. In addition, although the E-value was used to assess the effect of unknown confounding factors on the results, the possibility of residual confounding cannot be excluded due to the observational nature of the studies. Finally, this study focused on the effect of dietary diversity on frailty without considering people's different dietary preferences, even though their DVS may be the same. In the upcoming research, the inclusion of dietary intake is planned to provide a more comprehensive exploration of the relationship between dietary factors and frailty.

5. Conclusions

In conclusion, the latent class trajectory analysis revealed that in Chinese older adults, individuals with a DVS change trajectory characterized by "Moderate-Slow decline-Slow growth" were at an increased risk of frailty. The optimal time for intervention is in the early stages of aging. These novel findings had significant clinical and public health implications.

Supplementary Materials: The following supporting information can be downloaded at https://www.mdpi.com/article/10.3390/nu16101445/s1, Table S1: List of items included in the frailty index; Table S2: Food intake frequency and scoring criteria; Table S3: Fitting statistics for DVS trajectories; Table S4: The prevalence of frailty; Figure S1: Trajectories of DVS after eliminating the lowest 2% DVS in the baseline survey; Table S5: Logistic regression models for frailty and DVS trajectories after eliminating the lowest 2% DVS in the baseline survey; Figure S2: Trajectories of DVS after excluding persons older than 100 years in the 2018 survey; Table S6: Logistic regression models for frailty and DVS trajectories after excluding persons older than 100 years in the 2018 survey.

Author Contributions: X.S. and Y.Y.: conceptualization, visualization, resources, supervision; C.Z. and Y.W.: methodology, data curation, formal analysis, writing—original draft; X.J.: methodology, data curation, formal analysis; J.F. and N.W.: reviewing and editing. All authors have read and agreed to the published version of the manuscript.

Funding: This research received no external funding.

Institutional Review Board Statement: The study was conducted in accordance with the Declaration of Helsinki, and approved by the Institutional Review Board (or Ethics Committee) of Biomedical Ethics Committee of Peking University (Approval Code: IRB00001052-13074, Approval Date: 9 November 2021).

Informed Consent Statement: Informed consent was obtained from all subjects involved in the study.

Data Availability Statement: The data of CLHLS are available at https://opendata.pku.edu.cn/dataverse/CHADS (accessed on 15 March 2023).

Acknowledgments: We thank all investigators and participants in the CLHLS cohort for providing access to data.

Conflicts of Interest: The authors have no actual or potential conflicts of interest to declare.

References

1. Hoogendijk, E.O.; Afilalo, J.; Ensrud, K.E.; Kowal, P.; Onder, G.; Fried, L.P. Frailty: Implications for clinical practice and public health. *Lancet* **2019**, *394*, 1365–1375. [CrossRef] [PubMed]
2. O'Caoimh, R.; Sezgin, D.; O'Donovan, M.R.; Molloy, D.W.; Clegg, A.; Rockwood, K.; Liew, A. Prevalence of frailty in 62 countries across the world: A systematic review and meta-analysis of population-level studies. *Age Ageing* **2021**, *50*, 96–104. [CrossRef] [PubMed]
3. Ida, S.; Kaneko, R.; Imataka, K.; Murata, K. Relationship between frailty and mortality, hospitalization, and cardiovascular diseases in diabetes: A systematic review and meta-analysis. *Cardiovasc. Diabetol.* **2019**, *18*, 81. [CrossRef] [PubMed]
4. Fan, J.; Yu, C.; Guo, Y.; Bian, Z.; Sun, Z.; Yang, L.; Chen, Y.; Du, H.; Li, Z.; Lei, Y.; et al. Frailty index and all-cause and cause-specific mortality in Chinese adults: A prospective cohort study. *Lancet Public Health* **2020**, *5*, e650–e660. [CrossRef] [PubMed]
5. Angulo, J.; El Assar, M.; Rodriguez-Manas, L. Frailty and sarcopenia as the basis for the phenotypic manifestation of chronic diseases in older adults. *Mol. Aspects Med.* **2016**, *50*, 1–32. [CrossRef] [PubMed]
6. Arauna, D.; Cerda, A.; Garcia-Garcia, J.F.; Wehinger, S.; Castro, F.; Mendez, D.; Alarcon, M.; Fuentes, E.; Palomo, I. Polypharmacy Is Associated with Frailty, Nutritional Risk and Chronic Disease in Chilean Older Adults: Remarks from PIEI-ES Study. *Clin. Interv. Aging* **2020**, *15*, 1013–1022. [CrossRef] [PubMed]
7. Liperoti, R.; Vetrano, D.L.; Palmer, K.; Targowski, T.; Cipriani, M.C.; Lo Monaco, M.R.; Giovannini, S.; Acampora, N.; Villani, E.R.; Bernabei, R.; et al. Association between frailty and ischemic heart disease: A systematic review and meta-analysis. *BMC Geriatr.* **2021**, *21*, 357. [CrossRef] [PubMed]
8. Dent, E.; Martin, F.C.; Bergman, H.; Woo, J.; Romero-Ortuno, R.; Walston, J.D. Management of frailty: Opportunities, challenges, and future directions. *Lancet* **2019**, *394*, 1376–1386. [CrossRef] [PubMed]
9. Gale, C.R.; Westbury, L.; Cooper, C. Social isolation and loneliness as risk factors for the progression of frailty: The English Longitudinal Study of Ageing. *Age Ageing* **2018**, *47*, 392–397. [CrossRef] [PubMed]
10. Kehler, D.S.; Theou, O. The impact of physical activity and sedentary behaviors on frailty levels. *Mech. Ageing Dev.* **2019**, *180*, 29–41. [CrossRef] [PubMed]
11. Yang, G.; Cao, X.; Li, X.; Zhang, J.; Ma, C.; Zhang, N.; Lu, Q.; Crimmins, E.M.; Gill, T.M.; Chen, X.; et al. Association of Unhealthy Lifestyle and Childhood Adversity With Acceleration of Aging Among UK Biobank Participants. *JAMA Netw. Open* **2022**, *5*, e2230690. [CrossRef] [PubMed]
12. Otaki, N.; Yano, M.; Yokoro, M.; Tanino, N.; Fukuo, K.; Raymo, J. Relationship Between Dietary Variety and Frailty in Older Japanese Women During the Period of Restriction on Outings due to COVID-19. *J. Gerontol. Ser. B* **2021**, *76*, e256–e262. [CrossRef] [PubMed]
13. Hayakawa, M.; Motokawa, K.; Mikami, Y.; Yamamoto, K.; Shirobe, M.; Edahiro, A.; Iwasaki, M.; Ohara, Y.; Watanabe, Y.; Kawai, H.; et al. Low Dietary Variety and Diabetes Mellitus Are Associated with Frailty among Community-Dwelling Older Japanese Adults: A Cross-Sectional Study. *Nutrients* **2021**, *13*, 641. [CrossRef] [PubMed]
14. Kiuchi, Y.; Makizako, H.; Nakai, Y.; Tomioka, K.; Taniguchi, Y.; Kimura, M.; Kanouchi, H.; Takenaka, T.; Kubozono, T.; Ohishi, M. The Association between Dietary Variety and Physical Frailty in Community-Dwelling Older Adults. *Healthcare* **2021**, *9*, 32. [CrossRef] [PubMed]
15. Yokoro, M.; Otaki, N.; Yano, M.; Tani, M.; Tanino, N.; Fukuo, K. Associations between Dietary Variety and Frailty in Community-Dwelling Older People Who Live Alone: Gender Differences. *J. Frailty Aging* **2023**, *12*, 30–36. [CrossRef] [PubMed]

16. Yokoro, M.; Otaki, N.; Yano, M.; Imamura, T.; Tanino, N.; Fukuo, K. Low Dietary Variety Is Associated with Incident Frailty in Older Adults during the Coronavirus Disease 2019 Pandemic: A Prospective Cohort Study in Japan. *Nutrients* **2023**, *15*, 1145. [CrossRef]
17. Xu, X.; Inglis, S.C.; Parker, D. Sex differences in dietary consumption and its association with frailty among middle-aged and older Australians: A 10-year longitudinal survey. *BMC Geriatr.* **2021**, *21*, 217. [CrossRef] [PubMed]
18. Aihemaitijiang, S.; Zhang, L.; Ye, C.; Halimulati, M.; Huang, X.; Wang, R.; Zhang, Z. Long-Term High Dietary Diversity Maintains Good Physical Function in Chinese Elderly: A Cohort Study Based on CLHLS from 2011 to 2018. *Nutrients* **2022**, *14*, 1730. [CrossRef] [PubMed]
19. Wang, X.-M.; Zhong, W.-F.; Li, Z.-H.; Chen, P.-L.; Zhang, Y.-J.; Ren, J.-J.; Liu, D.; Shen, Q.-Q.; Yang, P.; Song, W.-Q.; et al. Dietary diversity and frailty among older Chinese people: Evidence from the Chinese Longitudinal Healthy Longevity Study. *Am. J. Clin. Nutr.* **2023**, *117*, 383–391. [CrossRef] [PubMed]
20. Lennon, H.; Kelly, S.; Sperrin, M.; Buchan, I.; Cross, A.J.; Leitzmann, M.; Cook, M.B.; Renehan, A.G. Framework to construct and interpret latent class trajectory modelling. *BMJ Open* **2018**, *8*, e020683. [CrossRef] [PubMed]
21. Lu, L.; Contrand, B.; Dupuy, M.; Ramiz, L.; Sztal-Kutas, C.; Lagarde, E. Mental and physical health among the French population before and during the first and second COVID-19 lockdowns: Latent class trajectory analyses using longitudinal data. *J. Affect. Disord.* **2022**, *309*, 95–104. [CrossRef] [PubMed]
22. Zeng, Y.; Feng, Q.; Hesketh, T.; Christensen, K.; Vaupel, J.W. Survival, disabilities in activities of daily living, and physical and cognitive functioning among the oldest-old in China: A cohort study. *Lancet* **2017**, *389*, 1619–1629. [CrossRef] [PubMed]
23. Searle, S.D.; Mitnitski, A.; Gahbauer, E.A.; Gill, T.M.; Rockwood, K. A standard procedure for creating a frailty index. *BMC Geriatr.* **2008**, *8*, 24. [CrossRef] [PubMed]
24. Gu, D.; Dupre, M.E.; Sautter, J.; Zhu, H.; Liu, Y.; Yi, Z. Frailty and Mortality Among Chinese at Advanced Ages. *J. Gerontol. Ser. B Psychol. Sci. Soc. Sci.* **2009**, *64B*, 279–289. [CrossRef] [PubMed]
25. Kulminski, A.; Yashin, A.; Ukraintseva, S.; Akushevich, I.; Arbeev, K.; Land, K.; Manton, K. Accumulation of health disorders as a systemic measure of aging: Findings from the NLTCS data. *Mech. Ageing Dev.* **2006**, *127*, 840–848. [CrossRef] [PubMed]
26. Mitnitski, A.; Song, X.; Skoog, I.; Broe, G.A.; Cox, J.L.; Grunfeld, E.; Rockwood, K. Relative fitness and frailty of elderly men and women in developed countries and their relationship with mortality. *J. Am. Geriatr. Soc.* **2005**, *53*, 2184–2189. [CrossRef] [PubMed]
27. Goggins, W.B.; Woo, J.; Sham, A.; Ho, S.C. Frailty index as a measure of biological age in a Chinese population. *J. Gerontol. A Biol. Sci. Med. Sci.* **2005**, *60*, 1046–1051. [CrossRef]
28. Song, Y.; Deng, Y.; Li, J.; Hao, B.; Cai, Y.; Chen, J.; Shi, H.; Xu, W. Associations of falls and severe falls with blood pressure and frailty among Chinese community-dwelling oldest olds: The Chinese Longitudinal Health and Longevity Study. *Aging* **2021**, *13*, 16527–16540. [CrossRef] [PubMed]
29. Zhang, J.; Wang, Q.; Hao, W.; Zhu, D. Long-Term Food Variety and Dietary Patterns Are Associated with Frailty among Chinese Older Adults: A Cohort Study Based on CLHLS from 2014 to 2018. *Nutrients* **2022**, *14*, 4279. [CrossRef] [PubMed]
30. Lv, Y.-B.; Gao, X.; Yin, Z.-X.; Chen, H.-S.; Luo, J.-S.; Brasher, M.S.; Kraus, V.B.; Li, T.-T.; Zeng, Y.; Shi, X.-M. Revisiting the association of blood pressure with mortality in oldest old people in China: Community based, longitudinal prospective study. *BMJ* **2018**, *361*, k2158. [CrossRef]
31. Rubin, D.B. *Multiple Imputation for Nonresponse in Surveys*; Wiley & Sons: New York, NY, USA, 1987. [CrossRef]
32. VanderWeele, T.J.; Ding, P. Sensitivity Analysis in Observational Research: Introducing the E-Value. *Ann. Intern. Med.* **2017**, *167*, 268–274. [CrossRef] [PubMed]
33. Zhou, Q.; Li, Y.; Gao, Q.; Yuan, H.; Sun, L.; Xi, H.; Wu, W. Prevalence of Frailty among Chinese Community-Dwelling Older Adults: A Systematic Review and Meta-Analysis. *Int. J. Public Health* **2023**, *68*, 1605964. [CrossRef] [PubMed]
34. Cesari, M.; Gambassi, G.; van Kan, G.A.; Vellas, B. The frailty phenotype and the frailty index: Different instruments for different purposes. *Age Ageing* **2014**, *43*, 10–12. [CrossRef] [PubMed]
35. Colizzi, M.; Lasalvia, A.; Ruggeri, M. Prevention and early intervention in youth mental health: Is it time for a multidisciplinary and trans-diagnostic model for care? *Int. J. Ment. Health Syst.* **2020**, *14*, 23. [CrossRef] [PubMed]
36. Claesson, M.J.; Jeffery, I.B.; Conde, S.; Power, S.E.; O'Connor, E.M.; Cusack, S.; Harris, H.M.B.; Coakley, M.; Lakshminarayanan, B.; O'Sullivan, O.; et al. Gut microbiota composition correlates with diet and health in the elderly. *Nature* **2012**, *488*, 178–184. [CrossRef] [PubMed]
37. Tiihonen, K.; Ouwehand, A.C.; Rautonen, N. Human intestinal microbiota and healthy ageing. *Ageing Res. Rev.* **2010**, *9*, 107–116. [CrossRef] [PubMed]
38. Schuette, L.K.; Song, W.O.; Hoerr, S.L. Quantitative use of the Food Guide Pyramid to evaluate dietary intake of college students. *J. Am. Diet. Assoc.* **1996**, *96*, 453–457. [CrossRef] [PubMed]
39. Mirmiran, P.; Azadbakht, L.; Esmaillzadeh, A.; Azizi, F. Dietary diversity score in adolescents—A good indicator of the nutritional adequacy of diets: Tehran lipid and glucose study. *Asia Pac. J. Clin. Nutr.* **2004**, *13*, 56–60.
40. DeLuca, H.F. Overview of general physiologic features and functions of vitamin D. *Am. J. Clin. Nutr.* **2004**, *80*, 1689S–1696S. [CrossRef] [PubMed]
41. Morcos, S.R.; El-Shobaki, F.A.; El-Hawary, Z.; Saleh, N. Effect of vitamin C and carotene on the absorption of calcium from the intestine. *Z. Ernährungswissenschaft* **1976**, *15*, 387–390. [CrossRef]

42. Gasmi, A.; Bjørklund, G.; Peana, M.; Mujawdiya, P.K.; Pivina, L.; Ongenae, A.; Piscopo, S.; Severin, B. Phosphocalcic metabolism and the role of vitamin D, vitamin K2, and nattokinase supplementation. *Crit. Rev. Food Sci. Nutr.* **2022**, *62*, 7062–7071. [CrossRef] [PubMed]
43. Cao, X.; Ma, C.; Zheng, Z.; He, L.; Hao, M.; Chen, X.; Crimmins, E.M.; Gill, T.M.; Levine, M.E.; Liu, Z. Contribution of life course circumstances to the acceleration of phenotypic and functional aging: A retrospective study. *eClinicalMedicine* **2022**, *51*, 101548. [CrossRef] [PubMed]
44. Tarasuk, V.S.; Brooker, A.S. Interpreting epidemiologic studies of diet-disease relationships. *J. Nutr.* **1997**, *127*, 1847–1852. [CrossRef] [PubMed]
45. Satija, A.; Yu, E.; Willett, W.C.; Hu, F.B. Understanding nutritional epidemiology and its role in policy. *Adv. Nutr.* **2015**, *6*, 5–18. [CrossRef] [PubMed]

nutrients

MDPI

Article

Targeting Aging and Longevity with Exogenous Nucleotides (TALENTs): Rationale, Design, and Baseline Characteristics from a Randomized Controlled Trial in Older Adults

Shuyue Wang [1,2], Lixia Song [1,2], Rui Fan [1,2], Qianqian Chen [1,2], Mei You [1,2], Meng Cai [1,2], Yuxiao Wu [1,2], Yong Li [1,2] and Meihong Xu [1,2,*]

1 Department of Nutrition and Food Hygiene, School of Public Health, Peking University, Beijing 100191, China; 2111210090@stu.pku.edu.cn (S.W.); 1710306221@pku.edu.cn (L.S.); fanruirf@bjmu.edu.cn (R.F.); 2211210120@stu.pku.edu.cn (Q.C.); 1610306135@pku.edu.cn (M.Y.); 1810306215@pku.edu.cn (M.C.); 2311210112@bjmu.edu.cn (Y.W.); liyong@bjmu.edu.cn (Y.L.)

2 Beijing Key Laboratory of Toxicological Research and Risk Assessment for Food Safety, Peking University, Beijing 100191, China

* Correspondence: xumeihong@bjmu.edu.cn; Tel.: +86-010-8280-5243

Abstract: Nucleotides (NTs), important biomolecules involved in numerous cellular processes, have been proposed as potential candidates for anti-aging interventions. However, whether nucleotides can act as an anti-aging supplement in older adults remains unclear. TALENTs is a randomized, double-blinded, placebo-controlled trial that evaluates the efficacy and safety of NTs as an anti-aging supplement in older adults by exploring the effects of NTs on multiple dimensions of aging in a rigorous scientific setting. Eligible community-dwelling adults aged 60–70 years were randomly assigned equally to two groups: nucleotides intervention group and placebo control group. Comprehensive geriatric health assessments were performed at baseline, 2-months, and 4-months of the intervention. Biological specimens were collected and stored for age-related biomarker testing and multi-omics sequencing. The primary outcome was the change from baseline to 4 months on leukocyte telomere length and DNA methylation age. The secondary aims were the changes in possible mechanisms underlying aging processes (immunity, inflammatory profile, oxidative stress, gene stability, endocrine, metabolism, and cardiovascular function). Other outcomes were changes in physical function, body composition and geriatric health assessment (including sleep quality, cognitive function, fatigue, frailty, and psychology). In the RCT, 301 participants were assessed for eligibility and 122 were enrolled. Participants averaged 65.65 years of age, and were predominately female (67.21%). All baseline characteristics were well-balanced between groups, as expected due to randomization. The majority of participants were pre-frailty and had at least one chronic condition. The mean scores for physical activity, psychological, fatigue and quality of life were within the normal range. However, nearly half of the participants still had room for improvement in cognitive level and sleep quality. This TALENTs trial will represent one of the most comprehensive experimental clinical trials in which supplements are administered to elderly participants. The findings of this study will contribute to our understanding of the anti-aging effects of NTs and provide insights into their potential applications in geriatric healthcare.

Keywords: healthy aging; nucleotides; randomized trial; aging biomarkers; older adults

1. Introduction

The World Demography Report shows that the world is experiencing an unprecedented rapid aging process, and all countries will face the economic, social, and political challenges, as well as other issues brought about by population aging. Aging in humans physically refers to a multidimensional process, in that all the changes are accumulated in a person over time [1,2]. These aging changes are responsible for the progressive increases in the chance of disease and death associated with them. The central hallmarks of age-related changes include genomic instability, telomere depletion, epigenetic changes, loss of protein homeostasis, deregulated nutrient sensing, mitochondrial dysfunction, cellular senescence, stem cell exhaustion, and altered intercellular communication [3]. The reported anti-aging star compounds/drugs that have been subject to preclinical and clinical trials targeting the above aging characteristics, achieving certain results [4–8], include rapamycin, anti-aging drugs (senolytics), metformin, acarbose, spermidine, NAD+ supplements, and lithium. However, the function of these compounds needs to be further confirmed, and there are still many adverse reactions reported. The intervention of food-derived nutritional active ingredients has the advantages of safety, multi-effect, economy and preventive long-term interventions, which is an important way to explore the intervention methods of aging.

Nucleotides (NTs), the building blocks of DNA and RNA, are the material basis of heredity, energy metabolism, and signal transduction, and play an important role in the growth and development of organisms, metabolism, reproduction, heredity, and aging [9,10]. In the field of food nutrition, nucleotides (NTs) are the main forms of nucleoside composition detection and functional ingredient application measurement, and nucleotides have important nutritional value for body health. As a food raw material approved by the State Food and Drug Administration, nucleotides have been widely used in infant formula milk powder/food, and health food/special medical use formula food [11–14]. Additionally, their safety was fully verified by a lifetime feeding experiment and four generation breeding experiment with high safety [15–17]. What is more, NTs have been found to have many biological functions, such as anti-aging [18], telomere length regulation [19,20], tumor inhibition [21], immune regulation [22–24], anti-inflammatory and antioxidant properties [25,26], learning and memory improvement [27,28], anti-fatigue [29], liver protection [30,31], intestinal protection [32–34], flora regulation [35,36], growth promotion [37,38], neuroprotective effect [39], and participation in one carbon unit metabolism.

The food sources of NTs are animal viscera (liver, kidney, brain, etc.) and seafood, lean meat, broth, fish, goose meat, yeast are also rich in NTs. Nucleic acids in food are mostly in the form of nuclear proteins. Nucleoproteins are broken down into nucleic acids and proteins by the action of gastric acid in the stomach. The nucleic acid is gradually hydrolyzed in the small intestine by various hydrolases in the pancreatic and intestinal fluids, and finally produce the bases and pentose [10]. Evidence suggests that NTs in the diet may be regarded as conditionally essential nutrients. Under normal circumstances, the body can synthesize nucleotides de novo and remediate them to maintain nucleotide levels in the tissue. However, when the body is in a special state (immunosuppression, recovery from injury, infection, specific disease states, and insufficient nutrient intake) or a special life stage (rapid growth and functional decline), the ability of NTs to synthesize from scratch is greatly reduced and the remedial capacity of nucleotide synthesis is insufficient or limited [22]. Therefore, obtaining NTs externally is a better mode of supplementation when the organism is under special circumstances, which plays an important role in saving energy consumption of de novo or remediation synthesis in the body, maintaining tissue nucleotide pool homeostasis, and improving and optimizing tissue function.

In the process of aging, the ability of NTs to synthesize from scratch is greatly reduced and the remedial capacity of nucleotide synthesis is insufficient or limited. Moreover, due to the increase of DNA repair and mitochondrial DNA synthesis and the depletion of metabolic enzymes, the content of deoxynucleotide triphosphates (dNTPs) decreases, leading to cell cycle arrest during aging. dNTPs supplementation can promote sustained cell proliferation [40]. NTs as synthetic raw material of dNTPs with potential anti-aging effects may have positive significance for delaying senescence. However, the comprehensive evaluation of exogenous nucleotides on the improvement of nutrition in the elderly population is still lacking. To address this knowledge gap, this randomized controlled trial aims to explore whether NTs have anti-aging effects and explore the possible pathways of the anti-aging effects of NTs. This manuscript describes the study protocol and baseline results for the TALENTs trial.

2. Methods

2.1. Trial Design

This was an exploratory study of participants from the TALENTs trial (Targeting Aging and Longevity with Exogenous Nucleotides), which was a pragmatic 4-month prospective single-center randomized control trial (Figure 1). Briefly, the TALENTs trial enrolled community-living people aged 60–70 years, to evaluate the efficacy and safety of nucleotides as an anti-aging supplement. Major exclusion criteria included those having terminal or serious illness, inability to communicate adequately with the researcher, and having already taken nucleotide supplements, etc. Following the provision of written informed consent, baseline assessments were conducted. The demographic factors, lifestyle, general clinical data, current medication or supplementation use, and diet of the subjects were registerd. Additionally, the subjects underwent comprehensive health assessments, including clinical health physical examination, questionnaire surveys, physical function assessment, and anthropometrics as part of the screening process. Eligible participants were randomly assigned by a 1:1 ratio to receive supplementation of nucleotides (1.2 g/day) or placebo daily for 4 months. Assessments and biological sample collection at baseline, 2 months, and 4 months post-intervention were executed by well-trained and skilled researchers.

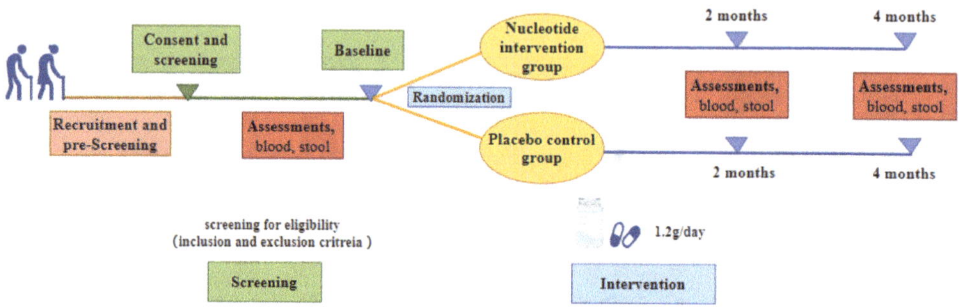

Figure 1. Overview of the TALENTs trial.

2.2. Ethic Committee Review and Approval

The study design and procedures have been conducted under the guidance of the Declaration of Helsinki and later amendments. Study approval has obtained from the Biomedical Ethics Committee of Peking University (ethics review approval number: IRB00001052-21114, date of approval: 17 November 2021). The TALENTs trial has been registered prospectively at Clinical Trials.gov (NCT05243018). The purpose, nature, and potential risks of participation in the trial were fully explained to the participants, and all participants provided written informed consent before participating in the trial.

2.3. Sample Size Calculation

A power analysis was not conducted due to the exploratory and preliminary nature of this study. Therefore, the sample size was based on the requirements of the human health food test, with 50 people in each group and an estimated loss of follow-up rate of 20%, so 60 people were required in each group; a 1:1 control group and intervention group totalled of 120 people.

2.4. Recruitment and Eligibility Criteria

In this study, subjects were recruited by the project leader and researchers through the method of publishing recruitment inspiration and posters on the Internet. As the trial site was set in Chengdu, potential participants were required to be local community residents and non-hospital patients. Research staff were to fully explain the purpose, nature, and potential risks of participation in this trial for prospective participants before trial participation (consent shown in Supplementary S1). Then, participants who provided written informed consent were assessed for their eligibility by comprehensive geriatric health assessments. Clinical physical examinations were conducted at Aikang Guobin Medical Examination Hospital (Chengdu, China) by clinical doctors with medical qualifications. Researchers evaluated participants' final eligibility based on the trial's inclusion and exclusion criteria.

Inclusion criteria: (1) age between 60 and 70 (2) no serious physical or mental illness; (3) never take nucleotide-related supplements/health food (4)Be able to follow the test protocol and sign the informed consent.

Exclusion criteria: (1) patients with related confirmed diseases, such as autoimmune system diseases, serious cardiovascular and cerebrovascular diseases, major organ complications such as in the liver and kidney, or complicated with other serious diseases such as malignant tumors, pancreatic diseases and mental diseases; (2) having abnormal screening laboratory test values or other lab test results that would preclude study participation in the judgment of the investigator; (3) severe visual or hearing loss affecting communication; (4) being participants in other clinical trials within the last 6 months; (5) having used food or medicine that is relevant to the function being tested.

2.5. Interventions

The intervention group was to receive the capsules of exogenous NTs, which were extracted from cane sugar by enzymatic hydrolysis, provided by Zhen-Ao Biotechnology Co., Ltd., (Dalian, China). The placebo capsules in the control group looked and tasted as similar as possible to those in the intervention group, making the participants indistinguishable. Both groups needed to take four capsules daily. Among them, each capsule in the nucleotides group contained 0.1 g starch excipients + 0.3 g nucleotides, and each capsule in the control group contained 0.4 g starch excipients. The specific nucleotide composition in this trial was 5′-AMP:5′-CMP:5′-GMPNa2:5′-UMPNa$_2$, prepared according to the ratio of 16:41:19:24, which is consistent with the ratio in breast milk and meets the requirements of national infant formula addition and special medical food formula. The dose of 1.2 g/day used is currently approved as the ingredient dose of nucleotides in conventional commercially available health food products. The experimental intervention continued for 4 months without interruption.

2.6. Randomization and Blinding

In this trial, participants were to be centrally randomized at a 1:1 allocation ratio into intervention and control groups using computer-generated random numbers. After grouping, the main factors that may affect the results and outcome indicators, such as gender and age, were tested for balance to ensure the comparability between groups. The random scheme was generated and known only by the experiment designer. Study participants and data collection staff including investigators, clinicians and follow-up staff were blinded. Pharmacists from Zhen-Ao Biotechnology Co., Ltd., (Dalian, China) manufactured and packed each subject's capsule into an opaque medicine bottle and numbered it on the outside. In the end, only the participant's name was marked on the bottle, and the bottle packaging and the capsules contained were identical in appearance. Blinding was fully explained when subjects sign the informed consent. If subjects did not accept the blinding method, they were not included in this study. The effectiveness of the study blinding was assessed with a short questionnaire completed by the study subjects when the trial was finished. The randomization code would not be broken without consent from the experiment designer under unusual circumstances, such as in the event of numerous serious adverse events before the end of the study. Neither the trial designer nor the pharmacist participated in the trial.

2.7. Follow-Up

After entering the follow-up period of intervention, trained follow-up staff were responsible for follow-ups and regular home visits to distribute capsules to reduce the burden on participants. In addition to handing out intervention capsules during each home visit to the subjects, the follow-up staff also presented practical daily activities to keep the participants upbeat. A total of 120 subjects were randomly divided into 12 follow-up groups with 10 people in each group and followed up by 12 trained follow-up staff respectively. Participants were required to report the time taken, and dosage of intervention capsules daily in WeChat groups and report special circumstances or adverse reactions. Researchers provided participants with an interventional follow-up manual, which explains the content and time points they needed to cooperate with during the study, as well as what they should do and who to contact if any new symptoms or side effects occurred. This study specifically hired a clinical doctor to provide relevant clinical consultation for participants to understand their participation and participate in clinical management when necessary.

2.8. Outcomes

The primary outcomes assessed changes in molecular biomarkers associated with aging, encompassing telomere length and DNA-methylation clock compared with baseline. Secondary outcomes included alterations in mechanisms underlying aging processes such as immune function, inflammatory cytokine levels, oxidative stress levels, DNA stability, endocrine, metabolic, and cardiovascular functions. Other outcomes comprised physical function, body composition and the Geriatric Assessment Questionnaire (including sleep quality, cognitive function, fatigue, frailty, and psychology). Figure 2 lists the outcome measures. Data were obtained at baseline, 2 months, and 4 months by teams of fieldworkers, who were trained in standard measurement and protocols of questionnaire administration.

2.8.1. Primary Outcomes

(1) Leukocyte telomere length: telomeres are repetitive nucleotide elements at the end of chromosomes that protect chromosomes from degradation and genetic information loss. In this study, leukocyte telomere length was measured using quantitative polymerase chain reaction assay and reported as T/S ratios (the relative ratio of telomere repeat copy number to single-copy gene) [41].

(2) DNAm clocks: the DNAm clock, an algorithm that combines DNAm measurement information in the genome to quantify biological age changes, is considered a highly accurate molecular correlation of the true age of humans and can detect DNA methylation age reversal up to three months ahead of the respective individual clocks, so it has the potential to quantify biological aging and test life expectancy or assess intervention effectiveness [42–44].

2.8.2. Secondary Outcomes

(1) Immune function: general parameters of immune health that are known to change with age were assessed as described, including IgA, IgG, IgM, IgE, and PBMCs (peripheral blood mononuclear cells) analyzed by FACS to assess the proportion and number of T lymphocyte subtypes ($CD3^+$, $CD3^+CD4^+$, $CD3^+CD8^+$, $CD4^+CD25^+$);

(2) Serum cytokine levels: SASP is a consequence of cell senescence and may occur in cells that, though undergoing cell cycle arrest, are still metabolically active and secrete proteins [45,46]. Luminex-based assays were be used, as these allow simultaneous measurement of different cytokines (TNF-α, IL-1β, IL-6, IL-18, TNF RI, TNF RII, ICAM-1, etc.);

(3) Oxidative stress: enzyme-linked immunosorbent assay (ELISA) kits were used to detect the level of serum oxidative (MDA, NOX, SOD, GSH-Px, etc.);

(4) Gene stability: γ-H2A.X phosphorylation is an initial and essential component of DNA damage foci and therefore a reliable marker of the extent of DNA damage [47,48];

(5) Glycolipid metabolic profile: glucose, insulin, total cholesterol, HDL-C, LDL-C, triglycerides, etc.;

(6) Endocrine and cardiovascular function: adiponectin, leptin, NT-proBNP, IGF-1, NO, carotid thickness of intima-media, spontaneous fluorescence of subcutaneous AGEs. etc.

2.8.3. Other Outcomes

Geriatric Assessment Questionnaire (CGAQ)

A self-designed questionnaire was used to evaluate the health status of all subjects before and after intervention (Full Questionnaire shown in Supplementary S2). The specific contents of the questionnaire are as follows:

(1) General information survey: demographic characteristics, including date of birth, occupation, education level, economic status, family status, etc.; lifestyle and living habits (smoking, drinking, tea, sports, etc.); the number of available teeth (including dentures) and the loss of teeth in the past one year; medical history (hypertension, coronary heart disease, obesity and other diseases); health food consumption, drug application history; expenditure on health, including physical examination, medicine, hospital expenses, etc.;
(2) Quality of life: assessment of physical and mental functioning (SF-12) [49];
(3) Frailty questionnaire: 28-item frailty index [50];
(4) Cognitive status: Montreal Cognitive Assessment Scale (MoCA) [51];
(5) Physical activity: Physical Activity Scale for the Elderly (PASE) [52];
(6) Sleep quality: Pittsburgh Sleep Index [53];
(7) Psychological function: Kessler Psychological Distress (K10) Scale [54] and Fatigue Scale-14 (FS-14) [55];
(8) Dietary survey: dietary data were collected at baseline and after intervention using a 3 day, 24 h photo-assisted dietary intake assessment method to assess dietary stability during the trial [56]. The Food Frequency Questionnaire (FFQ) was used to measure the intake of foods rich in nucleotides, including seasonings high in nucleotides, to assess daily dietary nucleotide intake.

Anthropometry and Physical Function

Waist circumference, hip circumference, calf circumference, mid-arm circumference and neck circumference were measured with a tape measure to the nearest 0.1 cm. We used bioelectrical impedance analysis (BIA) to measure the proportion of whole-body lean and fat mass, body fat, trunk fat, body muscle, trunk muscle, and other components in the human body.

The level of physical functioning was assessed with the short physical performance battery (SPPB), which consisted of standing balance tasks (side-by-side, semi-, and full-tandem stands for 10 s each), a 4 m walk to assess usual gait speed, and time to complete five repeated chair stands. Handgrip strength and a 30 s sarm curl test were to measure upper muscle strength. The Tinetti clinical scale including balance assessment items and gait assessment items, was used to evaluate fall risk.

Clinical Health Physical Examination and Safety

All blood measurements were performed in the morning after an overnight fast. Blood, urine, stool routine examination, and liver and kidney function were examined, and other safety indicators included chest X-ray, lung function, and electrocardiogram, etc.

The number and types of adverse events including changes in laboratory parameters (blood chemistry, lipid profile), monitored and documented. All adverse events, including falls, fractures, hospitalizations, and deaths were recorded in an "adverse events diary" during follow-up visits by the trained follow-up staff, and by self-report during the study period.

Multi-Omics Sequencing

The biospecimens of the TALENTs trial (shown in Supplementary S3) were transferred to the laboratory on dry ice within 24 h of collection and then stored at −80 °C. Blood samples were collected in three different types of collection tubes containing distinct stabilizers or coagulants, including ethylenediaminetetraacetic acid (EDTA) tubes, serum separation tubes (SST), and RNA preservation tubes. Whole blood, blood clots, and serum were taken for whole genome methylation, metabolomic and transcriptomic sequencing. Stool samples were taken for metagenomic sequencing.

Category	Assessment Domain	Item
Primary Outcome	Aging biomarker	(1) Telomere length of peripheral white blood cells. (2) DNA methylation age.
Secondary Outcome	Mechanism of aging	(1) Immune function: IgA, IgG, IgM, IgE, T lymphocyte subtypes($CD3^+$, $CD3^+CD4^+$, $CD3^+CD8^+$, $CD4^+CD25^+$). (2) Inflammatory cytokine: TNF-α, IL-1β,IL-6, IL-18, TNF RI,TNF RII, ICAM-1, etc. (3) Oxidative stress: MDA, NOX, SOD, GSH-Px, etc. (4) Gene stability: γ-H2A.X phosphorylation. (5) Glycolipid metabolic profile: HbA1c, FPG, insulin, HDL-C, LDL-C, TC, TG, etc. (6) Endocrine, metabolism and cardiovascular funtion:Adiponectin, leptin, NT-proBNP, IGF-1, NO, Carotid thickness of intima media, spontaneous fluorescence of subcutaneous AGEs, etc.
Other Outcome	Anthropometrics	(1) Physical motor function: SPPB, 6-meter walking speed test, chair standing test, grip strength, Tinetti scale, 30s sarm curl,etc. (2) Anthropometric measurement: waist circumference, hip circumference, middle arm circumference, neck circumference, calf circumference, etc. (3) Body composition: basic metabolism, body fat content, bone mass, muscle mass, etc.
	Clinical health physical examination	(1) Routine examination of blood: RBC, WBC, PLT, NEUT, LYMPH, Hb, ALB, etc. (2) Liver and kidney function: ALT, AST, ALP, TBA, Cys-C, Cr, β2-MG, etc. (3) Cancer marker screening: CEA, AFP. (4) Safety measurement: blood pressure, chest X-ray, lung function, electrocardiogram, abdominal B-ultrasonography.
	Comprehensive Geriatric Assessment Questionnaire (CGAQ)	(1) General information survey: general demographic characteristics, Lifestyle and living habits, Medical history, etc. (2) Quality of life: Physical and mental function assessment (SF-12). (3) Frailty questionnaire: 28-item frailty index. (4) Cognitive status: Montreal Cognitive Assessment Scale (MoCA). (5) Physical activity: Physical Activity Scale for the elderly(PASE). (6) Sleep quality: Pittsburgh Sleep Index. (7) Psychological function: Kessler 10 scale, Fatigue Scale-14 (FS-14) (8) Dietary survey: dietary frequency questionnaire (FFQ) and photo-assisted dietary intake assessment.
	Omics sequencing	Transcriptome, Methylomics, Metagenome, Metabolome.

Figure 2. Assessment outcomes of the TALENTs trial.

2.9. Statistical Analysis

The data in this study were analyzed by intention-to-treat (ITT), and no interpolation was performed for data missing from general data. The Shapiro-Wilk test was used for data normality and the Levene test for variance homogeneity. If the data conformed to normal distribution, the measurement data are expressed as mean ± SD, and the measurement data of non-normal distribution are expressed as median (P25, P75).

In the inter-group comparison, the changes between the outcomes after intervention and the baseline data were used as the outcome variable, and independent sample T test was used for the normal distribution data. Non-parametric Mann-Whitney U test was used to compare the non-normal distribution measurement data groups. Generalized estimating equations (Generalized Estimating Equation, GEE) were used to assess the effect of nucleotide intervention of outcome indicators. GEE can tolerate the presence of missing values in longitudinal data, and can perform unbiased estimation of research information, and GEE is relatively stable in various job-related matrices, so it is widely recommended in intervention studies. There are two levels of longitudinal data in this study; the assessment at each time point is level 1, nested in the study object (i.e., level 2). Continuous outcome indicators at each time point were used as dependent variables to evaluate the group main effect, the time main effect and the interaction effect of the time group. In addition, GEE model was used to adjust covariates including age, gender, BMI, and the baseline of the variables corresponding to the results. Cohens D was calculated using the mean difference between the two groups, with d = 0.2 as a small effect size, d = 0.5 as a medium effect size, and d = 0.8 as a large effect size. Analyses were performed using the SPSS 26.0 software and overseen by an experienced biostatistician; $p < 0.05$ indicates a statistically significant difference.

2.10. Procedures

Figure 3 shows the timeline for scheduled assessments of the TALENTs trial.

(1) Recruitment and screening period: physical examination in the partner hospital (blood and stool samples retained at the same time), questionnaire survey, and other items were completed. Questionnaires were completed by trained investigators and participants face-to-face. At the same time, participants were required to carry their past medical records and current medication and health supplements for on-site registration. According to the physical examination results, the subjects unfit to be enrolled were excluded;

(2) Initiation and follow-up period: during the intervention period, follow-up staffs are responsible for dispensing, reviewing, and documenting medications. Intervention capsules were distributed biweekly by follow-up staff during home visits. Old packaging products were collected for verification purposes and compliance is documented. Participants' physical activity levels, dietary patterns, changes in medication, and any adverse events experienced within the past two weeks were assessed and recorded by follow-up staff. At 2 and 4 months after the start of the intervention, physical examinations, questionnaires, and biological samples were obtained following the same procedures as at baseline.

(3) Final period: 4 months after the intervention, the project ended. Participants who successfully complete the project will receive compensation. At the end of the trial, we unblinded.

	STUDY PERIOD				
TIMEPOINT	**Enrolment**	**Allocation**	**Baseline**	**2-month**	**4-month**
ENROLMENT:					
Informed consent	✓				
Screening for eligibility	✓				
Socio-demographic and medical data collection	✓				
Allocation		✓			
INTERVENTIONS:					
Intervention arm: Nucleotides supplementation			⟵	—	⟶
Control arm: Placebo supplementation			⟵	—	⟶
ASSESSMENTS:					
Aging related biomarker			✓	✓	✓
Mechanism of aging			✓	✓	✓
Anthropometrics			✓		✓
Clinical health physical examination			✓	✓	✓
Comprehensive Geriatric Assessment			✓		✓
Omics sequencing			✓	✓	✓
adverse event monitoring			✓	✓	✓

Figure 3. Timeline for scheduled assessments of the TALENTs trial.

2.11. Quality Control

(1) Research design stage: review the relevant literature, understand the general situation of relevant research, and design the scheme according to the research purpose; simplify the study design process to minimize unnecessary inquiries and checks; For subjective data collection, standardized scales with high reliability, validity and responsiveness should were used as far as possible. The self-designed questionnaire and the medical records of the subjects were conducted in a small-scale pre-survey among the population before the formal experiment, and then discussed and evaluated with relevant experts and investigators. The final draft was revised several times according to the feedback;

(2) Investigator training: the sample collection and data measurement of the subjects was carried out by professional medical workers or trained qualified investigators in the physical examination hospital, and the measurements of each indicator were completed by the same personnel. The investigators were trained and passed the examination;
(3) Intervention follow-up stage: during the intervention period, we always communicated with the subjects, and subjects' medication and lifestyle remained unchanged, we strengthened education to improve the correct understanding and complianceof subjects to the research protocol; simplified study procedures and reduced the number of questions and tests to improve patient patience. Daily supervision was strengthened during the follow-up period, and the compliance of subjects was improved through the daily punch card exchange reward mechanism. We provided a high quality and free medical question answering service for the subjects;
(4) Data collection stage: we standardized the basic operation and process of implementation and organized the research subjects to carry out physical examination and questionnaire surveys on time. We strengthened the communication between investigators and research subjects, eliminated the concerns of survey subjects about this study, unified the measurement standards of indicators, reduced information bias, maintained a neutral investigation attitude, and improved the compliance of research subjects.
(5) Result analysis stage: Epidata and excel built databases were used to ensure parallel double entry and verification of data. We selected correct statistical analysis methods, and used blind methods for technical personnel, including subject sample processing and statistical analysis, to reduce measurement bias.
(6) Data management: according to the original observation records of the subjects, the researcher loaded the data into the physical examination data sheet and various questionnaires in a timely, complete, correct and clear manner. The survey form, which has been reviewed and signed by the Ombudsman, was sent to the research data manager in a timely manner. The corresponding database system was used for two-person and two-machine input, and then the database was compared twice. After all the physical examination data sheets and various questionnaires were entered and verified correctly, the data manager wrote the database inspection report, which included the study completion status (including the list of dropped subjects), inclusion/exclusion criteria, integrity checks, logical consistency checks, outlier data checks, time window checks, drug combination checks, and adverse event checks. After data entry and verification were completed as required, the physical examination data sheets and various questionnaires were filed and stored in numbered order and filled with search directories for reference. Electronic data files, including databases, inspection programs, analysis programs, analysis results, coding and explanatory files, should be classified and stored in different disks or recording media with multiple backups, properly stored to prevent damage. All original files were kept for the period specified accordingly.

3. Results

Between 23 August 2022 and 15 October 2022, we recruited 349 elderly people in the Chengdu community who were willing to participate and signed informed consent. Among them, 301 subjects underwent complete geriatric health assessments, including clinical health physical examination, questionnaire surveys, physical function assessment, and anthropometrics as part of the screening process. Blood and stool samples were collected for age-related biomarkers and omics monitoring. Based on inclusion and exclusion criteria, 123 were eligible to participate. The most common reasons for exclusion were not meeting inclusion criteria (n = 175) such as having abnormal screening laboratory test values or having cancer or other serious diseases. Of the eligible subjects, 123 subjects were randomized and 61were placed in the nucleotide intervention group and 62 in the placebo control group. However, after randomization, one participant in the nucleotide intervention group decided to withdraw from the trial before starting the intervention, so 122 participants were officially enrolled in the intervention. Figure 4 shows the flow chart of the study samples, including the number of participants who were screened and underwent randomization.

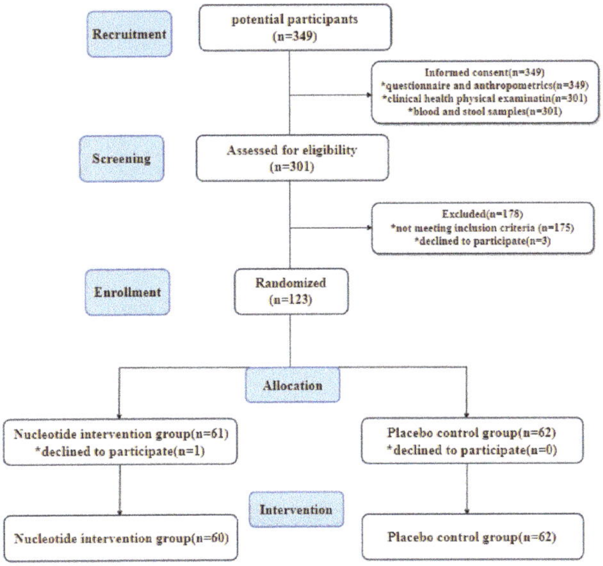

Figure 4. Participant flow of the TALENTs trial. The * represents a change after randomization.

Baseline data on sociodemographic characteristics, co-morbid conditions and health status of the enrolled participants appear in Table 1. All demographic characteristics were well-balanced between groups as expected due to randomization. The mean age was 65.65 ([SD] 2.59) years, 82 (67.21%) were women, and 121 (99.18%) were of Han nationality. Most participants were married (84.43%) and lived with others (91.80%). In terms of education level and monthly disposable income levels, a majority of the subjects received good education and were in good economic conditions. Only 4.10% of them received primary education or below, and only 14.75% had monthly disposable income level less than 2000 yuan. Analysis of smoking and drinking habits showed that only 6.56% and 13.93% of the subjects currently smoke or drink, respectively. The other subjects do not smoke or drink, or had quit smoking and drinking, suggesting that most of them have good living habits.

Table 1. Sociodemographic characteristics and health status of TALENTs participants.

	Overall (n = 122)	Intervention Group (n = 60)	Control Group (n = 62)	*p* for Diff
Age, years; mean ± SD	65.65 ± 2.59	65.55 ± 2.63	65.74 ± 2.58	0.685
Female, n (%)	82(67.21)	41(68.33)	41(66.13)	0.950
Nationality, n (%)				
Han nationality	121(99.18)	59(98.33)	62(100)	0.492
Marital status, n (%)				
single	2(1.64)	2(1.64)	0(0)	
married	103(84.43)	50(83.33)	53(85.48)	0.347
divorced/widowed	17(13.93)	8(13.33)	9(6.45)	

Table 1. *Cont.*

	Overall (n = 122)	Intervention Group (n = 60)	Control Group (n = 62)	*p* for Diff
Education level, n (%)				
primary school and below	5(4.10)	3(5.00)	2(3.23)	0.116
Junior high school	45(36.89)	25(41.67)	20(32.26)	
High school/technical secondary school	42(34.43)	23(38.33)	19(30.65)	
University/college or above	30(24.59)	9(15.00)	21(33.87)	
Living condition, n (%)				
Live alone	10(8.20)	5(8.33)	5(8.06)	0.957
Non-solitary	112(91.80)	55(91.67)	57(91.94)	
Monthly disposable income, n (%)				
<2000	18(14.75)	9(15.00)	9(14.52)	0.851
2000–3500	47(38.52)	25(41.67)	22(35.48)	
3501–5000	28(22.95)	14(23.33)	14(22.58)	
5001–6500	20(16.39)	9(15.00)	11(17.74)	
>6500	9(7.38)	3(5.00)	6(9.68)	
Smoking status, n (%)				
Never smoked	107(87.70)	49(81.67)	58(93.55)	0.135
Have quit smoking	7(5.74)	5(8.33)	2(3.23)	
smoking	8(6.56)	6(10.00)	2(3.23)	
Alcohol consumption, n (%)				
Never drank alcohol	99(81.15)	48(80.00)	51(82.26)	0.676
Have stopped drinking	6(4.92)	4(6.67)	2(3.23)	
drinking	17(13.93)	8(13.33)	9(14.52)	
Comorbidities, n (%)				
Hypertension	36(29.51)	20(33.33)	16(25.81)	0.362
Dyslipidemia	10(8.20)	2(3.33)	8(12.90)	0.095
Diabetes	17(13.93)	9(15.00)	8(12.90)	0.738
Cardiovascular disease	66(54.10)	37(61.67)	29(46.77)	0.099
Chronic respiratory diseases	40(32.79)	22(36.67)	18(29.03)	0.369
Fatty liver	45(36.89)	22(36.67)	23(37.10)	0.961
Renal disease	49(40.16)	20(33.33)	29(46.77)	0.130
Frailty index, mean ± SD	0.11 ± 0.06	0.11 ± 0.06	0.10 ± 0.06	0.528
Frailty statust, n (%)				
Robust	60(49.18)	27(45.00)	33(53.23)	0.547
Prefrail	57(46.72)	31(51.67)	26(41.94)	
Frail	5(4.10)	2(3.33)	3(4.84)	
PASE, mean ± SD	169.26 ± 51.93	161.98 ± 51.47	176.3 ± 51.82	0.128
SF-12, mean ± SD	108.00 ± 7.53	107.56 ± 7.26	108.42 ± 7.83	0.532
Physical health	52.26 ± 4.38	51.92 ± 3.94	52.58 ± 4.78	0.407
Mental health	55.75 ± 6.3	55.65 ± 6.00	55.84 ± 6.62	0.864
MoCA, mean ± SD	21.92 ± 3.86	21.5 ± 3.87	22.32 ± 3.84	0.241
PSQI, mean ± SD	4.89 ± 3.59	4.57 ± 3.3	5.21 ± 3.85	0.324
Kessler 10 scale, mean ± SD	11.61 ± 2.79	11.37 ± 2.03	11.84 ± 3.37	0.353
FS-14, mean ± SD	4.65 ± 3.46	5.1 ± 3.64	4.21 ± 3.24	0.156
Physical Fatigue	2.46 ± 2.29	2.75 ± 2.42	2.18 ± 2.14	0.168
Mental fatigue	2.19 ± 1.61	2.35 ± 1.69	2.03 ± 1.53	0.277

The baseline results indicate that both the NTs and control groups had similar levels of frail status, physical activity, cognitive function, sleep quality, depression, fatigue, and comorbidities at the start of the trial. Overall, the most common co-morbid conditions were cardiovascular disease (54.10%), renal disease (40.16%), fatty liver (36.89%), chronic respiratory disease (32.79%), hypertension (29.51%), diabetes (13.93%), and dyslipidemia (8.20%). The mean (SD) score of the PASE physical activity scale score was 169.26 (51.93), indicating that most participants were able to engage in moderate physical activity. The SF-12 Physical and Mental Functioning Assessment Scale showed that participants were in good physical and mental health, with a mean (SD) score of 108.00 (7.53). The MoCA

Montreal Cognitive Rating Scale score showed an average (SD) score of 21.92 (3.86) out of 30 and only 19.67% of participants had MoCA score $\geq$ 26, suggesting mild cognitive impairment (MCI) in participants accounting for 65.57%. The PSQI Pittsburgh Sleep Quality Index average (SD) score was 4.89 (3.59) with 40.98% of participants $\geq$ 5, indicating that nearly half of the participants had clinically significant sleep disturbances. The Kessler Anxiety and Depression Scale average (SD) score was 108.00 (7.53), indicating mild levels of anxiety and depression. The FS-14 Fatigue Scale had an average (SD) score of 4.65 (3.46), with most participants reporting low levels of fatigue. The frailty scale indicated that nearly half of the subjects were in the healthy stage (49.18%) or the pre-frailty stage (46.72%), except for a small number of participants (4.10%) who were in the frailty stage. In conclusion, the Comprehensive Geriatrics Health Assessment Questionnaire indicated that the health of the participants was relatively good overall, with good quality of life and good mental health, but the cognitive and sleep quality remained to be improved.

4. Discussions

Quantitative biomarkers of aging are valuable tools for measuring physiological age, assessing the degree of "healthy aging", and potentially predicting an individual's health span and lifespan. Given the complexity of the aging process, biomarkers of aging are multilayered and multifaceted. However, most studies so far concentrate on a single measure of biological aging or a single type of measure. Our study comprehensively evaluate the phenotype and molecular biomarkers of aging, placing particular emphasis on evaluating outcomes for markers of aging and possible mechanisms, and comprehensively evaluated the overall health status of older adults.

In our study, the primary outcome measures obtained information from single and multiple systems. For example, telomere length and epigenetic clock methods are cell-level measurements implemented only in a single-tissue blood, while age-related homeostasis, and aging rate measurements take information from multiple systems throughout the body. In addition, we use multi-faceted biomarkers of aging and multi-omics joint analysis to explore the specific mechanisms of targeting the aging process. What is more, the geriatric assessment questionnaire designed for evaluating aging phenotypes, as well as the items included in body composition and physical function measurements in this trial, are reliable and highly efficient. They can systematically, comprehensively, and holistically evaluate the anti-aging effect of nucleotides and their potential to prevent age-related diseases and extend healthy lifespan from a macroscopic.

The TALENTs trial successfully enrolled 122 older adults and completed the baseline assessments. What is more, the intervention and control groups were similar at the beginning of the trial, indicating that the randomization worked well. The results provide valuable information on the demographic and clinical characteristics of the study population and will serve as a reference point for evaluating the effects of NT supplementation on aging outcomes.

Clinical trials tailored to questions of importance to healthy older adults are urgently needed due to increasing numbers of people living into their 80s, 90s, and even 100s. In addition, relatively healthy elderly people have lower risks and higher benefits than those with more diseases, so the general research on anti-aging drugs or food active ingredients would first select relatively healthy elderly people for trial and exploration. Our baseline results of this trial suggest that the study population is a relatively healthy community of elderly aged 60–70 years, with good physical, mental and physical exercise abilities. However, the Montreal Scale indicated that this group of elderly people had mild cognitive impairment (mean 21.92) overall, thus were at increased risk of cognitive decline, which is consistent with the Montreal scores in those aged $\geq$60 years from a Chinese subsample drawn from a population-based study (mean 21.2) [57]. The overall disturbed sleep prevalence of household residents aged 60 and over recruited from a population-based random sample was 33.7%, which was basically similar to the rate of 40.98% in our study population [58]. Together, these data suggest that the baseline status in the current study are fairly representative and comparable to those reported in similar populations. Our intervention still had room for improvement in cognitive level and sleep quality.

This TALENTs trial also has many practical advantages during the implementation process. Firstly, all assessments and the data collection process are entirely in person rather than remote assessment. Secondly, the data collected in the survey are based on previous medical examination reports and available drug packaging, instead of being self- or proxy-reported, and participants were asked to look back in the past within the shortest feasible time to minimize the possibility of recall bias as much as possible. In addition, daily online WeChat online reminders for participants to take the capsules and biweekly home visits greatly reduced dropout rates, and information on daily physical activity, dietary patterns, and medication changes collected during follow-up provided a method for controlling confounding factors.

However, this trial still has certain limitations, such as having a small sample size and a short intervention period as an exploratory trial. Long-term follow-up in future studies will be necessary to assess the lasting effects of nucleotides on aging and longevity outcomes. Additionally, recruitment was performed through online channels, making it difficult for rural populations to participate. People who pay more attention to their health are more likely to learn about this project and participate, resulting in a certain degree of volunteer bias. Despite the limitations of this trial, the results are still of great clinical significance. Our study will reveal the complexity and multi-layered nature of the aging process, providing new directions and perspectives for future anti-aging research. Finally, our trial design and implementation methods also provide references for similar studies.

5. Conclusions

In conclusion, this RCT successfully recruited eligible participants and collected baseline data on their sociodemographic characteristics and health status. These data will serve as a reference for evaluating the efficacy and safety of nucleotides as an anti-aging supplement in older adults. The TALENTs trial combines the rigor of traditional double-blind randomized controlled trials with centralized and comprehensive systematic outcome determination, and will provide valuable insights into the anti-aging effects of NTs and contribute to our understanding of their underlying mechanisms. If proven effective, NT intervention may represent a promising strategy for counteracting aging-related decline and improving the quality of life for elderly individuals.

Supplementary Materials: The following supporting information can be downloaded at: https://www.mdpi.com/article/10.3390/nu16091343/s1, Supplementary S1: The informed consent of the TALENTs trial; Supplementary S2: The Comprehensive Geriatric Assessment Questionnaire of the TALENTs trial; Supplementary S3: The biospecimen collection of the TALENTs trial.

Author Contributions: S.W.: Investigation, Project administration, Methodology, Writing, Visualization. L.S.: Investigation, Project administration, Methodology. R.F.: Investigation, Methodology. Q.C.: Investigation, Validation. M.Y.: Investigation, Validation. M.C.: Investigation, Validation. Y.W.: Validation. Y.L.: Conceptualization. M.X.: Conceptualization, Funding acquisition, Resources, Supervision, Writing—review & editing. All authors have read and agreed to the published version of the manuscript.

Funding: This study is commissioned and funded by Zhen-Ao Biotechnology Co., Ltd., (Dalian, China). Funding number (2021220) was autonomous number. The sponsor provides assistance with subject recruitment and provides medical professional to participate in the entire health consultation process, but does not participate in the design, methodology, data collection, analysis, or paper writing.

Institutional Review Board Statement: Study approval has obtained from the Biomedical Ethics Committee of Peking University (ethics review approval number: IRB00001052-21114, date of approval: 17 November 2021).

Informed Consent Statement: Before data collection, the principal researcher obtained informed consent from all interested participants.

Data Availability Statement: The data presented in this study are available on request from the corresponding author. The data are not publicly available due to privacy.

Acknowledgments: The authors sincerely grateful to Xun Tang (Department of Epidemiology and Health Statistics, School of Public Health, Peking University) for his normative guidance in writing this article; The authors sincerely grateful to Lishi Zhang (West China School of Public Health, Sichuan University), Meiqin Cai (Shanghai Jiao Tong University School of Medicine), Haibin Xu (China National Center for Food Safety Risk Assessment), Weixing Yan (China National Center for Food Safety Risk Assessment), Ning Li (China National Center for Food Safety Risk Assessment) for their guidance on the design of the experiment; The authors sincerely grateful to Xianyu Guo (School of Medicine of Xi'an Jiao Tong University) for his medical consultation.

Conflicts of Interest: The authors don't have any conflict of interest.

References

1. Kennedy, B.K.; Berger, S.L.; Brunet, A.; Campisi, J.; Cuervo, A.M.; Epel, E.S.; Franceschi, C.; Lithgow, G.J.; Morimoto, R.I.; Pessin, J.E.; et al. Geroscience: Linking Aging to Chronic Disease. *Cell* **2014**, *159*, 708–712. [CrossRef]
2. Kirkwood, T.B.L. Understanding the odd science of aging. *Cell* **2005**, *120*, 437–447. [CrossRef] [PubMed]
3. Lopez-Otin, C.; Blasco, M.A.; Partridge, L.; Serrano, M.; Kroemer, G. Hallmarks of aging: An expanding universe. *Cell* **2023**, *186*, 243–278. [CrossRef]
4. Campisi, J.; Kapahi, P.; Lithgow, G.J.; Melov, S.; Newman, J.C.; Verdin, E. From discoveries in ageing research to therapeutics for healthy ageing. *Nature* **2019**, *571*, 183–192. [CrossRef] [PubMed]
5. Crimmins, E.; Vasunilashorn, S.; Kim, J.K.; Alley, D. Biomarkers related to aging in human populations. *Adv. Clin. Chem.* **2008**, *46*, 161–216. [PubMed]
6. Gonzalez-Freire, M.; Diaz-Ruiz, A.; Hauser, D.; Martinez-Romero, J.; Ferrucci, L.; Bernier, M.; de Cabo, R. The road ahead for health and lifespan interventions. *Ageing Res. Rev.* **2020**, *59*, 101037. [CrossRef]
7. Zhavoronkov, A.; Li, R.; Ma, C.; Mamoshina, P. Deep biomarkers of aging and longevity: From research to applications. *Aging* **2019**, *11*, 10771–10780. [CrossRef]
8. Xu, K.; Guo, Y.; Li, Z.; Wang, Z. Aging Biomarkers and Novel Targets for Anti-Aging Interventions. *Adv. Exp. Med. Biol.* **2019**, *1178*, 39–56. [PubMed]
9. Sánchez-Pozo, A.; Gil, A. Nucleotides as semiessential nutritional components. *Br. J. Nutr.* **2002**, *87*, S135–S137. [CrossRef]
10. Ding, T.; Song, G.; Liu, X.; Xu, M.; Li, Y. Nucleotides as optimal candidates for essential nutrients in living organisms: A review. *J. Funct. Foods* **2021**, *82*, 104498. [CrossRef]
11. Singhal, A.; Macfarlane, G.T.; Macfarlane, S.; Lanigan, J.; Kennedy, K.; Elias-Jones, A.; Stephenson, T.; Dudek, P.; Lucas, A. Dietary nucleotides and fecal microbiota in formula-fed infants: A randomized controlled trial. *Am. J. Clin. Nutr.* **2008**, *87*, 1785–1792. [CrossRef] [PubMed]
12. Singhal, A.; Kennedy, K.; Lanigan, J.; Clough, H.; Jenkins, W.; Elias-Jones, A.; Stephenson, T.; Dudek, P.; Lucas, A. Dietary Nucleotides and Early Growth in Formula-Fed Infants: A Randomized Controlled Trial. *Pediatrics* **2010**, *126*, E946–E953. [CrossRef] [PubMed]
13. Grimble, G.K.; Westwood, O.M. Nucleotides as immunomodulators in clinical nutrition. *Curr. Opin. Clin. Nutr. Metab. Care* **2001**, *4*, 57–64. [CrossRef] [PubMed]
14. Wang, L.; Mu, S.; Xu, X.; Shi, Z.; Shen, L. Effects of dietary nucleotide supplementation on growth in infants: A meta-analysis of randomized controlled trials. *Eur. J. Nutr.* **2019**, *58*, 1213–1221. [CrossRef] [PubMed]
15. Xu, M.; Ma, Y.; Xu, L.; Xu, Y.; Li, Y. Multigenerations Assessment of Dietary Nucleotides Consumption in Weaned Rats. *Birth Defects Res. Part. B-Dev. Reprod. Toxicol.* **2012**, *95*, 460–466. [CrossRef] [PubMed]
16. Meihong, X.; Zhao, Y.M.; Xu, L.; Xu, Y.; Li, Y. Multi-Generations Assessment of Dietary Nucleotides Consumption in Maternal Rats. *Birth Defects Res. Part. B-Dev. Reprod. Toxicol.* **2014**, *2*, 800–805. [CrossRef]
17. Xu, Y.; Li, Y. Safety assessment of exogenous nucleotides on embryonic and infant development: A multigeneration study in rats. *Ann. Nutr. Metab.* **2009**, *55*, 629.
18. Xu, M.; Liang, R.; Guo, Q.; Wang, S.; Zhao, M.; Zhang, Z.; Wang, J.; Li, Y. Dietary nucleotides extend the life span in Sprague-Dawley rats. *J. Nutr. Health Aging* **2013**, *17*, 223–229. [CrossRef]
19. Bryan, T.M. Nucleotide metabolism regulates human telomere length via telomerase activation. *Nat. Genet.* **2023**, *55*, 532–533. [CrossRef]
20. Mannherz, W.; Agarwal, S. Thymidine nucleotide metabolism controls human telomere length. *Nat. Genet.* **2023**, *55*, 568–585. [CrossRef]
21. Mao, D.; Xu, M.; Jiang, Q.; Sun, H.; Sun, F.; Yang, R.; Chai, Y.; Li, X.; Li, B.; Li, Y. A Single Nucleotide Mixture Enhances the Antitumor Activity of Molecular-Targeted Drugs Against Hepatocellular Carcinoma. *Front. Pharmacol.* **2022**, *13*, 951831. [CrossRef]
22. Yamamoto, S.; Wang, M.F.; Adjei, A.A.; Ameho, C.K. Role of nucleosides and nucleotides in the immune system, gut reparation after injury, and brain function. *Nutrition* **1997**, *13*, 372–374. [CrossRef] [PubMed]

23. Xu, M.; Zhao, M.; Yang, R.; Zhang, Z.; Li, Y.; Wang, J. Effect of dietary nucleotides on immune function in Balb/C mice. *Int. Immunopharmacol.* **2013**, *17*, 50–56. [CrossRef] [PubMed]
24. Sudo, N.; Aiba, Y.; Takaki, A.; Tanaka, K.; Yu, X.N.; Oyama, N.; Koga, Y.; Kubo, C. Dietary nucleic acids promote a shift in Th1/Th2 balance toward Th1-dominant immunity. *Clin. Exp. Allergy* **2000**, *30*, 979–987. [CrossRef] [PubMed]
25. Zhu, N.; Liu, X.; Xu, M.; Li, Y. Dietary Nucleotides Retard Oxidative Stress-Induced Senescence of Human Umbilical Vein Endothelial Cells. *Nutrients* **2021**, *13*, 3279. [CrossRef] [PubMed]
26. Wang, X.; Liu, R.; Wei, C.; Xu, M.; Li, Y. Exogenous Nucleotides Improved the Oxidative Stress and Sirt-1 Protein Level of Brown Adipose Tissue on Senescence-Accelerated Mouse Prone-8 (SAMP8) Mice. *Nutrients* **2022**, *14*, 2796. [CrossRef] [PubMed]
27. Dong, W.; Wu, Z.; Xu, L.; Fang, Y.; Xu, Y. Maternal supplementation of nucleotides improves the behavioral development of prenatal ethanol-exposed mice. *Cogn. Affect. Behav. Neurosci.* **2014**, *14*, 879–890. [CrossRef] [PubMed]
28. Chen, T.H.; Wang, M.F.; Liang, Y.F.; Komatsu, T.; Chan, Y.C.; Chung, S.Y.; Yamamoto, S. A nucleoside-nucleotide mixture may reduce memory deterioration in old senescence-accelerated mice. *J. Nutr.* **2000**, *130*, 3085–3089. [CrossRef] [PubMed]
29. Xu, M.; Liang, R.; Li, Y.; Wang, J. Anti-fatigue effects of dietary nucleotides in mice. *Food Nutr. Res.* **2017**, *61*, 1334485. [CrossRef]
30. Cai, X.; Bao, L.; Wang, N.; Ren, J.; Chen, Q.; Xu, M.; Li, D.; Mao, R.; Li, Y. Dietary nucleotides protect against alcoholic liver injury by attenuating inflammation and regulating gut microbiota in rats. *Food Funct.* **2016**, *7*, 2898–2908. [CrossRef]
31. Li, Y.; Xu, M.; Wang, J.; Mao, R. Long-term dietary nucleotides supplements and liver fibrosis in aging rats: A miRNA investigation. *Ann. Nutr. Metab.* **2017**, *71*, 1172.
32. You, M.; Liu, R.; Wei, C.; Wang, X.; Yu, X.; Li, Z.; Mao, R.; Hu, J.; Zhu, N.; Liu, X.; et al. Exogenous Nucleotides Ameliorate Ageing-Related Intestinal Inflammation in Senescence-Accelerated Mouse Prone-8 (SAMP8) Mice. *Nutrients* **2023**, *15*, 2533. [CrossRef]
33. Bao, L.; Xiao, Y.; Cai, X.; Wang, N.; Li, Y. Effects of 5′-nucleotides on the injury of colon function induced by chronic alcoholism in rats. *Sci. Technol. Food Ind.* **2017**, *38*, 299–303.
34. Bebawy, M.; Rasmussen, C.; Sambasivam, S.; Bao, S. Dietary Nucleotide Supplements in Infant Formula Modify the Expression of P-glycoprotein in the Intestinal Epithelium in vitro. *Int. J. Vitam. Nutr. Res.* **2009**, *79*, 381–387. [CrossRef] [PubMed]
35. Ding, T.; Li, Y. Promoting Effect and Mechanism of Exogenous Nucleotides on the Growth of Lactobacillus casei Q1. *Food Sci.* **2021**, *42*, 201–207.
36. Ding, T.; Xu, M.; Li, Y. An Overlooked Prebiotic: Beneficial Effect of Dietary Nucleotide Supplementation on Gut Microbiota and Metabolites in Senescence-Accelerated Mouse Prone-8 Mice. *Front. Nutr.* **2022**, *9*, 820799. [CrossRef]
37. Yi, M.A.; Linlin, X.U.; Yajun, X.U.; Yong, L.I. Effect of Exogenous Nucleotides on Development in Two-generation Rats. *Sci. Technol. Rev.* **2010**, *28*, 25–29.
38. Xu, M.; Ma, Y.; Xu, L.; Xu, Y.; Li, Y. Developmental Effects of Dietary Nucleotides in Second-Generation Weaned Rats. *J. Med. Food* **2013**, *16*, 1146–1152. [CrossRef]
39. Zhu, N.; Liu, R.; Xu, M.-H.; Li, Y. Neuroprotective Actions of Different Exogenous Nucleotides in H_2O_2-Induced Cell Death in Pc-12 Cells. *Molecules* **2023**, *28*, 1226. [CrossRef]
40. Aird, K.M.; Zhang, G.; Li, H.; Tu, Z.; Bitler, B.G.; Garipov, A.; Wu, H.; Wei, Z.; Wagner, S.N.; Herlyn, M.; et al. Suppression of Nucleotide Metabolism Underlies the Establishment and Maintenance of Oncogene-Induced Senescence. *Cell Rep.* **2013**, *3*, 1252–1265. [CrossRef]
41. Lin, J.; Epel, E.; Cheon, J.; Kroenke, C.; Sinclair, E.; Bigos, M.; Wolkowitz, O.; Mellon, S.; Blackburn, E. Analyses and comparisons of telomerase activity and telomere length in human T and B cells: Insights for epidemiology of telomere maintenance. *J. Immunol. Methods* **2010**, *352*, 71–80. [CrossRef] [PubMed]
42. Horvath, S.; Raj, K. DNA methylation-based biomarkers and the epigenetic clock theory of ageing. *Nat. Rev. Genet.* **2018**, *19*, 371–384. [CrossRef] [PubMed]
43. Weidner, C.I.; Lin, Q.; Koch, C.M.; Eisele, L.; Beier, F.; Ziegler, P.; Bauerschlag, D.O.; Joeckel, K.-H.; Erbel, R.; Muehleisen, T.W.; et al. Aging of blood can be tracked by DNA methylation changes at just three CpG sites. *Genome Biol.* **2014**, *15*, R24. [CrossRef] [PubMed]
44. Levine, M.E. Modeling the Rate of Senescence: Can Estimated Biological Age Predict Mortality More Accurately Than Chronological Age? *J. Gerontol. A Biol. Sci. Med. Sci.* **2013**, *68*, 667–674. [CrossRef] [PubMed]
45. Tchkonia, T.; Zhu, Y.; van Deursen, J.; Campisi, J.; Kirkland, J.L. Cellular senescence and the senescent secretory phenotype: Therapeutic opportunities. *J. Clin. Invest.* **2013**, *123*, 966–972. [CrossRef] [PubMed]
46. Franceschi, C.; Bonafè, M.; Valensin, S.; Olivieri, F.; De Luca, M.; Ottaviani, E.; De Benedictis, G. Inflamm-aging: An evolutionary perspective on immunosenescence. *Ann. N. Y Acad. Sci.* **2000**, *908*, 244–254. [CrossRef] [PubMed]
47. Wang, C.; Jurk, D.; Maddick, M.; Nelson, G.; Martin-Ruiz, C.; von Zglinicki, T. DNA damage response and cellular senescence in tissues of aging mice. *Aging Cell* **2009**, *8*, 311–323. [CrossRef] [PubMed]
48. Kuo, L.J.; Yang, L.-X. γ-H2AX: A novel biomarker for DNA double-strand breaks. *In Vivo* **2008**, *22*, 305–309. [PubMed]
49. Ware, J.E.; Kosinski, M.; Keller, S.D. A 12-item short-form health survey—Construction of scales and preliminary tests of reliability and validity. *Med. Care* **1996**, *34*, 220–233. [CrossRef]
50. Fan, J.; Yu, C.; Guo, Y.; Bian, Z.; Sun, Z.; Yang, L.; Chen, Y.; Du, H.; Li, Z.; Lei, Y.; et al. Frailty index and all-cause and cause-specific mortality in Chinese adults: A prospective cohort study. *Lancet Public. Health* **2020**, *5*, E650–E660. [CrossRef]

51. Nasreddine, Z.S.; Phillips, N.A.; Bédirian, V.; Charbonneau, S.; Whitehead, V.; Collin, I.; Cummings, J.L.; Chertkow, H. The montreal cognitive assessment, MoCA: A brief screening tool for mild cognitive impairment. *J. Am. Geriatr. Soc.* **2005**, *53*, 695–699. [CrossRef] [PubMed]
52. Washburn, R.A.; Smith, K.W.; Jette, A.M.; Janney, C.A. The physical activity scale for the elderly (PASE): Development and evaluation. J. *Clin. Epidemiol.* **1993**, *46*, 153–162. [CrossRef]
53. Buysse, D.J.; Reynolds, C.F.; Monk, T.H.; Berman, S.R.; Kupfer, D.J. The Pittsburgh Sleep Quality Index: A new instrument for psychiatric practice and research. *Psychiatry Res.* **1989**, *28*, 193–213. [CrossRef]
54. Brooks, R.T.; Beard, J.; Steel, Z. Factor structure and interpretation of the K10. *Psychol. Assess.* **2006**, *18*, 62–70. [CrossRef] [PubMed]
55. Chalder, T.; Berelowitz, G.; Pawlikowska, T.; Watts, L.; Wessely, S.; Wright, D.; Wallace, E.P. Development of a fatigue scale. *J. Psychosom. Res.* **1993**, *37*, 147–153. [CrossRef]
56. Fan, R.; Chen, Q.; Song, L.; Wang, S.; You, M.; Cai, M.; Wang, X.; Li, Y.; Xu, M. The Validity and Feasibility of Utilizing the Photo-Assisted Dietary Intake Assessment among College Students and Elderly Individuals in China. *Nutrients* **2024**, *16*, 211. [CrossRef]
57. Dong, Y.; Lee, W.Y.; Hilal, S.; Saini, M.; Wong, T.Y.; Chen, C.L.-H.; Venketasubramanian, N.; Ikram, M.K. Comparison of the Montreal Cognitive Assessment and the Mini-Mental State Examination in detecting multi-domain mild cognitive impairment in a Chinese sub-sample drawn from a population-based study. *Int. Psychogeriatr.* **2013**, *25*, 1831–1838. [CrossRef]
58. Blay, S.L.; Andreoli, S.B.; Gastal, F.L. Prevalence of self-reported sleep disturbance among older adults and the association of disturbed sleep with service demand and medical conditions. *Int. Psychogeriatr.* **2008**, *20*, 582–595. [CrossRef]

nutrients

Article

Age-Related Mucus Barrier Dysfunction in Mice Is Related to the Changes in Muc2 Mucin in the Colon

Xueqin Sang [1], Qingyu Wang [2], Yueyan Ning [3], Huihui Wang [1], Rui Zhang [1], Yixuan Li [4], Bing Fang [4], Cong Lv [4], Yan Zhang [1], Xiaoyu Wang [1,5,*] and Fazheng Ren [1,4,6,*]

1 College of Food Science and Engineering, Gansu Agricultural University, Lanzhou 730070, China
2 The Key Laboratory of Geriatrics, Beijing Institute of Geriatrics, Institute of Geriatric Medicine, Chinese Academy of Medical Sciences, Beijing Hospital/National Center of Gerontology of National Health Commission, Beijing 100730, China
3 State Key Laboratories for Agrobiotechnology, College of Biological Sciences, China Agricultural University, Beijing 100083, China
4 Key Laboratory of Precision Nutrition and Food Quality, Department of Nutrition and Health, China Agricultural University, Beijing 100193, China
5 Key Laboratory of Functional Dairy, Co-Constructed by Ministry of Education and Beijing Municipality, College of Food Science & Nutritional Engineering, China Agricultural University, Beijing 100083, China
6 Food Laboratory of Zhongyuan, Luohe 462000, China
* Correspondence: xy.wang@cau.edu.cn (X.W.); renfazheng@cau.edu.cn (F.R.); Tel./Fax: +86-0931-7631201 (F.R.)

Abstract: During aging, the protective function of mucus barrier is significantly reduced among which changes in colonic mucus barrier function received the most attention. Additionally, the incidence of colon-related diseases increases significantly in adulthood, posing a threat to the health of the elderly. However, the specific changes in colonic mucus barrier with aging and the underlying mechanisms have not been fully elucidated. To understand the effects of aging on the colonic mucus barrier, changes in the colonic mucus layer were evaluated in mice aged 2, 12, 18, and 24 months. Microbial invasion, thickness, and structure of colonic mucus in mice at different months of age were analyzed by in situ hybridization fluorescence staining, AB/PAS staining, and cryo-scanning electron microscopy. Results showed that the aged colon exhibited intestinal mucus barrier dys-function and altered mucus properties. During aging, microorganisms invaded the mucus layer to reach epithelial cells. Compared with young mice, the thickness of mucus layer in aged mice in-creased by 11.66 μm. And the contents of the main components and glycosylation structure of colon changed. Among them, the proportion of goblet cells decreased significantly in older mice, and the expression of spdef genes that regulate goblet cell differentiation decreased. Further, the expression of key enzymes involved in mucin core structure formation and glycan modification also changed with aging. The expression of core 1 β1,3-galactosyltransferase (C1GalT1) which is the key enzyme forming the main core structure increased by one time, while core 2 β1,6 N-acetylglucosaminyltransferase (C2GnT) and core 3 β1,3 N-acetylglucosaminyltransferase (C3GnT) decreased 2 to 6- and 2-fold, respectively. Also, the expression of sialyltransferase, one of the mucin-glycan modifying enzymes, was decreased by 1-fold. Overall, our results indicate that the goblet cells/glycosyltransferase/O-glycan axis plays an important role in maintaining the physicochemical properties of colonic mucus and the stability of intestinal environment.

Keywords: aging; mucus; microbiota; glycosyltransferase; goblet cells

1. Introduction

In recent years, the average human life expectancy has increased significantly. The 2019 revision of the United Nations' World Population Prospects predicts that the proportion of people aged 65 years and older will increase by almost 16% by 2050 [1]. As the population ages, the incidence of aging-related chronic diseases has increased year by year, among which the incidence of gastrointestinal chronic diseases has increased significantly [2]. Age-related changes in intestinal physiological function could lead to various gastrointestinal diseases including, polyps, inflammatory bowel dis-ease, and so on [3]. Recently, the incidence of inflammatory bowel disease has significantly increased with age, but the exact pathogenesis remains unclear. Ulcerative colitis is a common inflammatory bowel disease that is mainly caused by deterioration of colonic mucus layer [4]. It has been reported that mucus layer becomes thinner and the inner mucus layer becomes penetrable during the development of colitis [5]. Overall, it could be concluded that intestinal dynamic homeostasis depends on the protective function of intestinal mucus barrier, and its disruption is one of the main causes of inflammation [6].

The mucosal barriers that maintain a normal dynamic balance of intestinal environment mainly include mucus barrier, intestinal epithelial barrier and immune barrier [7]. Mucus barrier is the main defensive layer that separates microbiota from epithelial cells. It consists of mucus secreted by goblet cells and overlies intestinal epithelial cells [8,9]. The small intestine has only one mucus layer, whereas the colon has two mucus layers [10], an outer layer that allows bacterial colonization and an inner layer that does not allow luminal bacteria to penetrate [11]. Thus, mucus layer is not only a protector of intestinal epithelial cells but also a key mediator of host-bacterial interactions [12]. And the colonic mucus barrier is a major defender of intestinal pathogens [13]. The physiological importance of mucus layer has been demonstrated over the years by studying patients with ulcerative colitis. These studies reported several defects in mucus layer function, including a decrease in mucus layer thickness and mucus chemical composition, and an increase in bacterial penetration of the mucus barrier [12]. Similar results were obtained in the literature related to changes in the mucus layer during aging in mice. It was speculated that such changes were related to alterations in mucus composition, but the mechanism of interaction between mucus composition and mucus functional defects has not been elucidated yet [14].

Muc2 mucin is a major component of the mucus layer [15]. The colonic mucus layer formed by Muc2 mucin builds a functional barrier between the host and bacterial microbiota [16]. Studies have shown that Muc2-deficient mice spontaneously developed inflammation and were more susceptible to infection by pathogens [17]. Furthermore, Muc2 is a high molecular weight glycoprotein formed by glycosyltransferases. Mucin glycosylation is mainly responsible for the formation of four key O-glycan core structures [18]. The major polysaccharides in colonic mucus layer are mucin O-glycans derived from core 1 and core 3 [19]. The key enzymes involved in the formation of core 1 and core 3 O-glycan chains are core 1 β1,3-galactosyltransferase (C1GalT1) and core 3 β1,3 N-acetylglucosaminyltransferase (C3GnT), respectively [20–22]. Core 1 O-glycans form core 2 O-glycan with the participation of core 2 β1,6 N-acetylglucosaminyltransferases (C2GnT) [18]. C2GnT1 and C2GnT3 are normally expressed in intestinal epithelial cells, and C2GnT2 is also known as mucus-type C2GnT2 because it is highly restricted to mucus-producing tissues which highly ex-presses in stomach and colon. Studies using a C2GnT2-deficient mouse model found that C2GnT2 deficiency impaired the mucosal barrier and increased mucosal barrier permeability and susceptibility to colitis. Previous studies using C1galt1 and C3GnT deficient mouse models (IEC C1galt1-/-, C3GnT-/- mice) have demonstrated that mucin-type O-glycans are key factors in protecting the intestinal mucus barrier from damage [20,22]. It has also been shown that defect in O-glycan chains lead to alterations in the composition of intestinal flora, resulting in bacteria-dependent intestinal inflammation [23]. Besides, recent studies have reported distal colonic O-glycosylation impaired during aging, and protein expression of the rate-limiting enzyme in O-glycosylation, T-synthetase, decreased with age [1]. However, mechanisms under-lying the interaction between mucin content, mucin glycosylation structure and mucus layer function during aging need to be further investigated.

Here, naturally aged mice were used to investigate how the physiological, physical and chemical properties of mucus altered during aging. This study focused on compositional and structural changes in mucus during aging and the underlying mechanisms.

2. Materials and Methods

2.1. Animals

Male C57BL/6J mice used in the experiment were purchased from the Laboratory Animal Center of Weitong Lihua Laboratory Animal Technology Company (Beijing, China). Mice of the same age (months) were purchased in chronological order of 2, 12, 18, and 24 months of age. Mice of each month of age (n = 12) were placed in individually ventilated cages with ad libitum access to mouse growth and reproduction diet (12.95% fat) (Beijing Keao Xieli Feed Co., Ltd., Beijing, China) and autoclave-sterilized water. All mice were sacrificed using appropriate anesthesia and/or cervical dislocation before sample collection. The experiments and protocols were approved by the Ethics Committee of China Agricultural University (AW50102202-5).

2.2. Mucus and Tissue Collection

Colon tissue was incised longitudinally and the mucus was gently scraped with a smooth slide after careful removal of the fecal contents, and the mucus samples were frozen rapidly in liquid nitrogen. In addition, colon tissues were immediately frozen in liquid nitrogen and then stored at −80 °C for subsequent RNA and protein assays.

2.3. pH Measurements

The colon was cut longitudinally, and mucus was gently scraped from the entire colon with a slide after the feces were gently removed. About 100 to 120 µL of colonic mucus was scraped from each mouse, and pH was measured using a portable pen acidity meter (Yanlin Laboratories (Shenzhen) Co., Shenzhen, China). The samples were immediately frozen in liquid nitrogen after the measurements and stored at −80 °C for further experiments [24].

2.4. Water Content and Mucus Weight Determination

This method is modified based on methods from the literature. Briefly, the collected mucus samples were quickly placed into a freeze-dryer at a cold trap temperature of −50 °C for 48 h [24]. At the end of freeze-drying, the samples were placed in a desiccator for 30 min and weighed. The weight per unit length of mucus was calculated by the weight of collected mucus divided by the length of collected colonic tissue [24].

2.5. In Situ Hybridization Fluorescent Staining

The methods refer to the literature and are briefly described as follows [1,19,25]. Colon sections were stained with universal bacterial probe EUB338 (*50-GCTGCCTCCCGTAGGAGT-30*; bp 337–354 in bacteria EU622773) at 50 °C overnight. A nonspecific probe (*NON338 50-ACTCCTACGGGAGGCAGC-30*) was used as negative control. For Muc2-Fish double labeling experiments, Muc2 immunofluorescence staining was performed on sample sections before Fish labeling. Sample sections were incubated with anti-Muc2 (Service bio, GB11344, Wuhan Sevier Biotechnology Co., Ltd., Wuhan, China, incubated at 4 °C for 12 h) and goat anti-mouse IgG (H + L) cross-adsorbed secondary antibody, Alexa fluor™ 488 (Invitrogen, A-11001, Thermo Fisher Scientific, Beijing, China, incubated at room temperature for 1 h).

2.6. Cryo-Scanning Electron Microscopy and Image Analysis

The preparation of samples, Cryo-SEM, and image analysis were taken from the literature and modified [26]. In brief, samples were thawed at room temperature, and a single drop of each sample was cast onto a metal holder and then frozen in liquid nitrogen. Frozen samples were transferred into the sample preparation bin for freeze-fracture and sublimated under vacuum for 15 min at −80 °C. The Cryo-SEM images were counted for hole size and shape using ImageJ software, with at least 100 holes involved in each field of view. The minimum diameter of the FERET, which refers to the shortest distance between any two parallel tangents of the hole, was used to describe the size of the hole. The Aspect Ratio (AR) was used to describe the pore shape, which was described as the ratio of the largest ferret diameter to the smallest ferret diameter.

2.7. Immunostaining

Colon tissue was collected and the mesentery was gently removed with forceps. Then, the tissue was ligated at both ends with surgical thread and fixed in Carnoy's fixative solution for no more than 12 h. After fixation, the tissues were dehydrated and embedded in paraffin. Mucus thickness staining was performed using the AB-PAS Stain Kit (Solarbio, G1285, Beijing Solarbio Science & Technology Co., Ltd., Beijing, China) following manufacturer instructions [27]. PS software (Adobe Photoshop 2021 22.0.0, Adobe Systems Incorporated, San Jose, CA, USA) was used to measure the thickness of the mucus layer [19,28].

For IHC, the sections were subjected to antigen retrieval in 0.01 M sodium citrate for 20 min. The nonspecific antibody binding was blocked by goat serum working solution (Zhongshan Jinqiao, SP-9002, Beijing ZhongShan JinQiao Biotechnology Co., Beijing, China) for 1 h. Following blocking, rabbit anti-Muc2 (Service bio, GB11344, Wuhan Sevier Biotechnology Co., Ltd., Wuhan, China) antibody was used to incubate the sections at 4 °C overnight. After 30 min, sections were stained with the corresponding immune secondary antibody for 30 min. Finally, the slices were stained with DAB chromogenic kit (ZSGB-BIO, ZLI-9019, Beijing ZhongShan JinQiao Biotechnology Co., Beijing, China), counterstained stained with hematoxylin, dehydrated, and sealed.

For immunofluorescence [29], colon tissue sections (4 μm) were dehydrated, antigen-repaired, and blocked for 1 h in blocking reagent at room temperature. Staining was performed with rabbit anti-Muc2 (Service bio, GB11344, Wuhan Sevier Biotechnology Co., Ltd., Wuhan, China) antibody and placed at 4 °C overnight. Primary antibody identification was performed with a fluorescent secondary antibody and incubated at room temperature for 1 h. Finally, the nuclei were stained with DAPI (Beyotime Biotechnology, A-11008, Shanghai Biyuntian Biotechnology Co., Shanghai, China) for 10 min.

2.8. Real-Time PCR

To determine the effect of age on gene expression, total RNA was extracted from colon tissues using a specialized kit (Qiagen, Germantown, MD, USA) [3]. The RNA was then reverse transcribed into cDNA using a reverse transcription kit (abm, G592, Zhenjiang Ebimont Biotechnology Co., Jiangsu, China). The mRNA expression of O-glycosylase (C1galt1, C2gnt, C3gnt, and St6galnac-I) and goblet cell precursor markers (spdef, klf4, Beijing Liuhe Huada Gene Technology Co., Beijing, China, and muc2) was quantified on a real-time PCR system using the primers shown in Table 1.

Table 1. Primer sequences of qRT-PCR.

Gene Name	Forward 5′-3′	Reverse 5′-3′
Muc2	GGGAATGTTGCAAGAAGTGC	TTTTGTGAATCTCCCCAGGC
Spdef	ACGTTGGATGAGCACTCG	CCATAAAAGCCACTTCTGCAC
Klf4	CGAACTCACACAGGCGAGAA	GAGCGGGCGAATTTCCA
C1galt1	TGGAATTACAACTATTATCCTCCCATA	CAACATAGTGAAAAGAAACTGCGATA
C1galt2	TGGAGCCGTTCTAGATGCGGAAAA	GGGGCTTGCAGATGGTGATGCT
C2gnt	GCAGCCAAGAAGGTACCAAA	ACAGGCGAGGACCATCAA
C2gnt1	GCTTGATAGGAACTTGGCAGCAC	CACCTTCTGGATTTCTTCTGGGTC
C2gnt2	ACCTTCACTCCACATCACTCACGG	TTATTCAGCAGAGCCTGGGTCACC
C2gnt3	GCCGCTGTTCTTGCTGTTTTG	AGTCACTTGTCATCGCCACGA
C3gnt	GGCCAGATTCTCCTCTCTCAAACG	AGTGCTCCGCTGTCCAGTCCA
St6galnac-I	TGTTAGGGACCAGCCATCCA	ATGAACTGGCACCTGGAATC
St6galnac-II	CGGATGTTGTTGCTCGTTGC	AGTCGGCTCTTTCTGTTTTCC
Gapdh	AGGTCGGTGTGAACGGATTTG	TGTAGACCATGTAGTTGAGGTCA

2.9. Statistical Analysis

Data were expressed as mean ± SD. GraphPad Prism 8.0.2, (GraphPad Software, Inc. San Diego, CA, USA) was used for all statistical analyses. All samples had biological replicates. Data were statistically analyzed using one-way ANOVA with Tukey's post hoc test [30].

3. Results

3.1. Mucus Barrier Function

The mucus layer is the first protective layer of mucosa overlying the intestinal epithelial cells, and its integrity is essential for the maintenance of health [31]. As shown in Figure 1A, colonic mucus layers at 2, 12, and 18 months of age were intact and not invaded by microorganisms. In contrast, the mucus layer of 24-month-old mice was more loosely structured and colonized by a certain number of microorganisms. To investigate the effect of age on mucus thickness, AB-PAS staining on colonic tissue samples was performed. Surprisingly, results showed a continuous increase in mucus layer thickness in aged mice (Figure 1B,C). The thickness of the mucus layer in mice aged 12 to 24 months was significantly higher than that in mice aged 2 months. These results indicate that the function and thickness of the mucus layer are affected during aging, and the mucus barrier deteriorates with aging.

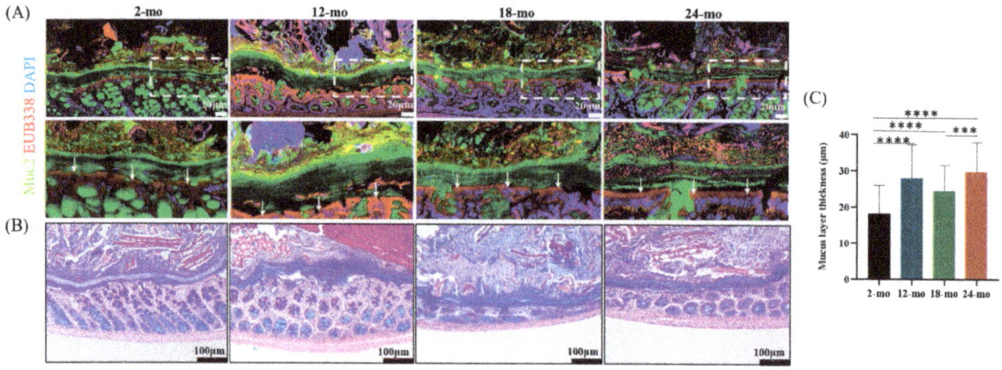

Figure 1. Aged colon showed impaired mucus barrier integrity and increased mucus thickness in mice. (**A**) FISH staining of distal colon sections using EUB338 probes and Muc2 antibodies showed bacterial penetration of the inner mucus layer. The magnified images showed the boxed region. White arrows indicated the apical surface of the epithelial cells. Scale bars: 20 μm. (**B**) AB/PAS staining images of the distal colons of mice aged 2, 12, 18, and 24 months. Scale bars: 100 μm. (**C**) Mucus thickness was measured in three colon tissues of mice aged 2, 12, 18, and 24 months (9 measurements per section, 3 sections per mouse). Data were presented as mean ± SD. Data were statistically analyzed using one-way ANOVA with Tukey's post hoc test. Asterisk indicates significant difference, *** $p < 0.001$, **** $p < 0.0001$, $n = 3$ per group.

3.2. Microscopic Characterization

The disulfide bonds between non-glycosylated region and multiple non-covalent bonds form a highly entangled mucus gel network. These networks are important structural components in the mucus layer that can interact with microbes and protect intestinal epithelial cells. The Cryo-SEM results of different ages were displayed in Figure 2A–C. Results showed that there were extended gel networks in the colonic mucus samples of all mice, and the size of the pores of these gel networks was statistically different in different age groups (Figure 2A,B). However, the shape of the pores did not change significantly (Figure 2C). Cryo-SEM images of the mucus at 2 months of age showed relatively large pores, while the colonic mucus network at 12, 18, and 24 months of age contained smaller pores. Compared with 2-month-old mice, the mini-mum diameter of the colonic mucus network was significantly reduced in 12-month-old, 18-month-old, and 24-month-old mice, especially at 12 months old. This suggests that the pore size of the mucus microscopic network changes during aging, but the network shape does not.

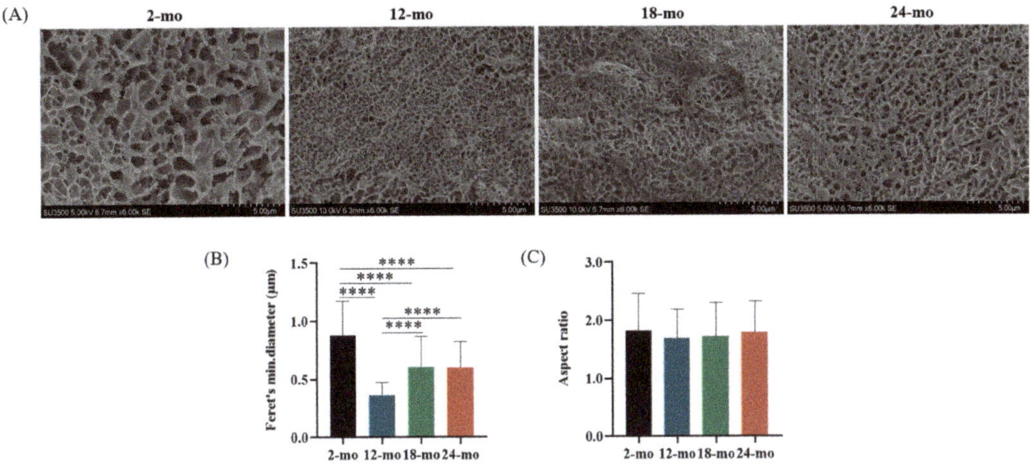

Figure 2. Microstructural characteristics of colonic mucus in mice. (**A**) Representative cryo-scanning electron micrographs of colon mucus at different ages (scale bars: 5 μm). (**B**,**C**) showed the Feret's minimum diameter and aspect ratio distribution of mucus pores, respectively. At least 100 pores per micrograph were identified and used in the analysis. Data were presented as mean ± SD. Data were statistically analyzed using one-way ANOVA with Tukey's post hoc test. Asterisk indicates significant difference, **** $p < 0.0001$, $n = 3$ per group.

3.3. Physiological Properties

pH is one of the influencing factors in the formation and release of mucus [32,33]. The changes in mucus pH with increasing age were shown in Figure 3A. Compared with 2 months of age, pH did not change significantly at 12 months of age while decreased significantly at 18 and 24 months of age. Changes in mucus pH indicate that mucus secretion and stratification may also be altered.

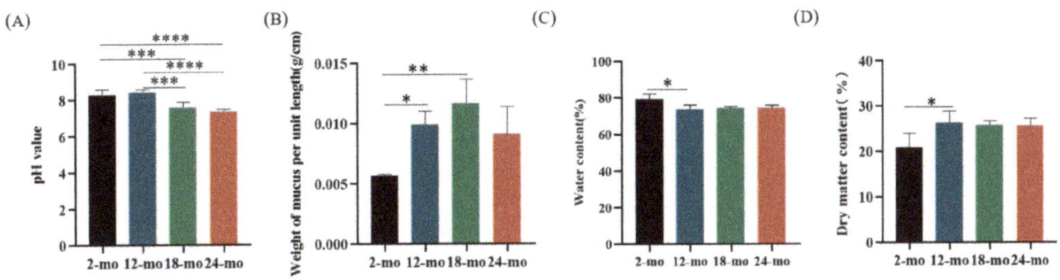

Figure 3. Physiological characteristics of colonic mucus in aging mice. The pH (**A**), mucus weight per unit length (**B**), water content (**C**), and dry matter content (**D**) of colonic mucus in aging mice were determined. Data were presented as mean ± SD. Data were statistically analyzed using one-way ANOVA with Tukey's post hoc test. Asterisk indicates significant difference, * $p < 0.05$, ** $p < 0.01$, *** $p < 0.001$, **** $p < 0.0001$, $n = 3$ per group.

The results of mucus weight and water content are shown in Figure 3B,C. The mucus weight per unit length of mice at 12, 18, and 24 months was significantly increased compared with that of mice at 2 months. The average water content of colonic mucus in

2-month-old mice was 79.12%. Compared with 2 months of age, there was a significant decrease at 12, 18, and 24 months of age. In contrast, the trend of mucus dry matter content was opposite to that of water content, with a significant increase at 24 months of age compared with 2 months of age (Figure 3D). These results indicate that the mucus layer becomes more viscous due to decreased water content and increased dry matter content during aging.

3.4. Muc2 Mucin

The gel-forming mucin Muc2 is a major structural and functional component of the mucus barrier [34]. Immunofluorescence staining of colon sections at different months of age was performed, and Muc2 gene expression in the colon was quantified. Compared with mice at 2 months of age, Muc2 protein content in the colon decreased significantly in mice at 24 months of age (Figure 4A,B). Interestingly, this decrease occurred mainly in the crypt base, which may have an impact on the activity of stem cells. Similarly, the Muc2 mRNA level reduced in 24-month-old mice compared with 2-month-old mice (Figure 4C).

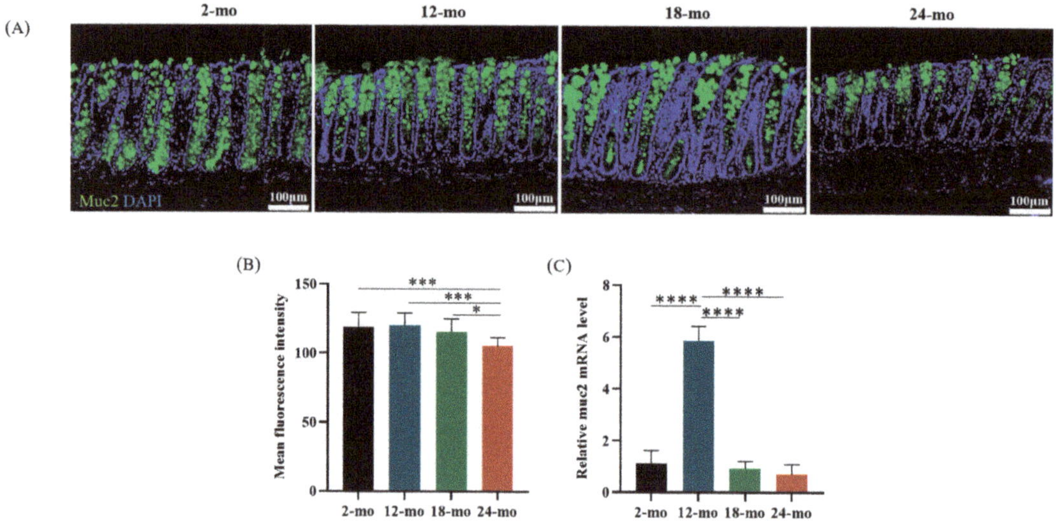

Figure 4. Muc2 mucin content of the mucus layer of colon in aging mice. (**A**) Immunofluorescence staining of 2-, 12-, 18- and 24-month-old mouse colon sections with anti-Muc2 (in green; cell nuclei in blue). Scale bars: 100 μm. Immunofluorescence results of Muc2 mucin (**B**) and relative mRNA expression (**C**) of Muc2 in the mouse colon. Data were presented as mean ± SD. Data were statistically analyzed using one-way ANOVA with Tukey's post hoc test. Asterisk indicates significant difference, * $p < 0.05$, *** $p < 0.001$, **** $p < 0.0001$, $n = 3$ per group.

3.5. Goblet Cells and Genes Regulating Goblet Cell Formation

Goblet cells are a cell type specializing in secreting mucus [35]. The number and proportion of goblet cells are shown in Figure 5A–C. The average number of goblet cells in the colon tissue of mice aged 2, 12, 18, and 24 months was 12, 14, 14, and 12, respectively (Figure 5B). Results showed that the number of goblet cells in aging mice did not change significantly compared with young mice, but the proportion of goblet cells decreased significantly at 24 months of age (29%) (Figure 5C), suggesting that the ability of goblet cells secreting mucus was inhibited with age.

Spdef is a goblet cell marker gene, which regulates the differentiation of intestinal goblet cells. Klf4 is also involved in the terminal differentiation of goblet cells. Compared with 2-month-old mice, the mRNA level of spdef was significantly reduced in 12-month-old,

18-month-old, and 24-month-old mice (Figure 5D). In contrast, the expression of klf4 was significantly higher in 12-, 18-, and 24-month-old mice than in 2-month-old mice (Figure 5E).

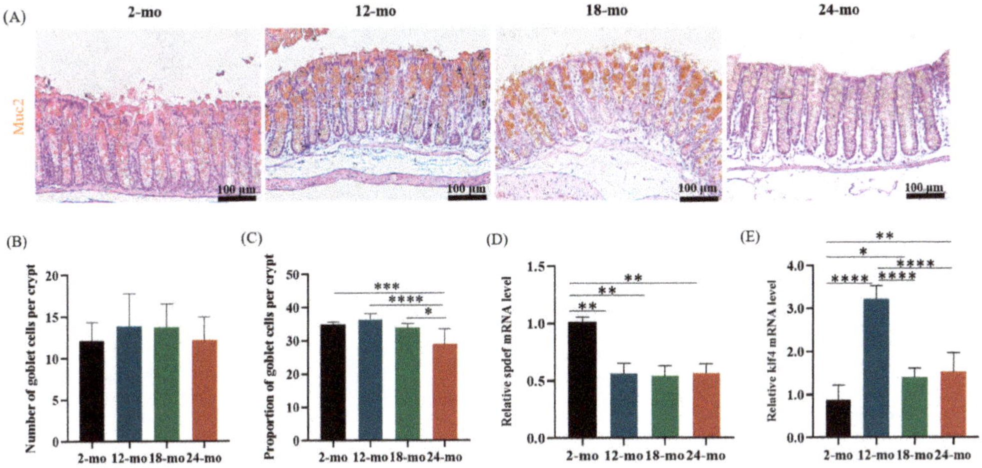

Figure 5. The number and proportion of goblet cells and the expression of genes regulating goblet cell formation changed in aging mice. Immunohistochemical staining of 2-, 12-, 18- and 24-month-old mouse colon sections with anti-Muc2 (in brown; cell nuclei in blue) (**A**), Scale bars: 100 μm. Number (**B**) and proportion (**C**) of goblet cells were measured. Relative mRNA expression of spdef (**D**) and klf4 (**E**) in the colon of mice. Data were presented as mean ± SD. Data were statistically analyzed using one-way ANOVA with Tukey's post hoc test. Asterisk indicates significant difference, * $p < 0.05$, ** $p < 0.01$, *** $p < 0.001$, **** $p < 0.0001$, $n = 3$ per group.

3.6. Mucin Glycosylation

Core enzymes are known to be directly involved in mucin glycosylation, forming important core structures. C1GalT1 and C1GalT2 are key enzymes in core 1 O-glycan biosynthesis and are widely expressed in epithelial cells. Compared with 2-month-old mice, the expression of C1galt1 in the colon of 12-, 18- and 24-month-old mice significantly increased, but there was no significant change in the expression of C1galt1 in the colon of 18- and 24-month-old mice (Figure 6A). The expression of C1galt2 was shown in Figure 6B. Compared with 2-month-old mice, the expression of C1galt2 was significantly increased in the colon of 12-month-old mice, while no significant changes were observed in the colon of 18-month-old and 24-month-old mice. The core 2 O-glycan is an important component of Muc2 O-glycans which is regulated by the core 2 transferase family (C2gnts). Compared with 2-month-old mice, C2gnt1 expression was significantly increased in the colon of 12- and 18-month-old mice and decreased in the colon of 24-month-old mice (Figure 6C). Meanwhile, the expression of both C2gnt2 and C2gnt3 decreased significantly with age compared with 2 months old mice (Figure 6D,E). C3gnt is the only enzyme catalyzing the biosynthesis of core 3-derived O-glycan chains [22]. Results showed that the expression of C3gnt in mice at 12 and 24 months was significantly reduced compared with that of mice at 2 months, and the expression of mice at 24 months was about half that of mice at 2 months (Figure 6F).

ST6GALNAC1 is a major sialyltransferase that specifically expresses in goblet cells and is sialylated at the ends of glycans on intestinal mucus [36]. Compared with younger mice, the expression of St6galnac-I in colon tissues of mice aged 12, 18, and 24 months was significantly decreased, and the expression level of St6galnac-I in mice aged 18 and 24 months was about 50% of that of mice aged 2 months (Figure 6G). Compared with 2-month-old mice, the expression of St6galnac-II was significantly increased in 18-month-old mice, and decreased in 12-month-old and 24-month-old mice (Figure 6H).

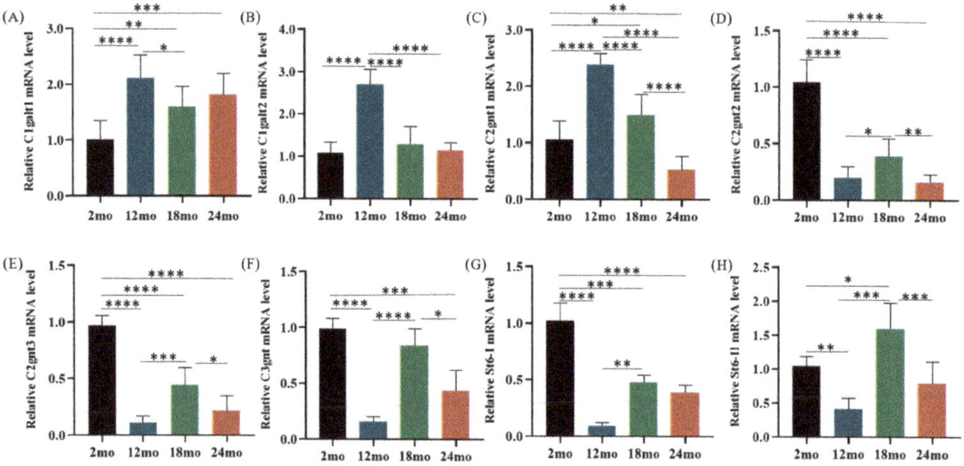

Figure 6. Altered mRNA expression of glycosylated transferase during aging. mRNA level of key enzymes in core 1 (**A**,**B**), core 2 (**C**–**E**), and core3 (**F**) formation in colon. Gene expression of the mucin-type O-glycan modifying enzymes (**G**,**H**) in colon tissue were measured. Data were presented as mean ± SD. Data were statistically analyzed using one-way ANOVA with Tukey's post hoc test. Asterisk indicates significant difference, * $p < 0.05$, ** $p < 0.01$, *** $p < 0.001$, **** $p < 0.0001$, $n = 3$ per group.

4. Discussion

Gut is one of the organs that undergoes significant changes during the aging process and the incidence of gut-related diseases in the elderly is quite high [37]. Mucus layer is one of the intestinal barriers, which prevents microorganisms from directly contacting intestinal epithelial cells to maintain host health [29]. The protective function of mucus barrier mainly depends on the mutual relationship between commensal microorganisms and mucus layer [38]. Our study showed that the mucus layer of 2-month-old mice was intact and free of microorganisms, whereas 24-month-old mice were invaded by bacteria. This observation is consistent with a previous study in which the aged (24 months) colonic mucus layer allowed bacterial penetration, but the inner mucus layer was intact in 2-month-old and 16-month-old mice [1]. This implies that with aging the protective function of mucus barrier is compromised and cannot prevent harmful microorganisms from reaching intestinal epithelial cells. Interestingly, the thickness of mucus layer increased during aging, which is different from previous studies [28]. In the present study, we found that the thickness of the colonic mucus layer increased at 12 and 24 months of age, whereas at 18 months of age, it increased compared with 2 months of age but significantly decreased compared with 12 and 24 months of age. A decreasing trend of mucus thickness was observed at 18 months compared with 12 months, which may be due to the beginning of aging in mice starting from 18 months of age. Gut may respond to changes in the body with a transient decrease in the thickness of the mucus layer. The thickness of mucus at 24 months of age was significantly higher than that at 18 months of age, which may be due to the slow peristalsis of the intestine in old mice and the slow mucus elimination. However, mucus thickness was not necessarily a useful indicator of mucus barrier function. Previous studies have shown that all mouse models of colitis had bacteria in contact with epithelial cells, but IL-$10^{-/-}$ mice showed a thicker mucus layer than wild-type mice [39]. Based on the results of microbial invasion into the colonic mucus layer and the increased mucus thickness, we hypothesized that this might be due to the changes in mucus composition and mucin structure during aging. Although thickness of the mucus layer increases during aging, it is also possible that changes in mucus composition make the mucus layer easier for microbial attachment, and that changes in mucin structure make the mucus layer loose and easier for microbial invasion.

As mentioned earlier, the major component of intestinal mucus layer is gel-forming mucin, which acts together with other components to protect and lubricate GI tract [40]. The species of mucins varied in different GI segments. Muc2 is essential for colon protection, and Muc2-deficient mice spontaneously develop colitis [17]. It has been reported that aged mice (19 months) exhibited impaired mucus barrier without colitis. However, $Muc2^{-/-}$ mice lacking the secretory mucus layer developed spontaneous colitis around 6 weeks of age. This further suggests that muc2 plays an important role in the colonic mucus layer [28]. The importance of muc2 in the mucus of aged mice was also demonstrated by experiments in a metformin-treated model which found that metformin reduced intestinal leakage by increasing Muc2 expression in the colon of aged mice. In our study, it was also found that Muc2 protein content decreased at 24 months of age. However, Muc2 gene expression increased at 12 months of age but did not differ significantly at 2, 18, and 24 months of age. The reason for this inconsistent may be that Muc2 gene is subject to many different regulations during transcription and translation, and these transcriptional, post-transcriptional, and translational regulations all play roles in protein expression. Since goblet cells are major sources of Muc2, the changes in goblet cell number and proportion with age were further examined. There was no change in the number of colonic goblet cells but a reduction in proportion. This result is consistent with a previous study showing that mucin expression decreased in aged mice (24 months) but the number of goblet cells in colon did not change [1]. These results indicate that the ability of goblet cells secreting mucin decreases during aging. Studies using Muc2 deficient mouse models have re-ported that goblet cells appear to be absent in the entire colon of $muc2^{-/-}$ mice and have smaller goblet cell morphology compared to $muc2^{+/+}$ mice [17]. This further suggests that goblet cells play a crucial role in mucus secretion. To explore the mechanism by which goblet cells are affected during aging, the expression of genes regulating goblet cell formation was determined. It was found that the expression of gene spdef was significantly decreased with aging which is consistent with previous studies [3]. These results indicate that the expression of regulatory factors controlling goblet cell formation is inhibited during aging, which prevents the formation of goblet cells, thereby reducing mucin secretion and leading to weakened mucus barrier function.

In addition to changes in the content of the main components of mucus, changes in the microstructure of mucus are also associated with mucus barrier dysfunction. Mucins are divided into two major subfamilies, including transmembrane mucins and gel-forming mucins [41]. The monomer structure of gel-forming mucin Muc2 has been extensively studied, and the monomer creates a highly entwined gel network through disulfide bonds [42]. Network structures were observed in all cryo-electron microscopy images of colonic mucus in this study. Results showed that the pore size of the network structure decreased with aging, and the pore shape was round or nearly round in all ages of mice. The same shape of colonic mucus was also reported in another study in which mucus microstructure was examined in different segments of porcine GI tract. At the same time, it is also speculated that the size of the network structure may be related to bacteria, and small pores can hinder the diffusion of bacteria without impeding the diffusion of other substances [26]. The results of bacterial invasion during aging could also be explained by the hypothesis that mucus blocks bacterial invasion by changing the size of the network during aging. In this study, we also found that the minimum pore size of the colonic mucus network was significantly reduced at 12 months of age compared with 18 and 24 months of age. The microstructure of the colonic mucus network has been reported to be dependent on factors other than mucus type. It has been established that mucus networks are formed by the supramolecular assemblies and that mucin gel assemblies is governed by the collective action of non-mucins, disulfide bridges, Ca^{2+}-mediated links and hydrogen bonds, so that disruption of any single element results in alterations of the network [42].

Based on the results of mucus microstructure changes, the underlying mechanism was further investigated. It was found that the change in the microstructure of mucus may be related to the glycosylation of mucin-type O-glycan. Several studies in mouse models lacking core 1 or core 3-derived O-glycans have found disruption of mucus layer and have shown that O-polysaccharides stabilize the structure of mucus layer [20]. The biosynthesis of core structure O-glycan is regulated by different transferases including core enzymes and modifying enzymes [43]. Our study found an increase in core-1 glycosyltransferase expression and a decrease in core-2 and core-3 transferase expression during aging. A previous study in which immature rats fed polyamines found that polyamines were involved in the regulation of mucin glycosylation during postnatal intestinal development by promoting the synthesis of glycoprotein galactosyltransferase [44]. This may explain the increased expression of galactosyltransferase during aging. We hypothesized that intestinal microbiota-derived polyamines increased during aging, thus leading to increased mucin galactosylation. Integrity of the core structure plays an important role in mucin molecules. Previous studies using mouse models of intestinal core 1- and core 3-derived O-glycans deficiency revealed that Muc2 mucin required both O-glycans to protect it from degradation by microbial-derived proteolytic enzymes [45]. Reduced expression of core 2 and core 3 transferases during aging makes it difficult to form core 2 and core 3 structures, thus truncating the O-glycan chains. Low expression of St6I and St6II transferase results in reduced mucin glycan modification. Short and less modified O-glycan chains may expose the protein core to proteases that degrade mucins. Overall, these results suggest that decreased expression of mucin glycosylases during aging leads to truncation of O-glycans and degradation of protein core, which disrupts mucin microstructure and leads to decreased mucus barrier function.

5. Conclusions

In conclusion, this study explored the changes in the physical and chemical prop-erties of mucus layer with increasing age and the underlying mechanisms. We demon-strated that mucin content and structural changes in aging mice lead to impaired mucus barrier function. Furthermore, we found that low expression of genes regulating goblet cell formation reduced the percentage of goblet cells and mucin content, and low expression of mucin glycosylase resulted in shorter O-glycan chains, thereby affecting mucus microstructure with aging. Taken together, this study is important for better understanding the role of the intestinal mucus layer in the aging process and may provide potential targets for the treatment of intestinal diseases in the elderly.

Author Contributions: X.S., Q.W. and F.R.: Conceptualization; X.S. and Y.N.: Investigation; X.S. and R.Z.: Methodology; H.W.: Software. X.S. and Y.L.: Data curation; B.F.: Resources; X.S. and C.L.: Formal analysis; Y.Z.: Visualization; X.S.: Writing—original draft; X.S., X.W. and F.R.: Writing—Review and Editing; X.W.: Project Administration. F.R.: Supervision. All authors have read and agreed to the published version of the manuscript.

Funding: This research was funded by the National Natural Science Foundation of China (Grant No. 32001676).

Institutional Review Board Statement: The animal study protocol was approved by the Animal Experimentation Ethics Committee of the China Agricultural University (approval number: AW50102202-5).

Informed Consent Statement: This paper is not concerned with the study of human beings.

Data Availability Statement: Data will be sent upon request from the corresponding author.

Conflicts of Interest: The authors declare no conflict of interest.

References

1. Huang, L.; Sun, T.Y.; Hu, L.J.; Hu, S.L.; Sun, H.M.; Zhao, F.Q.; Wu, B.; Yang, S.; Ji, F.Q.; Zhou, D.S. Elevated miR-124-3p in the aging colon disrupts mucus barrier and increases susceptibility to colitis by targeting T-synthase. *Aging Cell* **2020**, *19*, e13252. [CrossRef] [PubMed]
2. Merchant, H.A.; Rabbie, S.C.; Varum, F.J.; Afonso-Pereira, F.; Basit, A.W. Influence of ageing on the gastrointestinal environment of the rat and its implications for drug delivery. *Eur. J. Pharm. Sci.* **2014**, *62*, 76–85. [CrossRef]
3. Ahmadi, S.; Razazan, A.; Nagpal, R.; Jain, S.; Wang, B.; Mishra, S.P.; Wang, S.; Justice, J.; Ding, J.; McClain, D.A.; et al. Metformin Reduces Aging-Related Leaky Gut and Improves Cognitive Function by Beneficially Modulating Gut Microbiome/Goblet Cell/Mucin Axis. *J. Gerontol. Ser. A* **2020**, *75*, e9–e21. [CrossRef] [PubMed]
4. van der Post, S.; Jabbar, K.S.; Birchenough, G.; Arike, L.; Akhtar, N.; Sjovall, H.; Johansson, M.E.V.; Hansson, G.C. Structural weakening of the colonic mucus barrier is an early event in ulcerative colitis pathogenesis. *Gut* **2019**, *68*, 2142–2151. [CrossRef]
5. Anthony, P.C.; Corfield, A.P.; Carroll, D.; Myerscough, N.; Probert, C.S. Mucins in the gastrointestinal tract in health and disease. *Front. Biosci.* **2001**, *6*, 1321–1357. [CrossRef]
6. Sartor, R.B.; Wu, G.D. Roles for Intestinal Bacteria, Viruses, and Fungi in Pathogenesis of Inflammatory Bowel Diseases and Therapeutic Approaches. *Gastroenterology* **2017**, *152*, 327–339. [CrossRef] [PubMed]
7. Camilleri, M. Leaky gut: Mechanisms, measurement and clinical implications in humans. *Gut* **2019**, *68*, 1516–1526. [CrossRef]
8. Bansil, R.; Turner, B.S. Mucin structure, aggregation, physiological functions and biomedical applications. *Curr. Opin. Colloid Interface Sci.* **2006**, *11*, 164–170. [CrossRef]
9. Johansson, M.E.V.; Hansson, G.C. Immunological aspects of intestinal mucus and mucins. *Nat. Rev. Immunol.* **2016**, *16*, 639–649. [CrossRef]
10. Johansson, M.E.V.; Ambort, D.; Pelaseyed, T.; Schütte, A.; Gustafsson, J.K.; Ermund, A.; Subramani, D.B.; Holmén-Larsson, J.M.; Thomsson, K.A.; Bergström, J.H.; et al. Composition and functional role of the mucus layers in the intestine. *Cell. Mol. Life Sci.* **2011**, *68*, 3635–3641. [CrossRef]
11. Hansson, G.C.; Johansson, M.E. The inner of the two Muc2 mucin-dependent mucus layers in colon is devoid of bacteria. *Gut Microbes* **2010**, *1*, 51–54. [CrossRef] [PubMed]
12. Swidsinski, A.; Loening-Baucke, V.; Theissig, F.; Engelhardt, H.; Bengmark, S.; Koch, S.; Lochs, H.; Dörffel, Y. Comparative study of the intestinal mucus barrier in normal and inflamed colon. *Gut* **2007**, *56*, 343–350. [CrossRef] [PubMed]
13. Desai, M.S.; Seekatz, A.M.; Koropatkin, N.M.; Kamada, N.; Hickey, C.A.; Wolter, M.; Pudlo, N.A.; Kitamoto, S.; Terrapon, N.; Muller, A.; et al. A Dietary Fiber-Deprived Gut Microbiota Degrades the Colonic Mucus Barrier and Enhances Pathogen Susceptibility. *Cell* **2016**, *167*, 1339–1353.e21. [CrossRef] [PubMed]
14. Elderman, M.; Sovran, B.; Hugenholtz, F.; Graversen, K.; Huijskes, M.; Houtsma, E.; Belzer, C.; Boekschoten, M.; De Vos, P.; Dekker, J.; et al. The effect of age on the intestinal mucus thickness, microbiota composition and immunity in relation to sex in mice. *PLoS ONE* **2017**, *12*, e0184274. [CrossRef]
15. Paassen, N.B.-V.; Van Der Sluis, M.; Bouma, J.; Male, A.M.K.-V.; Lu, P.; Van Seuningen, I.; Boehm, G.; van Goudoever, J.; Renes, I.B. Colitis development during the suckling-weaning transition in mucin Muc2-deficient mice. *Am. J. Physiol. Liver Physiol.* **2011**, *301*, G667–G678. [CrossRef]
16. Johansson, M.E.V.; Phillipson, M.; Petersson, J.; Velcich, A.; Holm, L.; Hansson, G.C. The inner of the two Muc2 mu-cin-dependent mucus layers in colon is devoid of bacteria. *Proc. Natl. Acad. Sci. USA* **2008**, *105*, 15064–15069. [CrossRef]
17. Van der Sluis, M.; De Koning, B.A.E.; De Bruijn, A.C.J.M.; Velcich, A.; Meijerink, J.P.P.; van Goudoever, J.B.; Büller, H.A.; Dekker, J.; VAN Seuningen, I.; Renes, I.B.; et al. Muc2-Deficient Mice Spontaneously Develop Colitis, Indicating That MUC2 Is Critical for Colonic Protection. *Gastroenterology* **2006**, *131*, 117–129. [CrossRef]
18. Bergstrom, K.S.B.; Xia, L. Mucin-type O-glycans and their roles in intestinal homeostasis. *Glycobiology* **2013**, *23*, 1026–1037. [CrossRef]
19. Bergstrom, K.; Fu, J.; Johansson, M.E.V.; Liu, X.; Gao, N.; Wu, Q.; Song, J.; McDaniel, J.; McGee, S.; Chen, W.; et al. Core 1– and 3–derived O-glycans collectively maintain the colonic mucus barrier and protect against spontaneous colitis in mice. *Mucosal Immunol.* **2017**, *10*, 91–103. [CrossRef]
20. Fu, J.; Wei, B.; Wen, T.; Johansson, M.E.; Liu, X.; Bradford, E.; Thomsson, K.A.; McGee, S.; Mansour, L.; Tong, M.; et al. Loss of intestinal core 1–derived O-glycans causes spontaneous colitis in mice. *J. Clin. Investig.* **2011**, *121*, 1657–1666. [CrossRef]
21. Stotter, B.R.; Talbot, B.E.; Capen, D.E.; Artelt, N.; Zeng, J.; Matsumoto, Y.; Endlich, N.; Cummings, R.D.; Schlondorff, J.S. Cosmc-dependent mucin-type *O*-linked glycosylation is essential for podocyte function. *Am. J. Physiol. Physiol.* **2020**, *318*, F518–F530. [CrossRef] [PubMed]
22. An, G.; Wei, B.; Xia, B.; McDaniel, J.M.; Ju, T.; Cummings, R.D.; Braun, J.; Xia, L. Increased susceptibility to colitis and colorectal tumors in mice lacking core 3–derived O-glycans. *J. Exp. Med.* **2007**, *204*, 1417–1429. [CrossRef] [PubMed]
23. Sommer, F.; Adam, N.; Johansson, M.E.V.; Xia, L.; Hansson, G.C.; Bäckhed, F. Altered Mucus Glycosylation in Core 1 O-Glycan-Deficient Mice Affects Microbiota Composition and Intestinal Architecture. *PLoS ONE* **2014**, *9*, e85254. [CrossRef]
24. Dubbelboer, I.R.; Barmpatsalou, V.; Rodler, A.; Karlsson, E.; Nunes, S.F.; Holmberg, J.; Häggström, J.; Bergström, C.A.S. Gastrointestinal mucus in dog: Physiological characteristics, composition, and structural properties. *Eur. J. Pharm. Biopharm.* **2022**, *173*, 92–102. [CrossRef] [PubMed]

25. Kamphuis, J.B.J.; Mercier-Bonin, M.; Eutamène, H.; Theodorou, V. Mucus organisation is shaped by colonic content; a new view. *Sci. Rep.* **2017**, *7*, 1–13. [CrossRef]
26. Barmpatsalou, V.; Dubbelboer, I.R.; Rodler, A.; Jacobson, M.; Karlsson, E.; Pedersen, B.L.; Bergström, C.A. Physiological properties, composition and structural profiling of porcine gastrointestinal mucus. *Eur. J. Pharm. Biopharm.* **2021**, *169*, 156–167. [CrossRef]
27. Varum, F.J.O.; Veiga, F.; Sousa, J.; Basit, A.W. Mucus thickness in the gastrointestinal tract of laboratory animals. *J. Pharm. Pharmacol.* **2012**, *64*, 218–227. [CrossRef]
28. Sovran, B.; Hugenholtz, F.; Elderman, M.; Van Beek, A.A.; Graversen, K.; Huijskes, M.; Boekschoten, M.V.; Savelkoul, H.F.J.; De Vos, P.; Dekker, J.; et al. Age-associated Impairment of the Mucus Barrier Function is Associated with Profound Changes in Microbiota and Immunity. *Sci. Rep.* **2019**, *9*, 1437. [CrossRef]
29. Engevik, M.A.; Luk, B.; Chang-Graham, A.L.; Hall, A.; Herrmann, B.; Ruan, W.; Endres, B.T.; Shi, Z.; Garey, K.W.; Hyser, J.M.; et al. Bifidobacterium dentium Fortifies the Intestinal Mucus Layer via Autophagy and Calcium Signaling Pathways. *MBio* **2019**, *10*, e01087-19. [CrossRef]
30. Wang, Q.; Lin, Y.; Sheng, X.; Xu, J.; Hou, X.; Li, Y.; Zhang, H.; Guo, H.; Yu, Z.; Ren, F. Arachidonic Acid Promotes Intestinal Regeneration by Activating WNT Signaling. *Stem Cell Rep.* **2020**, *15*, 374–388. [CrossRef]
31. Guzmán-Mejía, F.; Godínez-Victoria, M.; Vega-Bautista, A.; Pacheco-Yépez, J.; Drago-Serrano, M. Intestinal Homeostasis under Stress Siege. *Int. J. Mol. Sci.* **2021**, *22*, 5095. [CrossRef] [PubMed]
32. Johansson, M.E.V.; Hansson, G.C. The goblet cell: A key player in ischaemia-reperfusion injury. *Gut* **2013**, *62*, 188–189. [CrossRef]
33. Fang, J.; Wang, H.; Zhou, Y.; Zhang, H.; Zhou, H.; Zhang, X. Slimy partners: The mucus barrier and gut microbiome in ulcerative colitis. *Exp. Mol. Med.* **2021**, *53*, 772–787. [CrossRef] [PubMed]
34. Johansson, M.E.V. Fast Renewal of the Distal Colonic Mucus Layers by the Surface Goblet Cells as Measured by In Vivo Labeling of Mucin Glycoproteins. *PLoS ONE* **2012**, *7*, e41009. [CrossRef] [PubMed]
35. Li, C.; Ma, D.; Zhou, H.; Zhang, M.; An, L.; Wang, Y.; Wu, C. Effects of different doses lipopolysaccharides on the mucosal barrier in mouse intestine. *Res. Veter Sci.* **2020**, *133*, 75–84. [CrossRef] [PubMed]
36. Yao, Y.; Kim, G.; Shafer, S.; Chen, Z.; Kubo, S.; Ji, Y.; Luo, J.; Yang, W.; Perner, S.P.; Kanellopoulou, C.; et al. Mucus sialylation determines intestinal host-commensal homeostasis. *Cell* **2022**, *185*, 1172–1188.e28. [CrossRef]
37. O'Mahony, D.; O'Leary, P.; Quigley, E.M.M. Aging and Intestinal Motility. *Drugs Aging* **2002**, *19*, 515–527. [CrossRef] [PubMed]
38. Etienne-Mesmin, L.; Chassaing, B.; Desvaux, M.; De Paepe, K.; Gresse, R.; Sauvaitre, T.; Forano, E.; Van De Wiele, T.; Schüller, S.; Juge, N.; et al. Experimental models to study intestinal microbes–mucus interactions in health and disease. *FEMS Microbiol. Rev.* **2019**, *43*, 457–489. [CrossRef]
39. Johansson, M.E.V.; Gustafsson, J.K.; Holmén-Larsson, J.; Jabbar, K.S.; Xia, L.; Xu, H.; Ghishan, F.K.; Carvalho, F.A.; Gewirtz, A.T.; Sjövall, H.; et al. Bacteria penetrate the normally impenetrable inner colon mucus layer in both murine colitis models and patients with ulcerative colitis. *Gut* **2014**, *63*, 281–291. [CrossRef] [PubMed]
40. Caldara, M.; Friedlander, R.S.; Kavanaugh, N.L.; Aizenberg, J.; Foster, K.R.; Ribbeck, K. Mucin Biopolymers Prevent Bacterial Aggregation by Retaining Cells in the Free-Swimming State. *Curr. Biol.* **2012**, *22*, 2325–2330. [CrossRef]
41. Syed, Z.A.; Zhang, L.; Hagen, K.G.T. In vivo models of mucin biosynthesis and function. *Adv. Drug Deliv. Rev.* **2022**, *184*, 114182. [CrossRef] [PubMed]
42. Meldrum, O.W.; Yakubov, G.E.; Bonilla, M.R.; Deshmukh, O.; McGuckin, M.A.; Gidley, M.J. Mucin gel assembly is controlled by a collective action of non-mucin proteins, disulfide bridges, Ca^{2+}-mediated links, and hydrogen bonding. *Sci. Rep.* **2018**, *8*, 5802. [CrossRef] [PubMed]
43. Arike, L.; Hansson, G.C. The Densely O-Glycosylated MUC2 Mucin Protects the Intestine and Provides Food for the Commensal Bacteria. *J. Mol. Biol.* **2016**, *428*, 3221–3229. [CrossRef] [PubMed]
44. Gréco, S.; Niepceron, E.; Hugueny, I.; George, P.; Louisot, P.; Biol, M.-C. Dietary Spermidine and Spermine Participate in the Maturation of Galactosyltransferase Activity and Glycoprotein Galactosylation in Rat Small Intestine. *J. Nutr.* **2001**, *131*, 1890–1897. [CrossRef]
45. Johansson, M.E.V. Mucus Layers in Inflammatory Bowel Disease. *Inflamm. Bowel Dis.* **2014**, *20*, 2124–2131. [CrossRef]

nutrients

MDPI

Review

Dietary Restriction and Lipid Metabolism: Unveiling Pathways to Extended Healthspan

Hye-Yeon Lee and Kyung-Jin Min *

Department of Biological Sciences and Bioengineering, Inha University, Incheon 22212, Republic of Korea; lee.hy@inha.ac.kr
* Correspondence: minkj@inha.ac.kr; Tel.: +82-32-860-8193

Abstract: Dietary restriction (DR) has been reported to be a significant intervention that influences lipid metabolism and potentially modulates the aging process in a wide range of organisms. Lipid metabolism plays a pivotal role in the regulation of aging and longevity. In this review, we summarize studies on the significant role of lipid metabolism in aging in relation to DR. As a potent intervention to slow down aging, DR has demonstrated promising effects on lipid metabolism, influencing the aging processes across various species. The current review focuses on the relationships among DR-related molecular signaling proteins such as the sirtuins, signaling pathways such as the target of rapamycin and the insulin/insulin-like growth factor (IGF)-1, lipid metabolism, and aging. Furthermore, the review presents research results on diet-associated changes in cell membrane lipids and alterations in lipid metabolism caused by commensal bacteria, highlighting the importance of lipid metabolism in aging. Overall, the review explores the interplay between diet, lipid metabolism, and aging, while presenting untapped areas for further understanding of the aging process.

Keywords: dietary restriction; lipid profile; healthspan; sirtuin; target of rapamycin; Ins/IGF-1; cell membrane; commensal bacteria

1. Introduction

Dietary restriction (DR), a dietary regimen characterized by reduced calorie intake without causing malnutrition, has emerged as a significant intervention that influences lipid metabolism and potentially modulates the aging process in a wide range of organisms including yeast, nematodes, fruit flies, rodents, and primates [1]. Many studies have demonstrated that DR delays aging via the sirtuin (SIRT), target of rapamycin (TOR), and insulin/insulin-like growth factor-1 (ins/IGF-1) signaling pathways [2,3]. These pathways are well known to be involved in lipid metabolism, thereby suggesting the role of lipid metabolism in aging [4,5]. Evidence corroborating this notion is discernible when examining aging-related research on compounds identified as potential calorie restriction mimetics, such as resveratrol, rapamycin, and metformin, which reveal an association between energy metabolism and the aging process [4–6]. Resveratrol, a typical DR mimetic, is known to not only regulate the biosynthesis of the adipocyte life cycle and fat [7] but also to enhance fat degradation in adipocytes cultured in vitro [8]. In addition, studies on mistletoe, which has been reported to increase the lifespan in yeast, nematodes, fruit flies, and silkworms, have shown that mistletoe increased sensitivity to starvation in fruit flies [9], reduced body weight in the silkworm moth [10], and inhibited fatty acids biosynthesis [11]. This suggests that the lifespan-extending mechanism of DR mimetics is correlated with energy storage.

It is currently suggested that the human lifespan is about 120 years [12]. Jeanne Calment (1875–1997), officially the world's longest-living person, was recorded as living for 122 years and 164 days (The Guinness Book of World Records 1999). The World Population Ageing 2009 report by the United Nations Department of Economic and Social Affairs (UN DESA) used the term 'Homo hundreds' to indicate that Homo sapiens (humans) can live for over 100 years. In addition, improvements in medicine, economics, and technology have improved the quality of life of human beings, and accordingly, 'food intake' has been associated with improved life quality in people today. Predominantly carbohydrate-based diets have now been replaced by high-fat-based diets in recent times. Also, the excessive intake of proteins and carbohydrates leads to excess energy, which is ultimately converted to the lipid form and stored in the body. This excessive intake of energy, including fat, has been reported to increase the incidence of metabolic diseases such as obesity, diabetes, cardiovascular disease, and cancer [13–16]. Thus, diet-based lipid metabolism in the body is closely related to the aging of organisms. This report reviews perspectives on the relationship between DR, lipid metabolism, and aging.

2. Molecular Mechanisms of Dietary Restriction Regulating Lipid Metabolism

DR is a proven intervention that extends the lifespan and enhances metabolic health across species, from yeast to mammals [3]. The interest in DR research is its influence on conserved cellular pathways such as the sirtuin, TOR, and insulin/IGF-1 signaling pathways, which regulate metabolism, energy homeostasis, stress responses, and longevity (Figure 1). Under DR conditions, elevated nicotinamide adenine dinucleotide (NAD^+) levels enhance sirtuin activity, particularly SIRT1. NAD^+-dependent deacetylation is a process crucial for sensing the energy status of the cell. NAD^+ levels are indicative of cellular energy balance, and the consumption of NAD^+ influences directly the cellular survival. This activation promotes metabolic adaptation through the deacetylation of key proteins, including AMP-activated protein kinase (AMPK), leading to increased energy efficiency and lipid oxidation. Sirtuins also modulate the activity of transcription factors such as Forkhead box-O and p53, enhancing stress resistance, DNA repair, and apoptosis regulation. TOR pathway, a critical nutrient-sensing mechanism that regulates growth, metabolism, and lifespan, is inhibited under the conditions of DR. Reduced nutrient availability leads to decreased activation of Akt, which in turn diminishes mammalian TOR complex 1 (mTORC1) activity. mTORC1, which includes TOR and raptor, is negatively regulated by the TSC1/TSC2 complex. Since TORC1 activation phosphorylates downstream effectors such as S6 kinase (S6K) and 4E-binding protein (4E-BP), lowered TORC1 activity downregulates S6K activation and upregulates 4E-BP, resulting in reduced protein synthesis and enhanced cellular stress resistance. These changes contribute to an increase in autophagy and a shift from anabolic processes to catabolic pathways, ultimately supporting cellular maintenance and promoting lifespan extension. The insulin/IGF-1 pathway is reduced under DR, leading to decreased signaling through phosphoinositide 3-kinase (PI3K) and phosphoinositide-dependent kinase (PDK). This cascade results in lower activation of Akt, which in turn prevents the phosphorylation and inactivation of FOXO transcription factors. Activated FOXO trans-localization to the nucleus, where it stimulates the expression of genes involved in stress resistance, lipid catabolism, and metabolic adaptation. These pathways play crucial roles in lipid metabolism, promoting fat oxidation, reducing fat storage, and modulating lipid synthesis. DR reduces the accumulation of fat in the liver, particularly under high-fat diet conditions, and improves mitochondrial efficiency in lipid metabolism. This suggests a protective role of DR in managing lipid disorders like non-alcoholic fatty liver disease [17]. In long-term DR diets during a 2-year period in Biosphere 2—a closed ecological system designed to study interactions between humans, agriculture, and the environment—significant reductions in total plasma cholesterol and triglyceride levels have been observed, demonstrating that DR prevents hyperlipidemia and improve lipid profiles, which are key factors in preventing cardiovascular diseases [18]. Interestingly, restricting specific dietary components, even without limiting overall diet, can also impact lipid profile. For instance, carbohydrate restriction can improve lipid metabolism by reducing triglycerides, low-density lipoprotein, and total cholesterol levels. It promotes lipid oxidation, improves insulin sensitivity, and reduces fat storage, contributing to improved cardiovascular health [19]. Also, restriction of dietary methionine increases fat oxidation and reduces liver lipid accumulation without affecting body weight, indicating a potential therapeutic role in lipid disorders independent of caloric intake [20]. In summary, DR beneficially impacts lipid profiles, reduces fat accumulation, enhances insulin sensitivity, and promotes fat oxidation, supporting its potential as a strategy to prevent and manage lipid disorders and metabolic diseases. However, long-term studies and further research are needed to fully understand its sustained impacts on lipid metabolism.

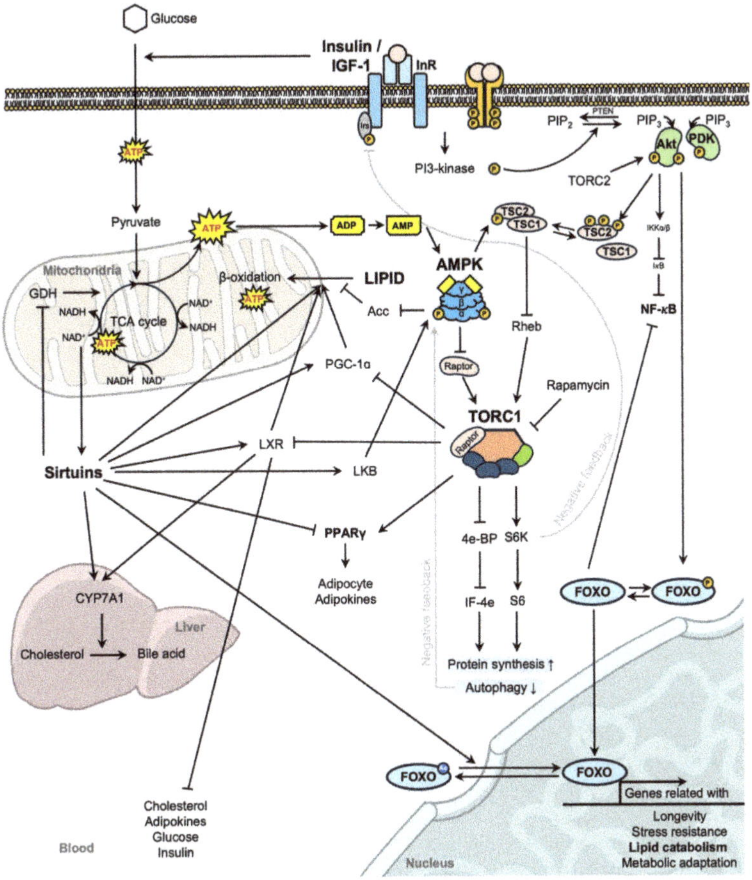

Figure 1. Molecular mechanisms linking dietary restriction to lipid metabolism and health benefits. The figure illustrates the molecular pathways influenced by dietary restriction (DR) and their effects on lipid metabolism and overall health. Key pathways include NAD^+-dependent sirtuin signaling, the TOR pathway, AMPK activation, and insulin/IGF-1 signaling, which collectively regulate metabolic processes. By reducing energy intake, DR enhances mitochondrial β-oxidation, promoting the breakdown of fatty acids to generate energy, while increasing NAD^+ levels to activate SIRT1, further driving lipid catabolism. DR also suppresses insulin/IGF-1 signaling by activating the TSC1/2 complex to inhibit mTORC1, shifting the cellular focus from anabolic processes to lipid utilization and autophagy. Additionally, FOXO transcription factors, activated by DR, upregulate genes involved in lipid breakdown and metabolic adaptation, supporting efficient fat utilization and stress resistance. In the liver, DR enhances cholesterol metabolism by activating CYP7A1, facilitating the conversion of cholesterol into bile acids to maintain lipid homeostasis. Furthermore, DR reduces circulating levels of cholesterol, adipokines, glucose, and insulin, promoting lipid catabolism, metabolic adaptation, stress resistance, and longevity. These coordinated changes highlight DR's critical role in optimizing energy balance, lowering the risk of metabolic diseases, and extending the healthspan.

2.1. Sirtuins and Lipid Profile

Silent information regulator 2 (Sir2) proteins, or sirtuins, are highly conserved NAD^+-dependent protein deacetylases present in a diverse range of organisms from yeasts to humans. Sir2 was first found to regulate replication-associated aging in the budding yeast, *Saccharomyces cerevisiae* [21]. In response to cellular energy states, it plays a role in activating the metabolism via metabolic enzymes, as well as histones and transcription factors [22]. Activation of sirtuins reduces blood cholesterol, adipokines, glucose, and insulin [23], and prolongs lifespan by managing energy homeostasis and stress [24,25]. The sirtuin-activating mechanism is involved in the lifespan-increasing effect of DR. In addition, sirtuin directly or indirectly affects lipid metabolism in white and brown adipose tissue (WAT and BAT), the liver, and skeletal muscle by modulating key metabolic pathways. The sirtuin family in mammals, encompassing Sirtuin 1 to Sirtuin 7, plays pivotal roles in numerous cellular functions, including cell survival, aging, genomic integrity, and metabolic regulation [26]. Each sirtuin is activated in distinct locations and performs unique functions, with varying responses to CR. For example, CR or fasting upregulates Sirtuin 1 [27], Sirtuin 2 [28], Sirtuin 3 [29], and Sirtuin 6 [30] while downregulating Sirtuin 4 [31] in adipose tissue, with no significant effect on Sirtuin 7 [32] expression. The sirtuin family not only acts as crucial sensors of cellular energy status but also confers protection against metabolic challenges. Sirtuin 1 (SIRT1) reduces adipocyte production in white adipocytes by inhibiting the transcriptional activity of nuclear receptor peroxisome proliferator-activated receptor γ (PPARγ), a major ligand-active transcription factor that regulates adipocyte differentiation [33]. Interestingly, SIRT1 causes the promoters of the PPARγ target gene to aggregate in starvation conditions [34] and regulates adipocyte production and function in response to starvation or DR [35]. In addition, PPARγ, PPARγ coactivator 1α (PGC-1α), and the liver X receptor (LXR) in the liver, which is the main organ responsible for fatty acid, cholesterol, and bile acid synthesis and lipoprotein absorption, synthesis, and secretion, increase the targeting of LXR and β-oxidation [36–38]. The skeletal muscle also promotes β-oxidation through the deacetylation of PGC-1α [39,40]. The promotion of the β-oxidation of free fatty acids plays an important role in producing adenosine triphosphate (ATP) and providing it to the brain under starvation conditions. ATP consumption increases the intracellular adenosine monophosphate (AMP)/ATP ratio and activates AMP-activated protein kinase (AMPK), which stimulates fatty acid oxidation, triglyceride accumulation, and mitochondrial biosynthesis [41]. SRIT1 also stimulates the activity of AMPK by deacetylating liver kinase B1 (LKB1), which is an activation factor of AMPK. In this way, SIRT1 and AMPK regulate the mitochondrial oxidation of fatty acids in a nutrient-limited manner. This effect is observed both in vitro and in vivo: for example, α-lipoic acid enhances SIRT1-mediated fat oxidation and reduces fat storage in cells and diabetic mice [42]. Sirtuin 2 (SIRT2) indirectly affects white adipose tissue by the deacetylation of the transcription factor Forkhead box O1 (FoxO1) protein. Even though the process of SIRT2 differs from that of SIRT1, it results in reducing the production of adipocytes in white adipose tissue and promoting lipolysis. Unlike SIRT2, sirtuin 3 (SIRT3) is involved in brown adipose tissue. The expression of SIRT3 in brown adipocytes not only increases the expression of PGC-1α, the main regulator of mitochondrial biosynthesis gene expression, but also increases the respiration rate through the activation of uncoupling protein 1 (UCP1) present in the mitochondrial inner membrane [29]. Also, SIRT3 enhances mitochondrial fatty acid oxidation, particularly in response to high fatty acid conditions, by regulating enzymes involved in the breakdown of fatty acids. This reduces lipid accumulation in liver cells and improves energy homeostasis [43]. Enhanced mitochondrial function by SIRT3, activated through DR, plays a crucial role in promoting fatty acid oxidation, thereby regulating brain lipid metabolism and supporting overall mitochondrial health [44]. Some studies predict that the stimulation of brown fat thermogenesis by the pharmacological activation of sirtuins could be used as a new strategy for the treatment of obesity and related diseases [45,46]. Sirtuin 4 (SIRT4) regulates lipid homeostasis and metabolism by controlling insulin secretion [23]. Insulin is generally known only as a hormone that synthesizes glucose into glycogen. However, research on the relationship between insulin and aging is actively underway, as it became known that FOXO is regulated by insulin and insulin-like growth hormones. SIRT4 is expressed in the islets of Langerhans that inhibit glutamate dehydrogenase (GDH) through NAD^+ dependent ADP-ribosylation [31]. GDH is a mitochondrial enzyme that promotes ATP production by inducing amino acids into the citric acid cycle by converting glutamate to α-ketoglutarate [31]. Islets of Langerhans isolated from SIRT4-deficient mice showed increased GDH activity and insulin secretion regardless of the glucose and amino acid levels [31]. The over-expression of SIRT4 in insulinoma cells has been shown to inhibit insulin secretion [47]. In addition, SIRT4 inhibits fatty acid oxidation by repressing malonyl CoA decarboxylase (MCD). This action increases malonyl CoA levels, which both promotes lipid synthesis and inhibits fat oxidation. Mice lacking SIRT4 showed elevated fat oxidation, highlighting its regulatory role in lipid balance [48]. Therefore, the modulation of SIRT4 activity can potentially provide a new therapeutic approach for the treatment of metabolic diseases. Sirtuin 5 (SIRT5), a mitochondria-localized sirtuin, plays a crucial role in regulating lipid metabolism and promoting fatty acid oxidation [49]. Studies using bovine preadipocytes and obese mice have shown that SIRT5 activates the AMPK pathway while repressing the MAPK pathway during adipocyte differentiation [50]. In Sirt5-knockout brown adipose tissue (BAT), reduced function of UCP1 and associated proteins leads to impaired mitochondrial respiration, defective mitophagy, and metabolic inflexibility [51]. Therefore, Sirt5 plays a critical role in maintaining energy homeostasis by regulating UCP1, a key thermogenic protein in BAT. Sirtuin 6 (SIRT6), in cooperation with SIRT5, reduces lipid deposition in bovine preadipocytes. It inhibits preadipocyte differentiation and lipid accumulation by activating the AMPKα pathway and suppresses adipose tissue differentiation and lipid synthesis in obese mice in vivo [52]. Fat-specific Sirt6-knockout mice are more susceptible to high-fat-diet-induced obesity and exhibit increased inflammation in adipose tissue. This effect is linked to the suppression of adipose triglyceride lipase (ATGL), a key lipolytic enzyme, caused by increased phosphorylation and acetylation of FoxO1, which impairs its transcriptional activation of ATGL. Similarly, in obese patients, reduced SIRT6 expression is associated with decreased ATGL expression [53]. A study investigating the role of SIRT6 in adipocytes found that fat-specific *Sirt6*-knockout mice showed impaired responses to intermittent fasting (IF), including glucose and insulin intolerance, reduced energy expenditure, impaired adipose tissue browning, and increased WAT inflammation. These effects were linked to reduced UCP1 expression and down-regulation of p38 MAPK/ATF2 signaling. Interestingly, SIRT6 deficiency in hepatocytes or in myeloid cells did not impair adaptation to IF. The findings highlight that SIRT6 in adipocytes is essential for the metabolic benefits of IF [54]. Additionally, sirtuin 7 (SIRT7) regulates lipid metabolism by influencing the degradation of key proteins involved in lipid storage. *Sirt7*-knockout mice showed resistance to fatty liver and obesity induced by a high-fat diet, suggesting that SIRT7 promotes lipid accumulation through regulation of fatty acid uptake and triglyceride synthesis [55]. Interestingly, *Sirt7*- knockout mice exhibit a significant reduction in white WAT due to elevated SIRT1 activity, and inhibition of SIRT1 restored WAT levels in these mice [56]. In summary, notably, SIRT1, SIRT3, and SIRT6 are central regulators of adipogenesis, lipid mobilization, and anti-inflammatory responses and play prominent roles in adipose tissue browning and thermogenesis by repressing PPARγ and PGC-1α and promoting AMPK activation. SIRT2 and SIRT5 modulate adipocyte function through pathways involving FOXO1, PGC-1α, and lipolysis regulation. SIRT4 increases adipocyte differentiation by repressing fatty acid oxidation. SIRT7 promotes adipogenesis by suppressing SIRT1 activity while lacking significant involvement in lipid mobilization, inflammation, fibrosis, or browning. The sirtuin family is intricately involved in the coordination of autophagic processes. The role of autophagy in lipid metabolism represents a novel and promising field of study, potentially offering new perspectives on the metabolic regulation of lipids [57,58]. Various natural compounds with a sirtuin modulation function, such as curcumin, resveratrol, and berberine, have been identified as influencing metabolism targeting sirtuins to regulate autophagy [26,59,60]. Overall, there is still a need to elucidate the diverse mechanisms by which sirtuins regulate aging through lipid metabolism. Particularly considering recent research trends, studies related to autophagy are anticipated to be vigorously pursued in the future. Thus, a study on the role of sirtuins in lipid metabolism can offer promising targets for treating metabolic diseases and extending the lifespan.

2.2. *Target of Rapamycin Pathway and Lipid Profile*

Target of rapamycin (TOR) is a major nutrient-sensing mechanism that regulates metabolism and is well conserved from yeasts to humans [61]. Mammalian TOR (mTOR) may be described as a serine/threonine kinase that forms two complexes, mTORC1 and complex 2 (mTORC2), which regulate growth, survival, autophagy, and metabolism [62]. Importantly, mTORC1 activity is subject to intricate feedback regulation to maintain cellular homeostasis. For example, the activation of mTORC1 inhibits insulin receptor substrate (IRS) signaling via the S6K1-mediated phosphorylation of IRS proteins, which creates a negative feedback loop that inhibits PI3K-Akt signaling [63]. Similarly, mTORC1 activation can suppress autophagy by phosphorylating Unc-51-like autophagy activating kinase 1 (ULK1), yet prolonged nutrient deprivation leads to upstream AMPK activation, which inhibits mTORC1 and reactivates autophagy to restore energy balance [64]. These feedback mechanisms highlight the dynamic and tightly regulated nature of the TOR pathway in cellular metabolism. When there is an anabolic stimulus, mTORC1 and mTORC2 are instrumental in increasing triglyceride accumulation through the enhancement of the adipogenic and lipogenic pathways [62]. Activation of mTORC1 promotes de novo lipogenesis by upregulating key enzymes like sterol regulatory element-binding protein 1c (SREBP1c) and fatty acid synthase (FAS). Moreover, TOR activation inhibits the expression of ATGL and hormone-sensitive lipase (HSL), thus reducing fat breakdown. This leads to the accumulation of triglycerides and fat storage in tissues like the liver and adipose tissue [65,66]. Concurrently, these complexes orchestrate the efficient sequestration of energy by attenuating catabolic pathways, notably lipolysis and β-oxidation, thereby ensuring the conservation of energy reserves within the organism. For instance, the inhibition of mTORC1 by rapamycin, an inhibitor of TORC1, increases β-oxidation and the expression of enzymes involved in fatty acid oxidation, such as acyl-CoA dehydrogenase, very long chain (ACADVL) and carnitine-palmitoyltransferase-1 [67,68]. Rapamycin reduces the expression of lipogenic genes like glucokinase and FAS in rainbow trout, resulting in decreased lipid accumulation [69]. In yeast, rapamycin treatment leads to the replenishment of lipid droplets, highlighting its role in lipid homeostasis [70]. Moreover, mTORC1 can control ketogenesis in mice in response to fasting [71]. Sengupta et al. demonstrated that the liver-specific loss of tuberous sclerosis complex 1 (TSC1), a known inhibitory regulator of mTORC1, markedly impairs the capacity of liver cells to initiate β-oxidation in response to fasting conditions and suggested a role for mTORC1 activity in promoting the aging of the liver by regulating the PPARα function and hepatic ketogenesis. Specifically, it is well-known that the functions of mTORC1 respond to various environmental inputs, like growth factors, amino acids, and stress, to regulate cell growth and proliferation [50]. In an experiment examining the relationship between a high-fat diet and TOR, the inhibition of the TOR signaling pathway prevented an increase in triglycerides and reduced insulin signaling in fruit flies fed on a high-fat diet [72]. In addition, the TOR signal mediates pancreatic β-cell growth and functional nutrient responses (such as insulin and glucose) [73–75]. The elimination of the TOR effector, S6 kinase 1 (S6K1), can prevent the accumulation of fat after consuming high-fat diets [76,77]. Although the inhibition of mTORC1 via rapamycin treatment can affect lipolysis [65], there is no evidence of an indirect effect by mTORC2 regulation. The mTORC1 function could also be inhibited by rapamycin in rats and some cell types [78,79]. However, most of the studies investigating the effect of rapamycin have focused only on the mTORC1 function, but not on the role of mTORC2 [65,80–83]. Despite considerable progress in unraveling the mechanisms linking the mTOR complexes to lipid metabolism, further research is essential to comprehensively understand the specific actions and contributions of each mTOR complex in the regulation of lipid metabolic processes.

2.3. Insulin/Insulin-like Growth Factor 1 and Lipid Profile

The insulin/insulin-like growth factor 1 (IGF-1) signaling pathway, which is evolutionarily conserved across various species ranging from yeast to humans, plays a pivotal role in a host of interconnected functions essential for metabolism, growth, and reproduction [84]. The insulin/IGF-1 signaling (IIS) pathway plays a pivotal role in the regulation of longevity and is represented as a pathway downregulated by DR [3]. Insulin transports blood glucose into the cells, regulates cellular growth, and controls metabolism-related gene transcription. Insulin signaling is closely related to carbohydrate metabolism but is also known to play an important role in lipid metabolism, affecting various aspects such as lipid synthesis, storage, and breakdown [85–87]. For instance, insulin activates lipogenic pathways, particularly in the liver and adipose tissue. It enhances the expression and activity of key enzymes involved in lipid synthesis, such as acetyl-CoA carboxylase (ACC) and FAS, which leads to increased fat production and storage [88]. Simultaneously, insulin suppresses the breakdown of stored triglycerides in adipocytes by inhibiting the action of hormone-sensitive lipase and adipose triglyceride lipase [89,90]. This prevents the release of free fatty acids (FFAs) into the bloodstream, promoting fat retention in adipose tissue. Elevated FFAs can impair insulin signaling, leading to metabolic insulin resistance by inhibiting insulin-stimulated glucose uptake and suppressing nitric oxide production, which can affect vascular function [91,92]. This process is crucial for converting excess glucose into stored energy, reducing fatty acid levels in the blood, and improving energy utilization efficiency. Furthermore, insulin plays a pivotal role in energy production by regulating the oxidation of fatty acids in muscles and other tissues, ensuring a balanced interplay between energy expenditure and storage. Also, a high proportion of fatty acids such as linoleic acid in the diet is associated with impaired insulin sensitivity independent of obesity [93]. In insulin-resistant conditions (e.g., type 2 diabetes), patients have altered fatty acid composition, indicating that the insulin pattern can alter the lipid composition in the body [93], and insulin fails to suppress hepatic glucose production but continues to promote lipogenesis, leading to excessive fat accumulation in the liver, a phenomenon known as selective insulin resistance [94]. Furthermore, skeletal muscle is a critical factor in determining insulin sensitivity. Research has consistently linked insulin resistance with higher saturation levels of phospholipids in the skeletal muscle cell membrane [95]. Additionally, insulin signaling regulates the expression of acyl Co-A synthetases (ACS), which are responsible for activating triglyceride catabolism in mammalian cells [96]. In *Drosophila*, increased ACS expression, transcriptionally regulated by FOXO activation, causes a decrease in triglyceride levels, indicating that ACS expression levels are important for lipid homeostasis [96]. Xu et al. also showed that insulin inhibits fatty acid oxidation by downregulating the activity of enzymes like carnitine palmitoyltransferase 1, reducing the transport of fatty acids into mitochondria for β-oxidation [96]. This process promotes lipid storage over lipid breakdown. Studies using long-lived mutant nematodes (*Caenorhabditis elegans*) have also demonstrated the relationship between lifespan and lipid metabolism. Long-lived mutant genes *age-1* and *daf-2* in these nematodes are closely related to the insulin signaling pathway and have been useful in aging studies [97,98]. Since insulin acts as a major anabolic agent that promotes the synthesis and storage of glucose, proteins, and lipids, it is imperative to research the interplay between DR, lipid metabolism, and the aging process with the regulation of insulin signaling.

2.4. Cell Membrane Lipid and Health

Lipid metabolism is very important in that it not only functions as a mechanism of metabolic energy storage but also has a close impact on cell membrane function. Lipid metabolism primarily involves the formation of membranes from non-lipid sources. Moreover, it was observed that a restricted diet increased the fluidity of erythrocyte membranes and had a protective effect on cell membranes in rats [99,100]. Considering the obvious evidence that aged individuals have reduced cell membrane fluidity [101], these results indicate that DR has beneficial effects on cell membranes. Also, the maximum lifespan of the mammals and the membrane lipid peroxidation index were inversely correlated [102]. Lipid peroxidation increases oxidative stress in cell membranes, causes cell damage, and can finally lead to cell death. At the organism level, the accumulation of oxidative stress promotes aging by inducing diseases such as cancer and cardiovascular disease and the production of mitochondrial reactive oxygen species (ROS). Hulbert used the phrase "membrane pacemaker theory" to emphasize that lipid peroxidation and its reactants determine the lifespan of organisms [103]. Hulbert showed this correlation not only in mammals having shortened lifespans due to feeding on toxic substances but also in mammals having an extended lifespan through DR. A study that identified genetic targets in long-lived mammals using a phylogenetic approach found that mammals with genes that bind to lipids selectively live longer [104]. Studies in birds have shown that parrots that live long produce more monounsaturated fatty acid (MUFA), less polyunsaturated fatty acid (PUFA), and lower cardiac phospholipid peroxidation index compared to short-lived birds [105]. Additionally, studies in birds and mice with the same body size and mass have shown that birds with larger, slower-beating hearts and membrane compositions are less susceptible to damage due to lipid peroxidation and live longer [106]. In the case of nematodes, it has been shown that the fatty acid composition in lipids is an important determinant of longevity in nematodes rather than the total lipid contents [107]. Knocking down the fat-4 gene (encoding the Δ5 desaturase enzyme that enables the conversion of 20:3*n*-6 → 20:4*n*-6 and 20:4*n*-3 → 20:5*n*-3) resulted not only in remarkable resistance to hydrogen peroxide exposure but also in significant lifespan extension, confirming that fat composition is involved in extending the lifespan [108]. A recent study using *C. elegans* fed *Escherichia coli* (*E. coli*) mutants depleted of intracellular glucose proved that glucose restriction increases the lifespan of worms by promoting membrane fluidity in the peripheral tissues [109]. Taken together, lipid metabolism plays an important role in both energy storage and cell membrane function, suggesting that DR may inhibit aging by regulating lipid metabolism and preventing cell damage in a variety of organisms.

2.5. Commensal Bacteria and Lipid Profile

Recent studies have shown evidence that interactions between host and gut bacteria affect the aging of the host [110–113]. Most commensal bacteria live in the intestines of the host, and the commensal gut microbiome changes dynamically depending on the host's diet type [114]. DR was found to have a beneficial effect on the host not only through changes in the individual's molecular mechanisms as mentioned above, but also through changes in the intestinal microbial flora [115–117]. For instance, in a study on obese and overweight adult subjects, DR, especially intermittent calorie restrictions, maintained stable gut microbiota involving *Akkermansiaceae*, *Christensenellaceae*, and *Tanerellaceae* and improved subjects' health status by inducing weight loss [116]. A study conducted among overweight and obese Chinese women demonstrated that both low-carbohydrate (LC) and CR diets effectively reduced weight and fat mass. However, only the LC diet significantly improved dyslipidemia. These improvements were associated with changes in gut microbiota, as the two diets altered microbial structure and co-abundance gene clusters to varying degrees. Notably, the *Bacteroidetes/Firmicutes* ratio, a marker of gut microbial health, increased significantly in the LC diet group but not in the CR group, highlighting the distinct impact of dietary composition on gut microbiota. These microbiota changes contributed to the improved lipid profiles observed in the participants [118]. Furthermore, the gut microbiota is different between mice fed diets that are high or low in fat [119–121]. Just as the host's diet affects the gut microbiota, many studies have shown that the commensal microbiota affects the lipid metabolism of the host by transforming, synthesizing, and breaking down the lipids. In mice, the levels of cholesterol, fatty acids, triglycerides, and phosphatidylcholine in the serum were reduced in a conventionally raised condition (with microbiota) compared to a germ-free condition (without microbiota), and these levels were affected by the presence of commensal bacteria [122]. Notably, the commensal bacterial flora plays pivotal roles in the regulation of host cholesterol and sphingolipid homeostasis [123]. Also, metabolites produced by commensal bacteria such as short-chain fatty acids (SCFA), secondary bile acids, and lipopolysaccharides affect host lipid metabolism [124]. Dietary methionine restriction, a form of dietary restriction, has been shown to extend lifespan, improve lipid metabolism, suppress inflammation, change the gut microbiota, and enhance gut immunity in a variety of animals [125–127]. A recent study revealed the interconnected role of dietary methionine restriction in regulating lipid metabolism, inflammatory responses, and gut microbiota, demonstrating that it reduces fat accumulation and systemic inflammation through gut microbial alterations in middle-aged mice fed high-fat diets [128]. Dysbiosis of commensal microbiota can induce disturbed homeostasis of the lipid metabolism, which leads to an increase in the risk of gut inflammation, which is a critical factor in the development of various age-related conditions and diseases [129]. Moreover, the modulation of gut microbiota by probiotics treatment resulted in lipid profile changes and enhanced insulin sensitivity in mice with Alzheimer's disease (AD), a representative age-related neurodegenerative disease. This result indicates that the symbiosis induced by probiotics positively changed the lipid composition in mice with AD [130]. Disruption of cerebral cholesterol metabolism is a hallmark of both aging and AD, with lipid-lowering agents being associated with a reduced risk of neurodegenerative disorders. Recent studies have shown that oral supplementation with multi-strain probiotics reduced amyloid beta aggregation and brain damage in AD transgenic mouse models, leading to improved cognitive function [131]. Recently, there has been active research on the host aging-commensal bacteria interaction. Despite the growing evidence linking gut microbiota, lipid metabolism, and aging, several knowledge gaps remain. Great progress is expected in the future on the elucidation of the relationship between host aging, commensal bacteria, and lipid metabolism. Future research should focus on elucidating the specific roles of individual microbial species and their metabolites in lipid regulation and age-related metabolic diseases. Long-term clinical studies are needed to understand the sustained effects of dietary modulation and probiotic interventions on gut microbiota and lipid homeostasis. Furthermore, exploring the interactions between diet, gut microbiota, and systemic inflammation may offer novel strategies for managing metabolic diseases such as obesity, diabetes, and cardiovascular disease.

2.6. Intermittent Fasting and Lipid Profile

Building on the insights from DR, intermittent fasting (IF) similarly exerts beneficial effects on lipid metabolism by modulating key pathways. IF is a dietary manipulation that involves cycling between periods of feeding and fasting [132]. There are different approaches to IF, such as alternate-day fasting (fasting every other day or 5 days of feeding followed by 2 days fasting) and time-restricted eating (16 h of fasting followed by 8 h feeding window). IF is distinct in that it focuses on the timing of food intake rather than the number of calories consumed. Both DR and IF alter the activities of common key metabolic pathways, such as sirtuins, TOR, and IIS pathways [133,134]. As discussed in the section on sirtuins, DR-induced SIRT1 activation plays a critical role in promoting β-oxidation and enhancing fat utilization. Similarly, IF has been shown to elevate *Sirt1* mRNA expression and activity in mouse liver and neuronal retina progenitor cells [135], driving lipid catabolism during fasting periods. Additionally, IF mirrors DR by inhibiting mTORC1, facilitating a shift from anabolic to catabolic processes and inducing autophagy. For instance, in a study using rats with experimentally induced NAFLD, IF significantly reduced liver injury markers, including serum triglycerides (TG), alanine transaminase (ALT), and aspartate transaminase (AST), compared to control groups [136]. Ma et al. demonstrated that IF could mitigate cardiomyopathy by normalizing desmin localization through autophagy-dependent mechanisms via reduced mTOR activation. Nevertheless, the inhibitory effects of IF on mTOR signaling in WAT have been little explored [137,138]. In the IIS pathway, IF promotes triacylglycerol deposition in white adipose tissue by regulating insulin and lipid-associated gene expression, such as *Fsp27/Cidec* [139]. Additionally, IF improves glycemia, lipid metabolism, and insulin resistance in mice subjected to dietary overload with high-fat or high-fructose diets [140]. Conversely, IF demonstrates dual effects by reducing fat mass while impairing hepatic insulin signaling and lipid metabolism in young rats fed high-protein and high-fat diets [141]. These findings underscore the complexity of IF's impact on IIS and lipid homeostasis. These shared molecular pathways highlight the convergence of DR and IF in their metabolic benefits, underscoring their potential to improve lipid metabolism and extend the healthspan. IF leads metabolic switching, stress resistance, and improved glucose metabolism and achieves many benefits of DR without sustained calorie reduction. Therefore, lipid changes are observed even through IF without calorie changes. IF promotes lipid oxidation, particularly during fasting periods, as it enhances the breakdown of fat stores, contributing to fat loss and reducing visceral fat accumulation [142]. IF may preferentially reduce ectopic fat, such as fat stored in the liver, which is beneficial for preventing conditions like fatty liver disease and improving overall metabolic health [143]. Also, IF reduce levels of total cholesterol, low-density lipoprotein, and triglycerides while increasing high-density lipoprotein levels. This improvement is observed in both healthy individuals, obesity, and dyslipidemia patients [144–146]. A recent study explored how IF affects lipid metabolism in obese mice. Mice on an alternate day fasting regimen showed weight loss, lower serum lipid levels, and reduced liver fat accumulation compared to those with continuous feeding. IF also decreased metabolic endotoxemia and altered gut microbiota, increasing diversity and shifting the balance toward beneficial bacteria (e.g., reduced the *Firmicutes/Bacteroidetes* ratio and increased the relative abundance of *Allobaculum* in the intestinal flora). Also, CR or IF reduces neuroinflammation in the hypothalamus and hippocampus of rodents exposed to bacterial lipopolysaccharide [147,148]. Overall, IF improved lipid metabolism, reduced fat storage, and supported healthier gut flora [149]. The clinical study involving 470 NAFLD patients with IF showed reductions in body weight, fat mass, and liver fat (hepatic steatosis) and alanine aminotransferase levels, all markers of liver damage. However, fat-free mass and HDL cholesterol levels showed no significant changes [150]. These findings suggest that the metabolic benefits of IF closely mirror those of DR, reinforcing the importance of shared molecular pathways in extending the healthspan. Given that the molecular mechanisms of IF closely resemble those of DR, it is hypothesized that the effects of IF on lipid profiles will be largely similar to those observed in DR. However, research specifically addressing the long-term effects of IF remains limited. Future studies should actively explore the molecular-level impacts of IF on lipid profiles, with an emphasis on elucidating its long-term outcomes.

3. Conclusions

In conclusion, the studies explored in this review describe the intricate and significant role of lipid metabolism in DR in relation to aging (Figure 2). DR, as a potent intervention for slowing aging, has shown promising effects in modulating lipid metabolism, thereby influencing aging across various species. Our review highlights that, despite significant progress in understanding the interplay between DR and lipid metabolism, several critical knowledge gaps persist, underscoring the need for further research. In conclusion, sirtuins play a pivotal role in lipid metabolism by promoting lipid oxidation, reducing fat storage, and enhancing energy homeostasis across various tissues through NAD^+-dependent deacetylation. Their activation, along with the regulation of autophagic pathways, positions sirtuins as promising therapeutic targets for metabolic diseases such as obesity and NAFLD. While the sirtuin family is central to lipid homeostasis, their precise roles, particularly in connection to autophagy, remain incompletely understood. Expanding autophagy-related studies is expected to provide valuable insights into how cellular degradation processes intersect with lipid regulation and aging. Additionally, the mTOR and insulin signaling pathways, which promote lipogenesis under anabolic conditions, can be modulated through IF or DR to shift metabolism toward fat utilization. Given the distinct contributions of mTORC2 to cellular signaling and metabolic processes, future research should aim to differentiate and define the specific actions of mTORC1 and mTORC2 in lipid regulation, providing a more comprehensive understanding of these pathways. The role of insulin signaling, a key anabolic regulator, also demands further exploration in the context of DR, lipid metabolism, and aging. Understanding how DR influences insulin-mediated lipid synthesis and storage processes, as well as its broader impact on aging, will be crucial for developing targeted interventions. IF, in particular, mirrors the benefits of DR by improving lipid profiles and reducing visceral fat without sustained calorie reduction. However, while IF has shown promise in modulating lipid profiles and improving metabolic health, studies addressing its long-term effects remain limited. Future research should focus on elucidating the molecular mechanisms underlying IF's benefits and assessing its sustained impact on lipid metabolism and aging-related processes. Moreover, the interaction between gut microbiota and host lipid metabolism further highlights the potential for future research into diet-microbiota modulation to improve metabolic health and combat age-related diseases. However, there are significant gaps in understanding the specific roles of microbial species and their metabolites in lipid regulation. Long-term clinical studies investigating the sustained effects of dietary modulation and probiotic interventions on gut microbiota and lipid homeostasis are needed. The gut–brain axis has emerged as a promising area of research with potential applications for treating neurological diseases. These findings suggest that modulating the gut microbiota through DR could positively influence lipid metabolism, thereby enhancing brain health. However, the specific role of lipid metabolism in the gut–brain axis under DR remains unexplored, highlighting the need for further studies to bridge this knowledge gap. By addressing these gaps, future studies can provide a more holistic understanding of how DR and related interventions influence lipid metabolism and aging. This will not only advance fundamental scientific knowledge but also pave the way for novel therapeutic strategies targeting age-related metabolic diseases. Overall, the interplay between diet, lipid metabolism, and aging represents a frontier with untapped potential, with the possibility of promising advances in understanding and managing the aging process.

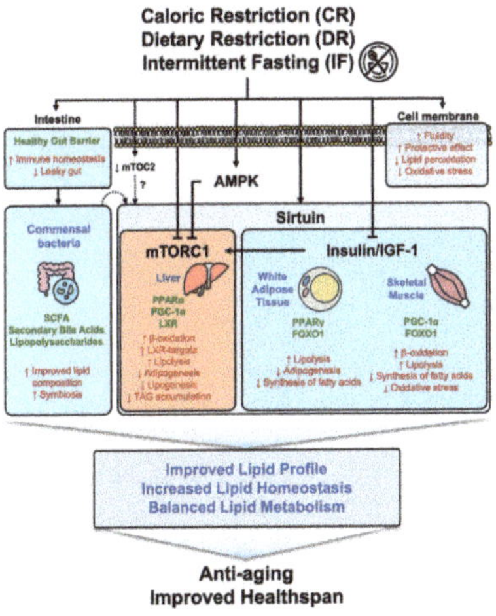

Figure 2. Effects of dietary restriction on lipid profile changes and healthspan. This figure summarizes the molecular mechanisms by which DR improves lipid profile and enhances the healthspan. DR impacts multiple pathways, including sirtuins, mTORC1, and the insulin/IGF-1 signaling pathway, which regulate lipid metabolism in the liver, adipose tissue, and skeletal muscle. DR also promotes a healthy gut barrier and modulates commensal microbiota, enhancing short-chain fatty acid (SCFA) production and immune homeostasis. Through increased β-oxidation, reduced lipogenesis, and enhanced oxidative stress resistance, DR maintains lipid homeostasis, reduces oxidative damage, and improves membrane fluidity. These coordinated effects result in balanced lipid metabolism, improved health outcomes, and anti-aging benefits.

Author Contributions: Study conception, design, and draft manuscript preparation: H.-Y.L.; manuscript supervision: K.-J.M. All authors have read and agreed to the published version of the manuscript.

Funding: This research was supported by Inha University (grant number 73030).

Conflicts of Interest: The authors declare no conflicts of interest.

References

1. Fontana, L.; Partridge, L.; Longo, V.D. Extending healthy life span--from yeast to humans. *Science* **2010**, *328*, 321–326. [CrossRef]
2. Lee, S.H.; Min, K.J. Caloric restriction and its mimetics. *BMB Rep.* **2013**, *46*, 181–187. [CrossRef]
3. Green, C.L.; Lamming, D.W.; Fontana, L. Molecular mechanisms of dietary restriction promoting health and longevity. *Nat. Rev. Mol. Cell Biol.* **2022**, *23*, 56–73. [CrossRef] [PubMed]
4. Ahn, J.; Cho, I.; Kim, S.; Kwon, D.; Ha, T. Dietary resveratrol alters lipid metabolism-related gene expression of mice on an atherogenic diet. *J. Hepatol.* **2008**, *49*, 1019–1028. [CrossRef]
5. Laplante, M.; Sabatini, D.M. An emerging role of mTOR in lipid biosynthesis. *Curr. Biol.* **2009**, *19*, R1046–R1052. [CrossRef]
6. Anurag, P.; Anuradha, C.V. Metformin improves lipid metabolism and attenuates lipid peroxidation in high fructose-fed rats. *Diabetes Obes. Metab.* **2002**, *4*, 36–42. [CrossRef] [PubMed]
7. Fischer-Posovszky, P.; Kukulus, V.; Tews, D.; Unterkircher, T.; Debatin, K.M.; Fulda, S.; Wabitsch, M. Resveratrol regulates human adipocyte number and function in a Sirt1-dependent manner. *Am. J. Clin. Nutr.* **2010**, *92*, 5–15. [CrossRef]
8. Szkudelska, K.; Nogowski, L.; Szkudelski, T. Resveratrol, a naturally occurring diphenolic compound, affects lipogenesis, lipolysis and the antilipolytic action of insulin in isolated rat adipocytes. *J. Steroid Biochem. Mol. Biol.* **2009**, *113*, 17–24. [CrossRef] [PubMed]
9. Lee, S.H.; An, H.S.; Jung, Y.W.; Lee, E.J.; Lee, H.Y.; Choi, E.S.; An, S.W.; Son, H.; Lee, S.J.; Kim, J.B.; et al. Korean mistletoe (*Viscum album coloratum*) extract extends the lifespan of nematodes and fruit flies. *Biogerontology* **2014**, *15*, 153–164. [CrossRef] [PubMed]
10. Jung, H.-Y.; Lee, A.-N.; Kim, S.-J.; Park, S.-M.; Song, T.-J.; Ko, B.-W.; Kim, Y.-H.; An, H.-S.; Yoo, Y.-C.; Min, B.-R.; et al. Effect of Korean Mistletoe (*Viscum album coloratum*) Extract (KME) on the Extensions of Life span in Silkworm Moth, Bombyx Mori. *Korean J. Pharmacogn.* **2008**, *39*, 310–315.
11. Wang, Y.; Deng, M.; Zhang, S.Y.; Zhou, Z.K.; Tian, W.X. Parasitic loranthus from Loranthaceae rather than Viscaceae potently inhibits fatty acid synthase and reduces body weight in mice. *J. Ethnopharmacol.* **2008**, *118*, 473–478. [CrossRef] [PubMed]
12. Carnes, B.A.; Olshansky, S.J.; Grahn, D. Biological evidence for limits to the duration of life. *Biogerontology* **2003**, *4*, 31–45. [CrossRef] [PubMed]
13. Wakil, S.J.; Abu-Elheiga, L.A. Fatty acid metabolism: Target for metabolic syndrome. *J. Lipid Res.* **2009**, *50*, S138–S143. [CrossRef] [PubMed]
14. Attie, A.D.; Scherer, P.E. Adipocyte metabolism and obesity. *J. Lipid Res.* **2009**, *50*, S395–S399. [CrossRef] [PubMed]
15. Ganguly, R.; Pierce, G.N. Trans fat involvement in cardiovascular disease. *Mol. Nutr. Food Res.* **2012**, *56*, 1090–1096. [CrossRef]
16. Santos, C.R.; Schulze, A. Lipid metabolism in cancer. *FEBS J.* **2012**, *279*, 2610–2623. [CrossRef]
17. Duivenvoorde, L.P.; van Schothorst, E.M.; Bunschoten, A.; Keijer, J. Dietary restriction of mice on a high-fat diet induces substrate efficiency and improves metabolic health. *J. Mol. Endocrinol.* **2011**, *47*, 81–97. [CrossRef] [PubMed]
18. Verdery, R.B.; Walford, R.L. Changes in plasma lipids and lipoproteins in humans during a 2-year period of dietary restriction in Biosphere 2. *Arch. Intern. Med.* **1998**, *158*, 900–906. [CrossRef] [PubMed]
19. Volek, J.S.; Fernandez, M.L.; Feinman, R.D.; Phinney, S.D. Dietary carbohydrate restriction induces a unique metabolic state positively affecting atherogenic dyslipidemia, fatty acid partitioning, and metabolic syndrome. *Prog. Lipid Res.* **2008**, *47*, 307–318. [CrossRef] [PubMed]
20. Plaisance, E.P.; Greenway, F.L.; Boudreau, A.; Hill, K.L.; Johnson, W.D.; Krajcik, R.A.; Perrone, C.E.; Orentreich, N.; Cefalu, W.T.; Gettys, T.W. Dietary methionine restriction increases fat oxidation in obese adults with metabolic syndrome. *J. Clin. Endocrinol. Metab.* **2011**, *96*, E836–E840. [CrossRef] [PubMed]
21. Landry, J.; Sutton, A.; Tafrov, S.T.; Heller, R.C.; Stebbins, J.; Pillus, L.; Sternglanz, R. The silencing protein SIR2 and its homologs are NAD-dependent protein deacetylases. *Proc. Natl. Acad. Sci. USA* **2000**, *97*, 5807–5811. [CrossRef] [PubMed]
22. Imai, S.; Armstrong, C.M.; Kaeberlein, M.; Guarente, L. Transcriptional silencing and longevity protein Sir2 is an NAD-dependent histone deacetylase. *Nature* **2000**, *403*, 795–800. [CrossRef] [PubMed]
23. Bordone, L.; Cohen, D.; Robinson, A.; Motta, M.C.; Veen, E.v.; Czopik, A.; Steele, A.D.; Crowe, H.; Marmor, S.; Luo, J.; et al. SIRT1 transgenic mice show phenotypes resembling calorie restriction. *Aging Cell* **2007**, *6*, 759–767. [CrossRef] [PubMed]
24. Blander, G.; Guarente, L. The Sir2 family of protein deacetylases. *Annu. Rev. Biochem.* **2004**, *73*, 417–435. [CrossRef] [PubMed]
25. Lin, S.J.; Defossez, P.A.; Guarente, L. Requirement of NAD and SIR2 for life-span extension by calorie restriction in Saccharomyces cerevisiae. *Science* **2000**, *289*, 2126–2128. [CrossRef] [PubMed]
26. Zheng, Y.; Kou, J.; Wang, P.; Ye, T.; Wang, Z.; Gao, Z.; Cong, L.; Li, M.; Dong, B.; Yang, W.; et al. Berberine-induced TFEB deacetylation by SIRT1 promotes autophagy in peritoneal macrophages. *Aging* **2021**, *13*, 7096–7119. [CrossRef]
27. Chen, D.; Bruno, J.; Easlon, E.; Lin, S.J.; Cheng, H.L.; Alt, F.W.; Guarente, L. Tissue-specific regulation of SIRT1 by calorie restriction. *Genes Dev.* **2008**, *22*, 1753–1757. [CrossRef] [PubMed]
28. Wang, F.; Nguyen, M.; Qin, F.X.; Tong, Q. SIRT2 deacetylates FOXO3a in response to oxidative stress and caloric restriction. *Aging Cell* **2007**, *6*, 505–514. [CrossRef] [PubMed]

29. Shi, T.; Wang, F.; Stieren, E.; Tong, Q. SIRT3, a mitochondrial sirtuin deacetylase, regulates mitochondrial function and thermogenesis in brown adipocytes. *J. Biol. Chem.* **2005**, *280*, 13560–13567. [CrossRef]
30. Kanfi, Y.; Shalman, R.; Peshti, V.; Pilosof, S.N.; Gozlan, Y.M.; Pearson, K.J.; Lerrer, B.; Moazed, D.; Marine, J.C.; de Cabo, R.; et al. Regulation of SIRT6 protein levels by nutrient availability. *FEBS Lett.* **2008**, *582*, 543–548. [CrossRef] [PubMed]
31. Haigis, M.C.; Mostoslavsky, R.; Haigis, K.M.; Fahie, K.; Christodoulou, D.C.; Murphy, A.J.; Valenzuela, D.M.; Yancopoulos, G.D.; Karow, M.; Blander, G.; et al. SIRT4 inhibits glutamate dehydrogenase and opposes the effects of calorie restriction in pancreatic beta cells. *Cell* **2006**, *126*, 941–954. [CrossRef] [PubMed]
32. Wronska, A.; Lawniczak, A.; Wierzbicki, P.M.; Kmiec, Z. Age-Related Changes in Sirtuin 7 Expression in Calorie-Restricted and Refed Rats. *Gerontology* **2016**, *62*, 304–310. [CrossRef] [PubMed]
33. Tontonoz, P.; Spiegelman, B.M. Fat and beyond: The diverse biology of PPARgamma. *Annu. Rev. Biochem.* **2008**, *77*, 289–312. [CrossRef] [PubMed]
34. Picard, F.; Kurtev, M.; Chung, N.; Topark-Ngarm, A.; Senawong, T.; Machado De Oliveira, R.; Leid, M.; McBurney, M.W.; Guarente, L. Sirt1 promotes fat mobilization in white adipocytes by repressing PPAR-gamma. *Nature* **2004**, *429*, 771–776. [CrossRef]
35. Lomb, D.J.; Laurent, G.; Haigis, M.C. Sirtuins regulate key aspects of lipid metabolism. *Biochim. Biophys. Acta* **2010**, *1804*, 1652–1657. [CrossRef] [PubMed]
36. van den Berghe, G. The role of the liver in metabolic homeostasis: Implications for inborn errors of metabolism. *J. Inherit. Metab. Dis.* **1991**, *14*, 407–420. [CrossRef]
37. Francis, G.A.; Fayard, E.; Picard, F.; Auwerx, J. Nuclear receptors and the control of metabolism. *Annu. Rev. Physiol.* **2003**, *65*, 261–311. [CrossRef]
38. Li, X.; Zhang, S.; Blander, G.; Tse, J.G.; Krieger, M.; Guarente, L. SIRT1 deacetylates and positively regulates the nuclear receptor LXR. *Mol. Cell* **2007**, *28*, 91–106. [CrossRef] [PubMed]
39. Gerhart-Hines, Z.; Rodgers, J.T.; Bare, O.; Lerin, C.; Kim, S.H.; Mostoslavsky, R.; Alt, F.W.; Wu, Z.; Puigserver, P. Metabolic control of muscle mitochondrial function and fatty acid oxidation through SIRT1/PGC-1alpha. *EMBO J.* **2007**, *26*, 1913–1923. [CrossRef]
40. Rodgers, J.T.; Lerin, C.; Haas, W.; Gygi, S.P.; Spiegelman, B.M.; Puigserver, P. Nutrient control of glucose homeostasis through a complex of PGC-1alpha and SIRT1. *Nature* **2005**, *434*, 113–118. [CrossRef]
41. Fulco, M.; Sartorelli, V. Comparing and contrasting the roles of AMPK and SIRT1 in metabolic tissues. *Cell Cycle* **2008**, *7*, 3669–3679. [CrossRef] [PubMed]
42. Chen, W.-L.; Kang, C.-H.; Wang, S.-G.; Lee, H.-M. α-Lipoic acid regulates lipid metabolism through induction of sirtuin 1 (SIRT1) and activation of AMP-activated protein kinase. *Diabetologia* **2012**, *55*, 1824–1835. [CrossRef]
43. Liu, L.; Xing, D.; Du, X.; Peng, T.; McFadden, J.W.; Wen, L.; Lei, H.; Dong, W.; Liu, G.; Wang, Z.; et al. Sirtuin 3 improves fatty acid metabolism in response to high nonesterified fatty acids in calf hepatocytes by modulating gene expression. *J. Dairy Sci.* **2020**, *103*, 6557–6568. [CrossRef] [PubMed]
44. Kincaid, B.; Bossy-Wetzel, E. Forever young: SIRT3 a shield against mitochondrial meltdown, aging, and neurodegeneration. *Front. Aging Neurosci.* **2013**, *5*, 48. [CrossRef]
45. Jin, L.; Galonek, H.; Israelian, K.; Choy, W.; Morrison, M.; Xia, Y.; Wang, X.; Xu, Y.; Yang, Y.; Smith, J.J.; et al. Biochemical characterization, localization, and tissue distribution of the longer form of mouse SIRT3. *Protein Sci.* **2009**, *18*, 514–525. [CrossRef] [PubMed]
46. Virtanen, K.A.; Lidell, M.E.; Orava, J.; Heglind, M.; Westergren, R.; Niemi, T.; Taittonen, M.; Laine, J.; Savisto, N.J.; Enerback, S.; et al. Functional brown adipose tissue in healthy adults. *N. Engl. J. Med.* **2009**, *360*, 1518–1525. [CrossRef]
47. Ahuja, N.; Schwer, B.; Carobbio, S.; Waltregny, D.; North, B.J.; Castronovo, V.; Maechler, P.; Verdin, E. Regulation of insulin secretion by SIRT4, a mitochondrial ADP-ribosyltransferase. *J. Biol. Chem.* **2007**, *282*, 33583–33592. [CrossRef] [PubMed]
48. Laurent, G.; German, N.J.; Saha, A.K.; Boer, V.C.J.d.; Davies, M.; Koves, T.R.; Dephoure, N.; Fischer, F.; Boanca, G.; Vaitheesvaran, B.; et al. SIRT4 coordinates the balance between lipid synthesis and catabolism by repressing malonyl CoA decarboxylase. *Mol. Cell* **2013**, *50*, 686–698. [CrossRef] [PubMed]
49. Michishita, E.; Park, J.Y.; Burneskis, J.M.; Barrett, J.C.; Horikawa, I. Evolutionarily conserved and nonconserved cellular localizations and functions of human SIRT proteins. *Mol. Biol. Cell* **2005**, *16*, 4623–4635. [CrossRef]
50. Hong, J.; Mei, C.; Raza, S.H.A.; Khan, R.; Cheng, G.; Zan, L. SIRT5 inhibits bovine preadipocyte differentiation and lipid deposition by activating AMPK and repressing MAPK signal pathways. *Genomics* **2020**, *112*, 1065–1076. [CrossRef]
51. Wang, G.; Meyer, J.G.; Cai, W.; Softic, S.; Li, M.E.; Verdin, E.; Newgard, C.; Schilling, B.; Kahn, C.R. Regulation of UCP1 and Mitochondrial Metabolism in Brown Adipose Tissue by Reversible Succinylation. *Mol. Cell* **2019**, *74*, 844–857.e847. [CrossRef]
52. Hong, J.; Mei, C.; Abbas Raza, S.H.; Khan, R.; Cheng, G.; Zan, L. SIRT6 cooperates with SIRT5 to regulate bovine preadipocyte differentiation and lipid metabolism via the AMPKalpha signaling pathway. *Arch. Biochem. Biophys.* **2020**, *681*, 108260. [CrossRef] [PubMed]
53. Kuang, J.; Zhang, Y.; Liu, Q.; Shen, J.; Pu, S.; Cheng, S.; Chen, L.; Li, H.; Wu, T.; Li, R.; et al. Fat-Specific Sirt6 Ablation Sensitizes Mice to High-Fat Diet-Induced Obesity and Insulin Resistance by Inhibiting Lipolysis. *Diabetes* **2017**, *66*, 1159–1171. [CrossRef] [PubMed]
54. Wu, D.; Bang, I.H.; Park, B.H.; Bae, E.J. Loss of Sirt6 in adipocytes impairs the ability of adipose tissue to adapt to intermittent fasting. *Exp. Mol. Med.* **2021**, *53*, 1298–1306. [CrossRef] [PubMed]

55. Yoshizawa, T.; Karim, M.F.; Sato, Y.; Senokuchi, T.; Miyata, K.; Fukuda, T.; Go, C.; Tasaki, M.; Uchimura, K.; Kadomatsu, T.; et al. SIRT7 controls hepatic lipid metabolism by regulating the ubiquitin-proteasome pathway. *Cell Metab.* **2014**, *19*, 712–721. [CrossRef]
56. Fang, J.; Ianni, A.; Smolka, C.; Vakhrusheva, O.; Nolte, H.; Kruger, M.; Wietelmann, A.; Simonet, N.G.; Adrian-Segarra, J.M.; Vaquero, A.; et al. Sirt7 promotes adipogenesis in the mouse by inhibiting autocatalytic activation of Sirt1. *Proc. Natl. Acad. Sci. USA* **2017**, *114*, E8352–E8361. [CrossRef] [PubMed]
57. Sun, X.; Yang, X.; Gui, W.; Liu, S.; Gui, Q. Sirtuins and autophagy in lipid metabolism. *Cell Biochem. Funct.* **2023**, *41*, 978–987. [CrossRef] [PubMed]
58. Baeken, M.W. Sirtuins and their influence on autophagy. *J. Cell. Biochem.* 2023; *early access*. [CrossRef]
59. Tang, F.; Ling, C. Curcumin ameliorates chronic obstructive pulmonary disease by modulating autophagy and endoplasmic reticulum stress through regulation of SIRT1 in a rat model. *J. Int. Med. Res.* **2019**, *47*, 4764–4774. [CrossRef] [PubMed]
60. Qin, H.; Zhang, H.; Zhang, X.; Zhang, S.; Zhu, S.; Wang, H. Resveratrol protects intestinal epithelial cells against radiation-induced damage by promoting autophagy and inhibiting apoptosis through SIRT1 activation. *J. Radiat. Res.* **2021**, *62*, 574–581. [CrossRef] [PubMed]
61. Weisman, R. Target of Rapamycin (TOR) Regulates Growth in Response to Nutritional Signals. *Microbiol. Spectr.* **2016**, *4*, 1–13. [CrossRef] [PubMed]
62. Szwed, A.; Kim, E.; Jacinto, E. Regulation and metabolic functions of mTORC1 and mTORC2. *Physiol. Rev.* **2021**, *101*, 1371–1426. [CrossRef] [PubMed]
63. Hsu, P.P.; Kang, S.A.; Rameseder, J.; Zhang, Y.; Ottina, K.A.; Lim, D.; Peterson, T.R.; Choi, Y.; Gray, N.S.; Yaffe, M.B.; et al. The mTOR-regulated phosphoproteome reveals a mechanism of mTORC1-mediated inhibition of growth factor signaling. *Science* **2011**, *332*, 1317–1322. [CrossRef] [PubMed]
64. Holczer, M.; Hajdu, B.; Lorincz, T.; Szarka, A.; Banhegyi, G.; Kapuy, O. Fine-tuning of AMPK-ULK1-mTORC1 regulatory triangle is crucial for autophagy oscillation. *Sci. Rep.* **2020**, *10*, 17803. [CrossRef]
65. Chakrabarti, P.; English, T.; Shi, J.; Smas, C.M.; Kandror, K.V. Mammalian target of rapamycin complex 1 suppresses lipolysis, stimulates lipogenesis, and promotes fat storage. *Diabetes* **2010**, *59*, 775–781. [CrossRef]
66. Laplante, M.; Sabatini, D.M. mTORC1 activates SREBP-1c and uncouples lipogenesis from gluconeogenesis. *Proc. Natl. Acad. Sci. USA* **2010**, *107*, 3281–3282. [CrossRef] [PubMed]
67. Brown, N.F.; Stefanovic-Racic, M.; Sipula, I.J.; Perdomo, G. The mammalian target of rapamycin regulates lipid metabolism in primary cultures of rat hepatocytes. *Metabolism* **2007**, *56*, 1500–1507. [CrossRef] [PubMed]
68. Peng, T.; Golub, T.R.; Sabatini, D.M. The immunosuppressant rapamycin mimics a starvation-like signal distinct from amino acid and glucose deprivation. *Mol. Cell. Biol.* **2002**, *22*, 5575–5584. [CrossRef]
69. Dai, W.; Panserat, S.; Mennigen, J.A.; Terrier, F.; Dias, K.; Seiliez, I.; Skiba-Cassy, S. Post-prandial regulation of hepatic glucokinase and lipogenesis requires the activation of TORC1 signalling in rainbow trout (*Oncorhynchus mykiss*). *J. Exp. Biol.* **2013**, *216*, 4483–4492. [CrossRef] [PubMed]
70. Madeira, J.B.; Masuda, C.A.; Maya-Monteiro, C.M.; Matos, G.S.; Montero-Lomeli, M.; Bozaquel-Morais, B.L. TORC1 inhibition induces lipid droplet replenishment in yeast. *Mol. Cell. Biol.* **2015**, *35*, 737–746. [CrossRef] [PubMed]
71. Sengupta, S.; Peterson, T.R.; Laplante, M.; Oh, S.; Sabatini, D.M. mTORC1 controls fasting-induced ketogenesis and its modulation by ageing. *Nature* **2010**, *468*, 1100–1104. [CrossRef] [PubMed]
72. Oldham, S. High Fat Diet Induced Obesity and Nutrient Sensing TOR Signaling. *Trends Endocrinol. Metab.* **2012**, *22*, 45–52. [CrossRef] [PubMed]
73. Kwon, G.; Marshall, C.A.; Pappan, K.L.; Remedi, M.S.; McDaniel, M.L. Signaling elements involved in the metabolic regulation of mTOR by nutrients, incretins, and growth factors in islets. *Diabetes* **2004**, *53* (Suppl. S3), S225–S232. [CrossRef] [PubMed]
74. McDaniel, M.L.; Marshall, C.A.; Pappan, K.L.; Kwon, G. Metabolic and autocrine regulation of the mammalian target of rapamycin by pancreatic beta-cells. *Diabetes* **2002**, *51*, 2877–2885. [CrossRef]
75. Xu, G.; Kwon, G.; Cruz, W.S.; Marshall, C.A.; McDaniel, M.L. Metabolic regulation by leucine of translation initiation through the mTOR-signaling pathway by pancreatic beta-cells. *Diabetes* **2001**, *50*, 353–360. [CrossRef]
76. Um, S.H.; Frigerio, F.; Watanabe, M.; Picard, F.; Joaquin, M.; Sticker, M.; Fumagalli, S.; Allegrini, P.R.; Kozma, S.C.; Auwerx, J.; et al. Absence of S6K1 protects against age- and diet-induced obesity while enhancing insulin sensitivity. *Nature* **2004**, *431*, 200–205. [CrossRef]
77. Polak, P.; Cybulski, N.; Feige, J.N.; Auwerx, J.; Ruegg, M.A.; Hall, M.N. Adipose-specific knockout of raptor results in lean mice with enhanced mitochondrial respiration. *Cell Metab.* **2008**, *8*, 399–410. [CrossRef] [PubMed]
78. Sarbassov, D.D.; Ali, S.M.; Sengupta, S.; Sheen, J.H.; Hsu, P.P.; Bagley, A.F.; Markhard, A.L.; Sabatini, D.M. Prolonged rapamycin treatment inhibits mTORC2 assembly and Akt/PKB. *Mol. Cell* **2006**, *22*, 159–168. [CrossRef] [PubMed]
79. Houde, V.P.; Brule, S.; Festuccia, W.T.; Blanchard, P.G.; Bellmann, K.; Deshaies, Y.; Marette, A. Chronic rapamycin treatment causes glucose intolerance and hyperlipidemia by upregulating hepatic gluconeogenesis and impairing lipid deposition in adipose tissue. *Diabetes* **2010**, *59*, 1338–1348. [CrossRef] [PubMed]
80. Aggarwal, D.; Fernandez, M.L.; Soliman, G.A. Rapamycin, an mTOR inhibitor, disrupts triglyceride metabolism in guinea pigs. *Metabolism* **2006**, *55*, 794–802. [CrossRef] [PubMed]

81. Lopes, P.C.; Fuhrmann, A.; Sereno, J.; Espinoza, D.O.; Pereira, M.J.; Eriksson, J.W.; Reis, F.; Carvalho, E. Short and long term in vivo effects of Cyclosporine A and Sirolimus on genes and proteins involved in lipid metabolism in Wistar rats. *Metabolism* **2014**, *63*, 702–715. [CrossRef]
82. Soliman, G.A.; Acosta-Jaquez, H.A.; Fingar, D.C. mTORC1 inhibition via rapamycin promotes triacylglycerol lipolysis and release of free fatty acids in 3T3-L1 adipocytes. *Lipids* **2010**, *45*, 1089–1100. [CrossRef]
83. Pereira, M.J.; Palming, J.; Rizell, M.; Aureliano, M.; Carvalho, E.; Svensson, M.K.; Eriksson, J.W. The immunosuppressive agents rapamycin, cyclosporin A and tacrolimus increase lipolysis, inhibit lipid storage and alter expression of genes involved in lipid metabolism in human adipose tissue. *Mol. Cell. Endocrinol.* **2013**, *365*, 260–269. [CrossRef]
84. Cheng, C.L.; Gao, T.Q.; Wang, Z.; Li, D.D. Role of insulin/insulin-like growth factor 1 signaling pathway in longevity. *World J. Gastroenterol.* **2005**, *11*, 1891–1895. [CrossRef] [PubMed]
85. Saltiel, A.R.; Kahn, C.R. Insulin signalling and the regulation of glucose and lipid metabolism. *Nature* **2001**, *414*, 799–806. [CrossRef] [PubMed]
86. Saltiel, A.R. Insulin Signaling in the Control of Glucose and Lipid Homeostasis. *Handb. Exp. Pharmacol.* **2016**, *233*, 51–71. [CrossRef]
87. Lee, S.; Dong, H.H. FoxO integration of insulin signaling with glucose and lipid metabolism. *J. Endocrinol.* **2017**, *233*, R67–R79. [CrossRef]
88. Leavens, K.F.; Birnbaum, M.J. Insulin signaling to hepatic lipid metabolism in health and disease. *Crit. Rev. Biochem. Mol. Biol.* **2011**, *46*, 200–215. [CrossRef] [PubMed]
89. Lan, Y.L.; Lou, J.C.; Lyu, W.; Zhang, B. Update on the synergistic effect of HSL and insulin in the treatment of metabolic disorders. *Ther. Adv. Endocrinol. Metab.* **2019**, *10*, 2042018819877300. [CrossRef] [PubMed]
90. Kershaw, E.E.; Hamm, J.K.; Verhagen, L.A.; Peroni, O.; Katic, M.; Flier, J.S. Adipose triglyceride lipase: Function, regulation by insulin, and comparison with adiponutrin. *Diabetes* **2006**, *55*, 148–157. [CrossRef]
91. Chueire, V.B.; Muscelli, E. Effect of free fatty acids on insulin secretion, insulin sensitivity and incretin effect—A narrative review. *Arch. Endocrinol. Metab.* **2021**, *65*, 24–31. [CrossRef] [PubMed]
92. Wang, J.; Wu, X.; Simonavicius, N.; Tian, H.; Ling, L. Medium-chain fatty acids as ligands for orphan G protein-coupled receptor GPR84. *J. Biol. Chem.* **2006**, *281*, 34457–34464. [CrossRef] [PubMed]
93. Vessby, B.; Tengblad, S.; Lithell, H. Insulin sensitivity is related to the fatty acid composition of serum lipids and skeletal muscle phospholipids in 70-year-old men. *Diabetologia* **1994**, *37*, 1044–1050. [CrossRef] [PubMed]
94. Titchenell, P.M.; Quinn, W.J.; Lu, M.; Chu, Q.; Lu, W.; Li, C.; Chen, H.; Monks, B.R.; Chen, J.; Rabinowitz, J.D.; et al. Direct Hepatocyte Insulin Signaling Is Required for Lipogenesis but Is Dispensable for the Suppression of Glucose Production. *Cell Metab.* **2016**, *23*, 1154–1166. [CrossRef] [PubMed]
95. Ozanne, S.E.; Martensz, N.D.; Petry, C.J.; Loizou, C.L.; Hales, C.N. Maternal low protein diet in rats programmes fatty acid desaturase activities in the offspring. *Diabetologia* **1998**, *41*, 1337–1342. [CrossRef]
96. Xu, X.; Gopalacharyulu, P.; Seppanen-Laakso, T.; Ruskeepaa, A.L.; Aye, C.C.; Carson, B.P.; Mora, S.; Oresic, M.; Teleman, A.A. Insulin signaling regulates fatty acid catabolism at the level of CoA activation. *PLoS Genet.* **2012**, *8*, e1002478. [CrossRef] [PubMed]
97. Klass, M.R. A method for the isolation of longevity mutants in the nematode Caenorhabditis elegans and initial results. *Mech. Ageing Dev.* **1983**, *22*, 279–286. [CrossRef]
98. Kenyon, C.; Chang, J.; Gensch, E.; Rudner, A.; Tabtiang, R. A *C. elegans* mutant that lives twice as long as wild type. *Nature* **1993**, *366*, 461–464. [CrossRef]
99. Pieri, C. Food restriction slows down age-related changes in cell membrane parameters. *Ann. N. Y. Acad. Sci.* **1991**, *621*, 353–362. [CrossRef] [PubMed]
100. Levin, G.; Cogan, U.; Mokady, S. Food restriction and membrane fluidity. *Mech. Ageing Dev.* **1992**, *62*, 137–141. [CrossRef] [PubMed]
101. Skowronska-Krawczyk, D.; Budin, I. Aging membranes: Unexplored functions for lipids in the lifespan of the central nervous system. *Exp. Gerontol.* **2020**, *131*, 110817. [CrossRef]
102. Pamplona, R.; Portero-Otin, M.; Riba, D.; Ruiz, C.; Prat, J.; Bellmunt, M.J.; Barja, G. Mitochondrial membrane peroxidizability index is inversely related to maximum life span in mammals. *J. Lipid Res.* **1998**, *39*, 1989–1994. [CrossRef]
103. Hulbert, A.J. On the importance of fatty acid composition of membranes for aging. *J. Theor. Biol.* **2005**, *234*, 277–288. [CrossRef]
104. Jobson, R.W.; Nabholz, B.; Galtier, N. An evolutionary genome scan for longevity-related natural selection in mammals. *Mol. Biol. Evol.* **2010**, *27*, 840–847. [CrossRef]
105. Buttemer, W.A.; Battam, H.; Hulbert, A.J. Fowl play and the price of petrel: Long-living Procellariiformes have peroxidation-resistant membrane composition compared with short-living Galliformes. *Biol. Lett.* **2008**, *4*, 351–354. [CrossRef]
106. Montgomery, M.K.; Hulbert, A.J.; Buttemer, W.A. The long life of birds: The rat-pigeon comparison revisited. *PLoS ONE* **2011**, *6*, e24138. [CrossRef]
107. Hou, N.S.; Taubert, S. Function and Regulation of Lipid Biology in Caenorhabditis elegans Aging. *Front. Physiol.* **2012**, *3*, 143. [CrossRef]
108. Watts, J.L.; Browse, J. Genetic dissection of polyunsaturated fatty acid synthesis in *Caenorhabditis elegans*. *Proc. Natl. Acad. Sci. USA* **2002**, *99*, 5854–5859. [CrossRef] [PubMed]

109. Jeong, J.H.; Han, J.S.; Jung, Y.; Lee, S.M.; Park, S.H.; Park, M.; Shin, M.G.; Kim, N.; Kang, M.S.; Kim, S.; et al. A new AMPK isoform mediates glucose-restriction induced longevity non-cell autonomously by promoting membrane fluidity. *Nat. Commun.* **2023**, *14*, 288. [CrossRef] [PubMed]
110. Park, M.R.; Ryu, S.; Maburutse, B.E.; Oh, N.S.; Kim, S.H.; Oh, S.; Jeong, S.Y.; Jeong, D.Y.; Oh, S.; Kim, Y. Probiotic *Lactobacillus fermentum* strain JDFM216 stimulates the longevity and immune response of *Caenorhabditis elegans* through a nuclear hormone receptor. *Sci. Rep.* **2018**, *8*, 7441. [CrossRef] [PubMed]
111. Lee, H.Y.; Lee, S.H.; Lee, J.H.; Lee, W.J.; Min, K.J. The role of commensal microbes in the lifespan of *Drosophila melanogaster*. *Aging* **2019**, *11*, 4611–4640. [CrossRef] [PubMed]
112. Smith, P.; Willemsen, D.; Popkes, M.; Metge, F.; Gandiwa, E.; Reichard, M.; Valenzano, D.R. Regulation of life span by the gut microbiota in the short-lived African turquoise killifish. *Elife* **2017**, *6*, e27014. [CrossRef]
113. Parker, A.; Romano, S.; Ansorge, R.; Aboelnour, A.; Le Gall, G.; Savva, G.M.; Pontifex, M.G.; Telatin, A.; Baker, D.; Jones, E.; et al. Fecal microbiota transfer between young and aged mice reverses hallmarks of the aging gut, eye, and brain. *Microbiome* **2022**, *10*, 68. [CrossRef] [PubMed]
114. Marchesi, J.R.; Adams, D.H.; Fava, F.; Hermes, G.D.; Hirschfield, G.M.; Hold, G.; Quraishi, M.N.; Kinross, J.; Smidt, H.; Tuohy, K.M.; et al. The gut microbiota and host health: A new clinical frontier. *Gut* **2016**, *65*, 330–339. [CrossRef] [PubMed]
115. Schmidt, N.S.; Lorentz, A. Dietary restrictions modulate the gut microbiota: Implications for health and disease. *Nutr. Res.* **2021**, *89*, 10–22. [CrossRef]
116. Sowah, S.A.; Milanese, A.; Schubel, R.; Wirbel, J.; Kartal, E.; Johnson, T.S.; Hirche, F.; Grafetstatter, M.; Nonnenmacher, T.; Kirsten, R.; et al. Calorie restriction improves metabolic state independently of gut microbiome composition: A randomized dietary intervention trial. *Genome Med.* **2022**, *14*, 30. [CrossRef] [PubMed]
117. Wang, L.; Wang, F.; Xiong, L.; Song, H.; Ren, B.; Shen, X. A nexus of dietary restriction and gut microbiota: Recent insights into metabolic health. *Crit. Rev. Food Sci. Nutr.* **2024**, *64*, 8649–8671. [CrossRef]
118. Ma, Y.; Sun, Y.; Sun, L.; Liu, X.; Zeng, R.; Lin, X.; Li, Y. Effects of gut microbiota and fatty acid metabolism on dyslipidemia following weight-loss diets in women: Results from a randomized controlled trial. *Clin. Nutr.* **2021**, *40*, 5511–5520. [CrossRef] [PubMed]
119. Caesar, R.; Tremaroli, V.; Kovatcheva-Datchary, P.; Cani, P.D.; Backhed, F. Crosstalk between Gut Microbiota and Dietary Lipids Aggravates WAT Inflammation through TLR Signaling. *Cell Metab.* **2015**, *22*, 658–668. [CrossRef] [PubMed]
120. Lam, Y.Y.; Ha, C.W.; Hoffmann, J.M.; Oscarsson, J.; Dinudom, A.; Mather, T.J.; Cook, D.I.; Hunt, N.H.; Caterson, I.D.; Holmes, A.J.; et al. Effects of dietary fat profile on gut permeability and microbiota and their relationships with metabolic changes in mice. *Obesity* **2015**, *23*, 1429–1439. [CrossRef] [PubMed]
121. Just, S.; Mondot, S.; Ecker, J.; Wegner, K.; Rath, E.; Gau, L.; Streidl, T.; Hery-Arnaud, G.; Schmidt, S.; Lesker, T.R.; et al. The gut microbiota drives the impact of bile acids and fat source in diet on mouse metabolism. *Microbiome* **2018**, *6*, 134. [CrossRef] [PubMed]
122. Velagapudi, V.R.; Hezaveh, R.; Reigstad, C.S.; Gopalacharyulu, P.; Yetukuri, L.; Islam, S.; Felin, J.; Perkins, R.; Boren, J.; Oresic, M.; et al. The gut microbiota modulates host energy and lipid metabolism in mice. *J. Lipid Res.* **2010**, *51*, 1101–1112. [CrossRef]
123. Lee, M.T.; Le, H.H.; Johnson, E.L. Dietary sphinganine is selectively assimilated by members of the mammalian gut microbiome. *J. Lipid Res.* **2021**, *62*, 100034. [CrossRef]
124. Besten, G.d.; Lange, K.; Havinga, R.; Dijk, T.H.v.; Gerding, A.; Eunen, K.v.; Müller, M.; Groen, A.K.; Hooiveld, G.J.; Bakker, B.M.; et al. Gut-derived short-chain fatty acids are vividly assimilated into host carbohydrates and lipids. *Am. J. Physiol. -Gastrointest. Liver Physiol.* **2013**, *305*, G900–G910. [CrossRef] [PubMed]
125. Orgeron, M.L.; Stone, K.P.; Wanders, D.; Cortez, C.C.; Van, N.T.; Gettys, T.W. The impact of dietary methionine restriction on biomarkers of metabolic health. *Prog. Mol. Biol. Transl. Sci.* **2014**, *121*, 351–376. [CrossRef]
126. Wallis, K.F.; Melnyk, S.B.; Miousse, I.R. Sex-Specific Effects of Dietary Methionine Restriction on the Intestinal Microbiome. *Nutrients* **2020**, *12*, 781. [CrossRef] [PubMed]
127. Ren, B.; Wang, L.; Mulati, A.; Liu, Y.; Liu, Z.; Liu, X. Methionine Restriction Improves Gut Barrier Function by Reshaping Diurnal Rhythms of Inflammation-Related Microbes in Aged Mice. *Front. Nutr.* **2021**, *8*, 746592. [CrossRef] [PubMed]
128. Wu, G.; Shi, Y.; Han, L.; Feng, C.; Ge, Y.; Yu, Y.; Tang, X.; Cheng, X.; Sun, J.; Le, G.W. Dietary Methionine Restriction Ameliorated Fat Accumulation, Systemic Inflammation, and Increased Energy Metabolism by Altering Gut Microbiota in Middle-Aged Mice Administered Different Fat Diets. *J. Agric. Food Chem.* **2020**, *68*, 7745–7756. [CrossRef]
129. Manaer, T.; Yu, L.; Nabi, X.H.; Dilidaxi, D.; Liu, L.; Sailike, J. The beneficial effects of the composite probiotics from camel milk on glucose and lipid metabolism, liver and renal function and gut microbiota in db/db mice. *BMC Complement. Med. Ther.* **2021**, *21*, 127. [CrossRef] [PubMed]
130. Bonfili, L.; Cuccioloni, M.; Gong, C.; Cecarini, V.; Spina, M.; Zheng, Y.; Angeletti, M.; Eleuteri, A.M. Gut microbiota modulation in Alzheimer's disease: Focus on lipid metabolism. *Clin. Nutr.* **2022**, *41*, 698–708. [CrossRef] [PubMed]
131. Bonfili, L.; Cecarini, V.; Berardi, S.; Scarpona, S.; Suchodolski, J.S.; Nasuti, C.; Fiorini, D.; Boarelli, M.C.; Rossi, G.; Eleuteri, A.M. Microbiota modulation counteracts Alzheimer's disease progression influencing neuronal proteolysis and gut hormones plasma levels. *Sci. Rep.* **2017**, *7*, 2426. [CrossRef] [PubMed]
132. Anson, R.M.; Guo, Z.; de Cabo, R.; Iyun, T.; Rios, M.; Hagepanos, A.; Ingram, D.K.; Lane, M.A.; Mattson, M.P. Intermittent fasting dissociates beneficial effects of dietary restriction on glucose metabolism and neuronal resistance to injury from calorie intake. *Proc. Natl. Acad. Sci. USA* **2003**, *100*, 6216–6220. [CrossRef] [PubMed]

133. Hwangbo, D.S.; Lee, H.Y.; Abozaid, L.S.; Min, K.J. Mechanisms of Lifespan Regulation by Calorie Restriction and Intermittent Fasting in Model Organisms. *Nutrients* **2020**, *12*, 1194. [CrossRef] [PubMed]
134. Vo, N.; Zhang, Q.; Sung, H.K. From fasting to fat reshaping: Exploring the molecular pathways of intermittent fasting-induced adipose tissue remodeling. *J. Pharm. Pharm. Sci.* **2024**, *27*, 13062. [CrossRef] [PubMed]
135. Hammer, S.S.; Vieira, C.P.; McFarland, D.; Sandler, M.; Levitsky, Y.; Dorweiler, T.F.; Lydic, T.A.; Asare-Bediako, B.; Adu-Agyeiwaah, Y.; Sielski, M.S.; et al. Fasting and fasting-mimicking treatment activate SIRT1/LXRalpha and alleviate diabetes-induced systemic and microvascular dysfunction. *Diabetologia* **2021**, *64*, 1674–1689. [CrossRef]
136. Hosny, S.A.; Moustafa, M.H.A.; Mehina, F.M.; Sabry, M.M. Therapeutic effect of autophagy induced by rapamycin versus intermittent fasting in animal model of fatty liver. *Folia Histochem. Cytobiol.* **2023**, *61*, 205–216. [CrossRef] [PubMed]
137. Ma, X.; Mani, K.; Liu, H.; Kovacs, A.; Murphy, J.T.; Foroughi, L.; French, B.A.; Weinheimer, C.J.; Kraja, A.; Benjamin, I.J.; et al. Transcription Factor EB Activation Rescues Advanced αB-Crystallin Mutation-Induced Cardiomyopathy by Normalizing Desmin Localization. *J. Am. Heart Assoc.* **2019**, *8*, e010866. [CrossRef] [PubMed]
138. Mukai, R.; Zablocki, D.; Sadoshima, J. Intermittent Fasting Reverses an Advanced Form of Cardiomyopathy. *J. Am. Heart Assoc.* **2019**, *8*, e011863. [CrossRef]
139. Karbowska, J.; Kochan, Z. Intermittent fasting up-regulates Fsp27/Cidec gene expression in white adipose tissue. *Nutrition* **2012**, *28*, 294–299. [CrossRef] [PubMed]
140. Spezani, R.; da Silva, R.R.; Martins, F.F.; de Souza Marinho, T.; Aguila, M.B.; Mandarim-de-Lacerda, C.A. Intermittent fasting, adipokines, insulin sensitivity, and hypothalamic neuropeptides in a dietary overload with high-fat or high-fructose diet in mice. *J. Nutr. Biochem.* **2020**, *83*, 108419. [CrossRef] [PubMed]
141. Park, S.; Yoo, K.M.; Hyun, J.S.; Kang, S. Intermittent fasting reduces body fat but exacerbates hepatic insulin resistance in young rats regardless of high protein and fat diets. *J. Nutr. Biochem.* **2017**, *40*, 14–22. [CrossRef] [PubMed]
142. Krizova, E.; Simek, V. Effect of intermittent feeding with high-fat diet on changes of glycogen, protein and fat content in liver and skeletal muscle in the laboratory mouse. *Physiol. Res.* **1996**, *45*, 379–383. [PubMed]
143. Antoni, R.; Johnston, K.L.; Collins, A.L.; Robertson, M.D. Effects of intermittent fasting on glucose and lipid metabolism. *Proc. Nutr. Soc.* **2017**, *76*, 361–368. [CrossRef] [PubMed]
144. Santos, H.O.; Macedo, R.C.O. Impact of intermittent fasting on the lipid profile: Assessment associated with diet and weight loss. *Clin. Nutr. ESPEN* **2018**, *24*, 14–21. [CrossRef] [PubMed]
145. Ahmed, N.; Farooq, J.; Siddiqi, H.S.; Meo, S.A.; Kulsoom, B.; Laghari, A.H.; Jamshed, H.; Pasha, F. Impact of Intermittent Fasting on Lipid Profile-A Quasi-Randomized Clinical Trial. *Front. Nutr.* **2020**, *7*, 596787. [CrossRef]
146. Mohamed, Y.A.; Abouelmagd, M.; Elbialy, A.; Elwassefy, M.; Kyrillos, F. Effect of intermittent fasting on lipid biokinetics in obese and overweight patients with type 2 diabetes mellitus: Prospective observational study. *Diabetol. Metab. Syndr.* **2024**, *16*, 4. [CrossRef]
147. Radler, M.E.; Hale, M.W.; Kent, S. Calorie restriction attenuates lipopolysaccharide (LPS)-induced microglial activation in discrete regions of the hypothalamus and the subfornical organ. *Brain Behav. Immun.* **2014**, *38*, 13–24. [CrossRef] [PubMed]
148. Vasconcelos, A.R.; Yshii, L.M.; Viel, T.A.; Buck, H.S.; Mattson, M.P.; Scavone, C.; Kawamoto, E.M. Intermittent fasting attenuates lipopolysaccharide-induced neuroinflammation and memory impairment. *J. Neuroinflamm.* **2014**, *11*, 85. [CrossRef]
149. Deng, Y.; Liu, W.; Wang, J.; Yu, J.; Yang, L.-Q. Intermittent fasting improves lipid metabolism through changes in gut microbiota in diet-induced obese mice. *Med. Sci. Monit.* **2020**, *26*, e926789. [CrossRef]
150. Castillo, M.F.; Salgado-Canales, D.; Arrese, M.; Barrera, F.; Mikhailidis, D.P. Effect of Intermittent Fasting on Lipid Profile, Anthropometric and Hepatic Markers in Non-Alcoholic Fatty Liver Disease (NAFLD): A Systematic Review. *Curr. Vasc. Pharmacol.* **2024**, *22*, 187–202. [CrossRef]

nutrients

MDPI

Review

Immune Alterations with Aging: Mechanisms and Intervention Strategies

Weiru Yu †, Yifei Yu †, Siyuan Sun, Chenxu Lu, Jianan Zhai, Yumei Lei, Feirong Bai, Ran Wang and Juan Chen *

Key Laboratory of Precision Nutrition and Food Quality, Department of Nutrition and Health, China Agricultural University, Beijing 100083, China; yuweiru0@163.com (W.Y.)
* Correspondence: chenjuan@cau.edu.cn
† These authors contributed equally to this work.

Abstract: Aging is the result of a complex interplay of physical, environmental, and social factors, leading to an increased prevalence of chronic age-related diseases that burden health and social care systems. As the global population ages, it is crucial to understand the aged immune system, which undergoes declines in both innate and adaptive immunity. This immune decline exacerbates the aging process, creating a feedback loop that accelerates the onset of diseases, including infectious diseases, autoimmune disorders, and cancer. Intervention strategies, including dietary adjustments, pharmacological treatments, and immunomodulatory therapies, represent promising approaches to counteract immunosenescence. These interventions aim to enhance immune function by improving the activity and interactions of aging-affected immune cells, or by modulating inflammatory responses through the suppression of excessive cytokine secretion and inflammatory pathway activation. Such strategies have the potential to restore immune homeostasis and mitigate age-related inflammation, thus reducing the risk of chronic diseases linked to aging. In summary, this review provides insights into the effects and underlying mechanisms of immunosenescence, as well as its potential interventions, with particular emphasis on the relationship between aging, immunity, and nutritional factors.

Keywords: aging; immune system; immunosenescence; immunotherapy; diet intervention

1. Introduction

The global trend of population aging has heightened the focus on age-related diseases [1], spurring investigations into delaying or reversing the detrimental effects of aging. Aging is a ubiquitous and complex physiological process driven by the progression of time and environmental interactions, resulting in structural and functional alterations at the molecular, cellular, and systemic levels throughout all biological systems [2]. The immune system, as a critical defense barrier, identifies antigens and mounts responses crucial for maintaining health. The intricate impact of aging on the immune system impairs physiological function and disrupts homeostatic equilibrium, markedly elevating the risk of degenerative diseases and mortality [3]. Immune cells are generally considered to act as monitors in the aging process and are primarily responsible for the removal of harmful senescent cells. The progressive decline in various immune functions during aging is termed "immunosenescence" [4]. This age-related deterioration of the immune system heightens vulnerability to diseases as a consequence of immunosenescence. The aging immune system can be targeted through various interventions, including lifestyle interventions, pharmacological interventions, and immunotherapies [5]. Among these, nutritional intervention stands out as the safe and promising approach for personalized treatment. In short, a thorough understanding of the interactions between aging and immune function is essential for developing targeted interventions. This review encapsulates the age-associated immune function changes, their health implications, the underlying biological mechanisms, and the potential strategies for intervention.

2. Immunological Changes During Aging

The replenishment of mature immune cells throughout an organism's life involves processes centered in the bone marrow and thymus, which are crucial for sustaining immune cell development and function. However, the functionality of these organs declines with age, which subsequently reduces their capacity to renew the pool of immune cells. The concept of 'immunosenescence' was introduced for the first time, and research into immunosenescence has since been initiated [6].

2.1. The Composition of the Immune System

The immune system is divided into the innate and adaptive components. Innate cells include phagocytes and natural killer (NK) cells, while adaptive cells consist of T and B cells, all originating from hematopoietic stem cells. Upon pathogen entry into the bloodstream, the complement system activates, tagging invaders for destruction. Phagocytes use pattern recognition receptors (PRRs) to detect and engulf pathogens. If innate defenses fail, adaptive immunity activates. Antigen-presenting cells (APCs) migrate to lymph nodes, presenting antigens to T cells, which recognize them through T cell receptors (TCRs). This triggers helper and cytotoxic T cells, as well as B cells that produce neutralizing antibodies. Memory T and B cells ensure a faster and stronger response upon re-exposure to the same antigens [7].

2.2. Impact of Aging on Innate Immune System

2.2.1. Neutrophils

Neutrophils are crucial immune cells that combat infections through phagocytosis, secretion of antimicrobials, and neutrophil extracellular traps (NETs) [8,9]. Some studies indicate that while neutrophil development and proliferation remain stable with age [10], their functional efficiency declines, as evidenced by reduced phagocytic capability, chemotaxis, cytokine production, and oxidative burst [11]. This reduction extends to decreased phagocytic and bactericidal activities, as observed in various aging models [12,13]. Moreover, a recent multi-omics analysis revealed significant transcriptomic and epigenomic alterations in neutrophils due to aging, implicating the aged bone marrow microenvironment in the dysregulation of these cells [14]. In vivo studies in mice have also established a direct correlation between neutrophil senescence and increased pro-inflammatory activity [15]. Such alterations in neutrophils heighten the susceptibility to infections and inflammatory diseases in the elderly [16,17]. Moreover, some research has demonstrated that aging interferes with transepithelial migration of neutrophils in injured tissues via a process regulated by CXC-chemokine ligand 1 (CXCL1), leading to improper neutrophil trafficking and consequent remote organ damage [18].

2.2.2. Macrophages

Macrophages, which are key phagocytic cells, rapidly migrate to infection or injury sites, where they recognize, engulf, and digest pathogens and cellular debris. They eliminate these pathogens through phagocytosis and the release of antimicrobial substances [19]. Aging mice showed increased CD38 in liver and adipose tissue, which was mainly derived from pro-inflammatory M1 macrophages, indicating that the differentiation balance of macrophages was destroyed [20]. Aging-related changes in tissue integrity, such as increased extracellular matrix stiffness [21] and structural loss [22], may alter macrophage niches and their effector signals. In aging mice, diminished DNA repair and increased senescence in macrophages promote inflammation and in turn accelerate aging [23]. Caloric restriction reverses the pro-inflammatory polarization of aged M1 macrophages, enhances anti-inflammatory phenotypes, and modifies macrophage–stromal interactions [24]. Furthermore, by restarting glucose metabolism in macrophages, inflammation can be reduced, immune function can be restored, and aging can be reversed. As a result, the cognitive performance in older individuals can be improved [25].

2.2.3. Dendritic Cells (DCs)

Dendritic cells (DCs) are key antigen-presenting cells that initiate and regulate immune responses. With high migratory capacity, immature DCs rapidly move to infection or injury sites, where they capture pathogens and present them to T cells, activating adaptive immunity [26]. Research indicates that aging differentially impacts dendritic cell subtypes in mice; conventional type 1 DCs diminish in the spleen with age, while type 2 DCs do not [27]. Aging impairs DCs' phagocytic functions, notably reducing their ability to perform efferocytosis of apoptotic cells [28]. Additionally, aging impairs DCs' migration and cytokine production (IL-1β and CCR7). The decline in the number, along with the deterioration of their phagocytic and migratory functions, leads to a reduced capacity for activating downstream immune cells, thereby impairing their antigen presentation capability. In response, a vaccine adjuvant, DC hyper-activators, was developed to restore DCs' migration, enhance anti-tumor $CD4^+$ T cell responses, and bolster tumor immunity in aged mice [29].

2.2.4. Natural Killer (NK) Cells

Natural killer (NK) cells are large granular lymphocytes that are distinct from T and B cells and possess both cytotoxic and immunomodulatory functions. They play a crucial role in controlling malignant tumors and infections. While aging typically reduces the number of lymphocytes, NK cells are an exception, showing an increase in number but a decline in function. An analysis of age-related immune cell data revealed a significant decrease in total lymphocyte count after the age of 60, whereas the NK and NKT cell numbers increased significantly [30]. This may be due to the fact that chemokines, such as CCL2 and CCL4, which rise during aging, can recruit macrophages and NK cells [31]. Despite this numerical increase, aging diminishes NK cell cytotoxicity and cytokine production [32]. There is a shift in NK cell subpopulations, with a decrease in the proportion of $CD56^{bright}$ cells and an increase in $CD56^{dim}$ cells. The increased $CD56^{dim}$ NK cells exhibit reduced proliferative capacity and impaired pathogen clearance, leading to diminished immunity [33]. Additionally, aged NK cells inadequately remove senescent fibroblasts, which release pro-inflammatory factors known as the senescence-associated secretory phenotype (SASP). These factors contribute to tissue degradation and propagate senescence, thereby accelerating skin aging [34].

2.3. *Impact of Aging on Adaptive Immune System*

2.3.1. T Cells

T cells, produced in the thymus, include helper ($CD4^+$) and cytotoxic ($CD8^+$) subtypes. $CD4^+$ T cells regulate immune responses by secreting cytokines, while $CD8^+$ T cells directly eliminate infected or cancerous cells. Aging is associated with a reduction in naive $CD8^+$ T cells, an expansion in memory subsets, and an increase in clonal expansions, reducing available TCR diversity [35,36]. As with $CD8^+$ T cells, the proportions of naive $CD4^+$ T cells decline, and the memory population expands with age in mice and humans [37,38]. This switch to a memory phenotype seems to be attributable to diminished thymic output and the cumulative effects of lifelong antigen exposure [35]. An increase in regulatory T cells (Tregs) in aging spleens and lymph nodes has also been observed [39,40]. A predictive model of immune aging in a Chinese cohort showed decreases in naive T cells, increases in central and effector T cells, and stability in programmed death protein (PD)-1-associated subpopulations [41]. In addition, it has been reported that $CD4^+$ T cells can stop the increase in senescent cells once a person is aging. The higher the number of $CD4^+$ T cells in a tissue sample, the lower the number of aging cells in aging skin [42]. Aging also leads to increased membrane rigidity, weakening the formation of the immune synapse and disrupting the signaling pathways essential for T cell activation, such as TCR and CD28. Additionally, aging causes the loss or reduced expression of CD28, altering T cell fate, reducing IL-2 production, and impairing the immunometabolic shift from oxidative phosphorylation (OXPHOS) to aerobic glycolysis. These metabolic changes, driven by altered mTOR signaling, result in insufficient energy production [43]. Overall, aging is associated with a decline in naive T cells and profound differentiation into memory phenotypes as co-stimulatory molecules (such as CD28 and CD27) are lost, leading to T cell senescence or exhaustion and impaired cytokine secretion. Aging T cells also exhibit characteristics such as mitochondrial dysfunction and epigenetic changes [44]. Additionally, CD57 and CD95 have been employed to identify senescent human T cells [45], and CD153 is commonly used to identify a minor population of senescent T cells in mice [46].

2.3.2. B Cells

B cells play crucial roles in synthesizing antibodies, presenting antigens, and secreting cytokines, all of which are essential for immune defense. Their development and function are critical to the body's capacity to respond to pathogens. However, aging significantly diminishes B cell production and functionality, contributing to decreased immunity in the elderly. Adaptive immune remodeling with age includes changes in B cell composition and functionality. Specifically, a unique subset termed 'age-associated B cells' (ABC) expands in the spleen and bone marrow of aged mice [47,48]. These cells, which are distinct from traditional naive and memory B cells, were identified as $CD43^-$ $CD21/CD35^-$ $CD23^-$ B cells by Hao et al. [47] and as $CD11b^+CD11c^+$ B cells by Rubtsov et al. [48]. Notably, they express Tbx21, linking them to autoimmunity in a similar manner to that observed in lupus-like autoimmunity in mice [49,50]. Aging is associated with a more mature B cell state and a reduction in naive B cells, indicating a compromised cellular immune response to new antigens [51]. Furthermore, aging disrupts DNA interactions within the topologically associating domain (TAD) of the immunoglobulin heavy chain locus (IgH locus), reducing the diversity of the primary BCR repertoire. These alterations impair B cell development and function, contributing significantly to immune senescence [52,53].

2.4. *Senescence-Associated Secretory Phenotype (SASP)*

Cells may enter a senescent state as a consequence of various stressors, which further drives organismal aging. Cell cycle arrest is a typical feature of cellular aging; it leads to a stable, terminal proliferative halt and the development of a SASP [54]. The SASP defines the secretion of various cytokines, chemokines, growth factors, proteases, and lipids by senescent cells. This concept was first introduced by Krtolica et al. in 2001 [55]; they proposed that senescent cells secrete factors into their microenvironment, thereby potentially modulating biological activities both locally and systemically. In 2008, the SASP, also known as the senescence-messaging secretome, was independently characterized by various laboratories as primarily consisting of pro-inflammatory and growth-stimulating proteins [56,57]. This secretion leads to chronic inflammation and tissue damage in organisms. The composition of the secretome varies depending on the trigger of senescence, and the proportion of senescent cells in very old primates is estimated to range from 5 to 20% [58], contributing to age-related diseases. The SASP can also interact with immune cells, creating feedback loops that exacerbate tissue damage [44].

In summary, there are fewer immune cells, weaker immune responses, less antibody production, and an increase in the immune aging marker in various immune cells with aging [59] (Figure 1).

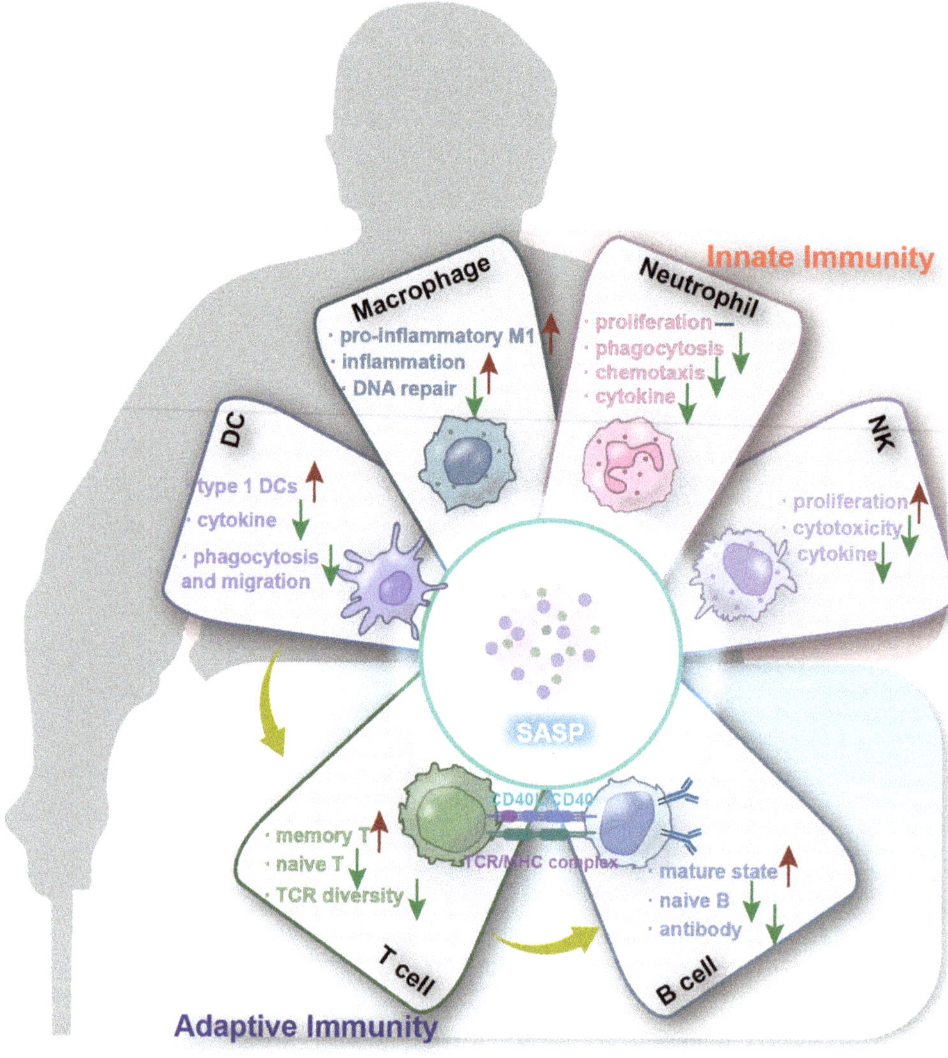

Figure 1. The effects of aging on different immune cells in the intrinsic and adaptive immune systems. Changes in the immune system involve the decline of both innate and adaptive immunity, along with the development of the senescence-associated secretory phenotype (SASP). Aging leads to alterations in the number and function of immune cells, with the adaptive immune system being more significantly affected. For innate immunity, the percentages of pro-inflammatory macrophages (M1) and type 1 DCs increase during aging; the phagocytosis and cytokine secretion decrease in both DCs and neutrophils; the cytotoxicity of NKs also decline in old individuals. For adaptive immunity, both T cell and B cell populations decline in the elderly, with a concomitant reduction in TCR diversity and impaired antibody production. The red upward arrow indicates an increase, the green downward arrow indicates a decrease, and "—" represents no change.

3. Mechanism of Aging Affecting Immunity

3.1. Mechanism at the Molecular Level

3.1.1. Thymic Involution

One of the most well-documented changes in the aging immune system is the involution of the thymus, which is crucial for developing a diverse and selectively refined T cell repertoire [60,61]. This process results in a reduced generation of naive T cells reaching the periphery, a compensatory expansion of memory T cells, and decreased diversity in the peripheral T cell repertoire, impairing pathogen detection. This aligns with observed trends in T cell aging. In murine thymic tissues, numerous nucleic acid-binding proteins (NABPs), which play key roles in immune regulation, immune cell activity maintenance, and cancer control, exhibit dynamic changes with age [62]. However, in murine spleen tissues, the binding activities of NABPs related to immune and defense functions remain relatively stable, possibly due to the spleen's inherent regenerative capacity and consistent phenotypic maintenance. Reversal of thymic atrophy has been demonstrated using intrathymic injections of thymic epithelial cells (TECs), a technique that enhances autoreactive thymocytes, reducing inflammation and T cell aging [63].

3.1.2. Inflammaging and Oxidative Stress

Originally perceived as a biomarker for aging, inflammaging is now recognized for its association with increased risks of age-related multimorbidity and mortality [64,65]. This condition results from a combination of age-related deficits, including enhanced gut permeability, chronic infections, and an accumulation of senescent cells, which collectively drive the overproduction of the pro-inflammatory cytokines characteristic of the SASP [66]. Elevated reactive oxygen species (ROS) levels induce HIF1-α expression, leading to enhanced glycolysis and TNF-α secretion [67]. For a rather small cohort of people, an increase in mitochondrial ROS production by $CD8^+$ T lymphocytes has been observed [68]. Additionally, in vitro studies have demonstrated that telomere shortening within this CD8+ subset can be mitigated by treatment with a ROS scavenger, underscoring a potential causal relationship between ROS accumulation and T cell aging [68]. In murine models, peritoneal leukocytes, particularly macrophages, exhibit an age-dependent increase in ROS levels, further implicating oxidative stress in the aging process [69].

3.1.3. DNA Damage

During aging, nuclear DNA in humans and model organisms accumulates somatic mutations, gene copy number variations, and chromosomal aneuploidies. These changes can affect functionally essential genes, leading to altered cells, tissue abnormalities, and organismal deficiencies, which contribute to aging and age-related pathologies [70,71]. In hematopoietic cells, this accumulation results in spontaneous DNA damage in immune cells and an increased expression of senescence and SASP markers in various cell types, including B cells, NK cells, $CD4^+$ and $CD8^+$ T cells, and monocytes/macrophages. This pattern resembles that observed in 2-year-old naturally aged mice [72,73]. These findings suggest that DNA damage may lead to increased cellular senescence in immune cells with age, potentially triggering secondary senescence through SASP factor secretion. Although all species have evolved complex DNA repair mechanisms to mitigate unavoidable genetic damage and maintain cellular homeostasis, these mechanisms become less effective with age. Consequently, the persistent accumulation of genomic damage increases susceptibility to cancer and other age-related diseases [74,75]. Moreover, deficiencies in DNA repair mechanisms have been implicated in aging and cancer. Disruptions in DNA repair processes can accelerate aging in experimental models [76].

3.1.4. Telomere Shortening

Telomere shortening prevents replicative DNA polymerase from fully replicating telomeric regions. Without intervention, successive cell divisions exacerbate telomere erosion, leading to genomic instability and ultimately resulting in permanent cell cycle arrest (senescence) or cell death [77].

3.1.5. Mitochondrial Function Changes

Mitochondria are crucial for bioenergetics, cellular metabolism (including NAD/NADH ratio maintenance), and various other functions. They act as signaling hubs, producing and responding to ROS and calcium spikes [78,79]. These signaling pathways become dysregulated in aged T cells [80]. As cells age, mitochondrial homeostasis is disrupted, leading to the release of mitochondrial DNA (mtDNA) into the cytoplasm. Age-related mutations in mtDNA impair mitochondrial oxidative phosphorylation (OXPHOS) functions [81]. In murine models, doxycycline-induced mutations result in decreased mtDNA levels, altered mitochondrial gene expression, and destabilization of OXPHOS complexes, contributing to skin aging and hair loss [82]. Studies on mice lacking DNA polymerase γ reveal that mtDNA mutations directly contribute to aging and related pathologies. These mice show accelerated aging and reduced lifespan, largely due to mtDNA deletions [83].

3.2. *The Role of Molecular Signaling Pathway*

Aging profoundly affects the immune system, increasing vulnerability to infections and chronic diseases. Key pathways, including mTOR, cGAS-STING, NF-κB, and JAK-STAT, regulate functions like cell growth, inflammation, and immune responses. Dysregulation of these pathways with age drives immune decline, emphasizing their role in immunosenescence.

3.2.1. mTOR Signaling

The mammalian target of rapamycin (mTOR), a member of the PI3K-related kinase (PIKK) family, regulates cellular growth and metabolism in response to nutrients and hormones. It is a highly conserved pathway influencing aging across species [84]. Dysregulation of mTOR signaling, often due to nutrient insufficiency, is closely associated with human aging and related diseases [85]. Impaired mTOR signaling affects T cell activation, differentiation, and immune response, underscoring its central role in aging [86]. In addition, mTOR is a key regulator of aging in organisms ranging from yeast to mammals. The extension of lifespan through the pharmacological inhibition of mTORC1 with rapamycin or genetic modification in S6K1 mutant mice highlights its significance [87]. Recent human studies have further demonstrated that mTOR inhibition with RAD001 enhances antibody titers to influenza vaccination and reduces PD-1 expression on $CD4^+$ and $CD8^+$ T cells [88].

3.2.2. cGAS-STING Signaling

Mitochondrial imbalances are significant in aging and neurodegenerative disorders, with mtDNA playing a central role in the activation of the cGAS-STING pathway. This pathway's activation during aging is a critical driver of inflammation, aging, and functional decline. Elevated phosphorylation levels of STING, TBK1, and IRF3 in Alzheimer's disease patients and aged mouse brains suggest the cGAS-STING pathway activation [89]. Administering H-151, a small molecule inhibitor of STING, to aging mice decreases the expression of age-associated immune genes (B2m, Il1b, Il6, Tnf and Ccl5) and inflammatory cell accumulation, enhancing muscle strength and cognitive functions. Moreover, similar results were obtained when the pathway was inhibited or the gene encoding the STING protein was knocked out [90].

3.2.3. NF-κB Signaling

The NF-κB pathway, a fundamental regulator of innate immunity found across various species from insects to vertebrates, plays a crucial role in inflammaging. This signaling pathway orchestrates the intracellular responses to age-related DNA and cellular stress, implicating it in aging and associated diseases [91]. NF-κB, along with C/EBP and p53, governs the SASP, influencing a range of biological processes, such as immune responses, inflammation, stress reactions, B cell development, and lymphoid organogenesis [58]. Additionally, in the adaptive immune system, similar regulatory axes involving Sirt1-HIF-1α and Sirt1-NF-κB modulate metabolism and inflammation in aging T cells [92,93].

3.2.4. JAK-STAT

The JAK-STAT pathway is crucial for the differentiation and development of immune cells and is modulated by a variety of cytokines. It plays a key role in T cell differentiation; IFN and IL-12 promote Th1 cell differentiation through STAT1 and STAT4, while IL-4 activates Th2 cells via STAT3 upregulation of GATA6. Similarly, IL-6 and TGF-β trigger Th17 cell differentiation through STAT17-mediated activation of RORt, and both IL-6 and IL-12 influence T follicular helper cell differentiation via STAT3-driven Bcl-6 expression. IL-2 supports Treg cell differentiation by coordinating STAT5A/B with Foxp3 [94]. The JAK-STAT signaling pathway also mediates IL-6 to activate the MAPK enzyme and transcription factor NF-κB and to regulate inflammation and immune function [95]. In addition, the JAK inhibitor ruxolitinib has been shown to decrease the size of the secretome from senescent cells, underscoring the pathway's relevance to aging [96] (Figure 2).

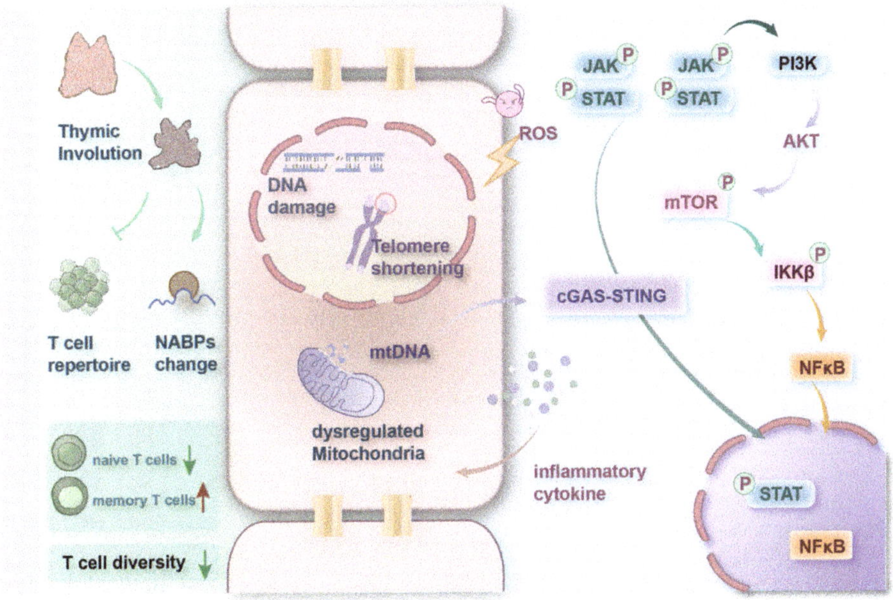

Figure 2. Multiple factors contribute to the aging of the immune system. These include thymic involution, inflammatory damage, DNA damage, telomere shortening, and mitochondrial dysfunction. The overall oxidation level increases with the higher level of inflammatory cytokines, activating pathways like NF-κB. The release of mtDNA from dysregulated mitochondria triggers the cGAS-STING pathway. The interaction of multiple pathways aggravates SASP. The red upward arrow indicates an increase, the green downward arrow indicates a decrease.

4. Aging Immune System and Aging-Related Diseases

Age-related changes in immune system function can have profound effects on the physiology of elderly organisms, contributing to the onset of infections, autoimmune diseases, and cancer. The decline in immune function weakens the body's defense against tumor cells and pathogens, while the chronic inflammatory state increases the risk of autoimmune diseases (Figure 3).

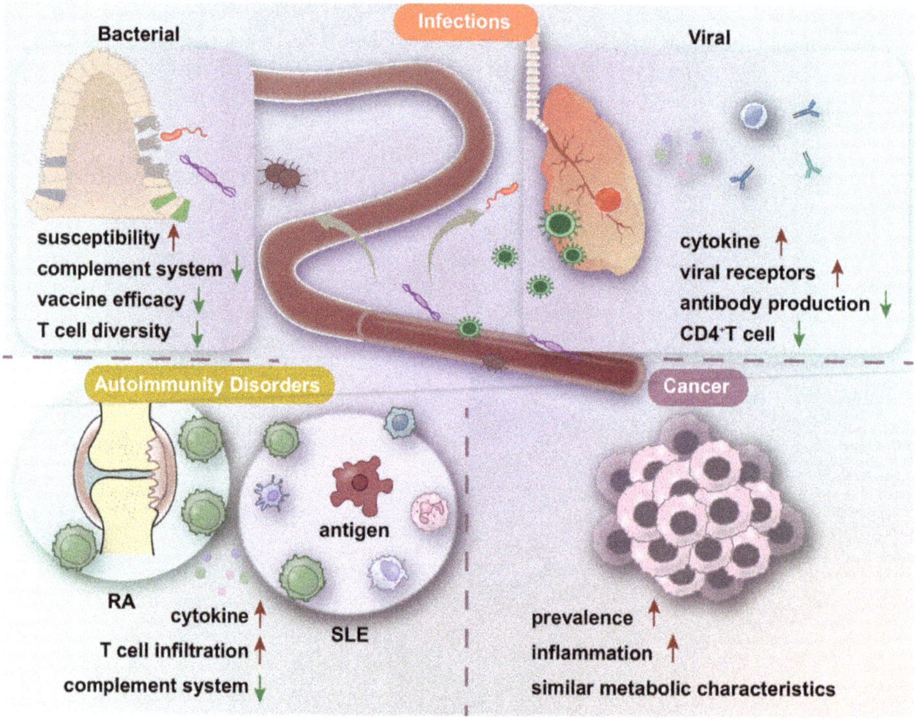

Figure 3. Aging affects the immune system and leads to age-related diseases. During aging, the likelihood of infection increases due to the deterioration of the immune system. This process is characterized by an enhanced inflammatory response, reduced T cell diversity, and diminished antibody production, which collectively increase vulnerability to bacterial and viral infections. Additionally, as aging progresses, the immune system's tolerance to self-antigens decreases, and immune cell infiltration occurs, contributing to the development of autoimmune diseases. Furthermore, compromised immune functions during aging provide a conducive environment for cancer development. The red upward arrow indicates an increase, the green downward arrow indicates a decrease.

4.1. Increased Susceptibility to Infections

4.1.1. Bacterial Infections

In addition to the previously noted decline in immune competence among older adults, the complement system, which is essential for rapid response to bacterial infections, is also compromised. Dysregulation of this system is frequently associated with and exacerbates age-related disease episodes [97]. Immune senescence increases the susceptibility of older adults to acute viral and bacterial infections, with mortality rates from these infections being threefold higher among elderly patients compared to younger adults [98]. During typical influenza seasons, approximately 90% of excess mortality occurs in individuals over the age of 65, with poor immune responses significantly reducing the efficacy of vaccines [99,100]. Throughout an individual's lifespan, continuous division, differentiation, and exposure to foreign infectious agents may diminish the diversity of T cells, thereby reducing their effectiveness. In the aging population, transformations in regulatory T cells occur; naive-like regulatory T cells decrease, whereas memory-like regulatory T cells increase with age. These changes in regulatory T cells, which help control the T cell compartment, promote anti-inflammation [101].

4.1.2. Viral Infections

Immunosenescence refers to the age-related alterations in both innate and adaptive immune systems, leading to diminished capacity to combat novel infections and fostering chronic inflammation [102]. This phenomenon also reactivates latent viruses such as the varicella-zoster virus, culminating in conditions like shingles and chronic neuralgia [103]. Infection with latent viruses such as cytomegalovirus (CMV) has also been shown to potentially play a role in immunosenescence by amplification of inflammation, and both CMV seropositivity and elevated CMV IgG antibodies are linked to phenotypic and functional changes in adaptive immunity, as well as an elevated risk of all-cause mortality [104]. These immunosenescent changes render older adults especially susceptible to SARS-CoV-2 and severe COVID-19 outcomes [105]. In the older individual cases of SARS-CoV, non-survivors exhibited heightened innate immune activation—specifically IFN-α and IFN-γ, CXCL10, CCL2, and downstream IFN-stimulated genes—yet showed deficient antibody production [105]. Early activation of the adaptive immune response by SARS-CoV-2 may disrupt the timely innate immune response necessary for rapid viral elimination, potentially leading to a detrimental cytokine storm [106]. In the immune defense against viruses, CD4 Th1 cells are crucial. During SARS-CoV-2 infections, the adaptive immune response is critical due to the virus's ability to evade innate immunity [107]; however, CD4 cell counts are typically reduced in aged individuals. In addition, the Angiotensin Converting Enzyme 2 (ACE2) [108] and the adipokine enzyme DPP4 [109] enzymes are more abundant in older people than in younger people, and after viruses enter the cells by binding to them, aging in turn induces further upregulation of the receptors in the context of a reduced immune response, resulting in older people being more susceptible to viral attack.

4.2. Autoimmunity Disorders and Inflammatory Disorders

Aging is strongly associated with the development of autoimmune diseases. Immunosenescence diminishes the immune system's pathogen defense while compromising self-tolerance, thereby increasing susceptibility to autoimmune reactions. Elderly individuals frequently exhibit chronic low-grade inflammation, or "inflammaging", which accelerates the progression of age-related conditions, such as rheumatoid arthritis (RA) and systemic lupus erythematosus (SLE) [110]. Altered immune cell functions, particularly in T and B lymphocytes, further heighten the risk of autoimmunity. Impaired self-/non-self-recognition and immune tolerance loss contribute to the increased incidence of autoimmune disorders in aging populations.

4.2.1. Rheumatoid Arthritis (RA)

Rheumatoid arthritis (RA) is a chronic, autoimmune inflammatory condition that can lead to disability and premature mortality. The age-specific prevalence of RA was highest among individuals aged 75–79, with 828.2 cases per 100,000 population in 2020 [111]. In RA patients, the levels of complement components C4d and C3bBbP were found to be 2–34 times higher than in the controls, aligning with data showing a correlation between complement C3 and C4 levels and age [112,113]. Longevity is negatively correlated with C3/C4 levels, suggesting that high C3 levels are detrimental to longevity. Furthermore, T cells in RA patients exhibit premature aging features, such as accumulation of CD28, telomere fragility, diminished DNA repair capacity, and excessive cytokine production [114]. These maladaptive T cells entering cellular senescence remain hyperactive and continuously secrete pro-inflammatory cytokines like TNF-α and IFN-γ, thereby perpetuating inflammation and accelerating aging due to deviation from homeostatic mechanisms.

4.2.2. Systemic Lupus Erythematosus (SLE)

Systemic lupus erythematosus (SLE) is a complex systemic autoimmune disease characterized by high heterogeneity and intricate pathogenesis. Dysfunction in both innate and adaptive immune responses is evident in SLE patients and involves activation of inflammatory pathways by dendritic cell type I IFN secretion to drive disease progression [115], neutrophilic leukocyte iron death [116], disruption of the balance of macrophage differentiation subtypes [117], and aberrant T and B cell proliferation and differentiation [118,119]. It was found that older age significantly increases the risk of neuropsychiatric events in systemic lupus erythematosus, demonstrating the complex interplay between aging and SLE progression [120]. Additionally, aging disrupts the expression of complement system proteins and impairs the clearance of apoptotic material due to deficiencies in C1q, C4, and C2, further increasing the SLE risk [121]. The increase in brain age is associated with neuronal damage and cognitive impairment in SLE patients [122]. Recent research suggests that T cells infiltrating the choroid plexus could drive neuropsychiatric symptoms in lupus, which could link these immune cells to the age-related cognitive decline observed in SLE patients [123].

4.3. Cancer

Cancer prevalence increases with age, which is attributed to heightened cancer risk factors and the expansion of the aging population [124]. Li et al. report that 60% of all cancer diagnoses and 70% of cancer-related deaths occur in individuals aged 65 and older, with projections suggesting that by 2030, this age group will account for 70% of all cancer diagnoses [125]. Older adults diagnosed with cancer (>65 years) are more likely to exhibit comorbidities and age-related conditions compared to those without cancer [126]. Reciprocally, the onset of malignancy can exacerbate the decline in general health due to the systemic effects of cancer on distant organs, including the intestinal microbiota [127,128]. Moreover, cancer treatments, whether localized (surgery or radiotherapy) or systemic (chemotherapy, immunotherapy, etc.), induce organismal stress and inflammatory responses, accelerating the aging process [129,130]. Therefore, the interaction between aging and cancer is bidirectional. Specifically, research indicates that several aging characteristics, such as genomic instability, epigenetic alterations, chronic inflammation, and dysbiosis, are intimately associated with cancer determinants. However, other aspects of aging, like telomerase attrition and stem cell exhaustion, exhibit tumor-suppressing properties [131]. Additionally, aging and cancer share similar metabolic characteristics. In short, these positively or negatively regulated factors work together and may explain the peculiar association between cancer and aging with an increment of malignancies.

5. Interventions and Strategies to Improve Immune Function in Aging

Immune aging contributes to a vicious cycle that exacerbates the aging process itself, thereby increasing susceptibility to a broad spectrum of age-related diseases. Developing interventions that target the mechanisms of immune aging may offer new strategies for preventing these conditions. By understanding the pathways and factors contributing to immune deterioration, targeted therapies can potentially disrupt this cycle and enhance overall health in the elderly population.

5.1. Lifestyle Interventions

5.1.1. Diet

With the advancement of research in the intersection of metabolism and immunology, there is increasing recognition of the critical role that nutrients play in modulating immune function. Dietary factors influence immune system functionality, thereby affecting the aging process and the onset and progression of diseases (Figure 4).

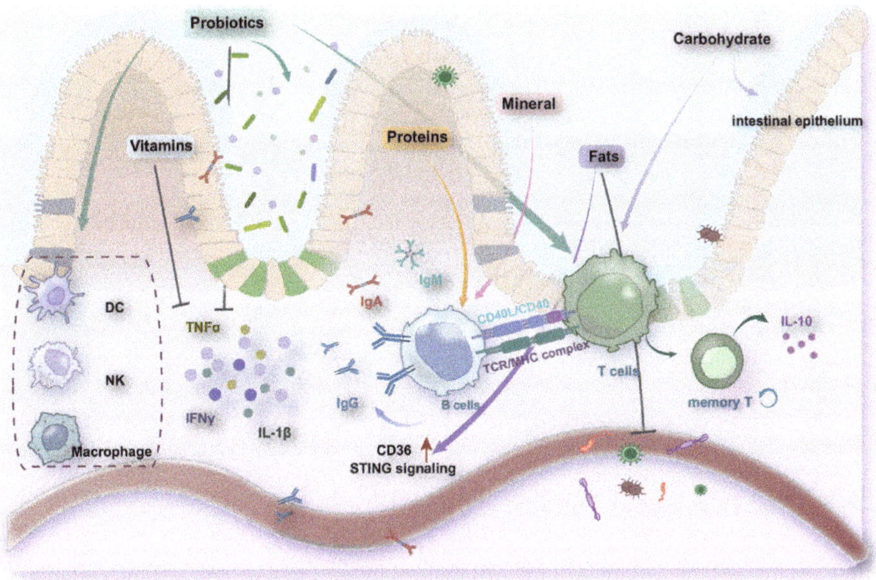

Figure 4. Dietary interventions, including proteins, carbohydrates, fats, vitamins, minerals, probiotics, and prebiotics, have significant effects on immune system function. Proteins enhance immunity by influencing T cell survival and differentiation, as well as by promoting B cell antibody production. Carbohydrates can strengthen innate immune responses and reduce cell apoptosis. The immunomodulatory effects of fats are primarily mediated by fatty acids, which enhance T cell antigen presentation, B cell proliferation, antibody production, and the antimicrobial activity of immune proteins. Vitamins regulate thymic and bone marrow homeostasis, enhancing the function of macrophages, dendritic cells, and T cells. Minerals play an essential role in modulating the production of immune regulators and antibodies. Probiotics and prebiotics modulate gut immunity by influencing the gut microbiota, reducing inflammation, and enhancing the function of various immune cells. The red upward arrow indicates an increase, the green downward arrow indicates a decrease.

Macronutrients

Macronutrients, including proteins, carbohydrates, and fats, support metabolic activities and energy balance through distinct pathways.

Proteins and amino acids are fundamental to the immune system, playing significant roles in the development and function of immune tissues and organs. They are essential for the production of various immune cells and the synthesis of antibodies. Amino acids are transported into the cytoplasm via solute carrier (SLC) transporters, where they activate sensor proteins, such as mTOR, GCN2, and Sestrin. These pathways can directly activate the TCR-CD3 complex or undergo further metabolism to influence T cell development and survival. For instance, an increase in arginine can restrict T cell differentiation, helping maintain T cells in a memory state under homeostatic conditions. Moreover, supplementation with L-arginine has been shown to significantly enhance the survival of activated $CD4^+$ and $CD8^+$ T cells in vitro and in mice, thereby improving antitumor activity [132]. During the late differentiation phase of T cell activation, asparagine (Asn) restriction enhances the proliferative and effector capacities of $CD8^+$ T cells by promoting metabolic fitness and antitumoral functionality through an NRF2-dependent stress response [133]. For B cells, threonine (Thr) further influences the differentiation and function of monocytes, while leucine (Leu) is transported into B cells via SLC7A5, targeting mTORC1 to promote B cell differentiation and support the production of IgG and cytokines. The inhibition of the glutamine transporter SLC1A5 or key enzymes in glutamine (Gln) metabolism results in reduced IgM production in B cells [134].

Research indicates that a high-carbohydrate diet can induce significant alterations in the transcriptional program of γδ T cells within just five days, leading to remodeling of the intestinal epithelium in mice [135]. While αβ T cells comprise over 95% of the T cell population, γδ T cells, though less abundant, are capable of rapidly initiating innate immune responses. Additionally, increased dietary carbohydrate intake elevates the proportion of B cells in the spleen, mesenteric lymph nodes, and Peyer's patches and enhances the production of antigen-specific IgG post-immunization. Compared to fructose, glucose serves as a superior substrate for B cell development by reducing apoptosis through the mTOR signaling pathway. Consequently, in autoimmune diseases such as rheumatoid arthritis, reducing carbohydrate intake may help mitigate excessive immune responses [136].

Research on the impact of dietary fats on immune system function has predominantly focused on specific types of fatty acids (FAs), which play crucial roles in the body as energy sources, components of biological membranes, and energy storage. As key metabolic products, FAs are integral to immune activation. Among known fatty acid transport proteins, CD36 is one of the most extensively studied. In mice, CD36 accounts for approximately 50% of fatty acid uptake in adipose and muscle tissues. Compared to wild-type mice, CD36-deficient mice exhibit reduced levels of specific IgM and IgG. He [137] et al. demonstrated that CD36-deficient B cells have significantly reduced plasma cell formation, proliferation, mitochondrial mobilization, and oxidative phosphorylation, resulting in weakened germinal center responses, impaired class switching, and decreased antibody production. In activated B cells of mice, the de novo fatty acid synthesis pathway is significantly upregulated, with marked increases in the expression of key genes involved in fatty acid biosynthesis [138]. Additionally, studies have shown that short-chain fatty acids (SCFAs) derived from the gut microbiota can modulate B cell class switching and autoantibody production, thereby ameliorating disease activity and extending survival in lupus mice [139]. Butyrate, a specific type of SCFA, has been shown by Hao et al. to be converted into butyryl-CoA (BCoA), which enhances the activity of CPT1A by antagonizing the binding of the metabolic intermediate malonyl-CoA (MCoA), thereby promoting fatty acid oxidation (FAO) and the differentiation of inducible regulatory T cells (iTregs) [140]. Long-chain fatty acids (LCFAs) require activation by acyl-CoA synthetase long-chain family members (ACSLs) for further metabolism and physiological function. ACSL5 enhances $CD8^+$ T cell-mediated cytotoxicity by modulating major histocompatibility complex class I (MHC-I) molecule-mediated antigen presentation in tumor cells [141]. Linoleic acid, a long-chain fatty acid, enhances the antitumor effects of T cells by promoting their differentiation into memory T cells rather than activating $CD8^+$ T cells through typical markers, such as GZMβ, IFN-γ, TNF-α, IL-2, and perforin [142]. Myristic acid, a long-chain saturated fatty acid, is significantly downregulated during viral infections and specifically inhibits STING-mediated type I interferon production while enhancing STING-dependent autophagy [143]. Intriguingly, intracellular lipid droplets, essentially reservoirs of 'oil', possess a variety of functions. As a major nutrient source, they not only support mitochondrial β-oxidation, providing energy for the body, but also recruit a range of innate immune proteins, contributing to antibacterial defense [144]. In summary, altering dietary fat content and the ratio of saturated to unsaturated fatty acids can significantly impact the lipid composition of lymphocyte membranes, thereby affecting lymphocyte function. The levels and proportions of polyunsaturated fatty acids, particularly essential fatty acids, play a crucial role in modulating both cellular and humoral immune responses [145].

Micronutrients

Vitamins, as essential dietary nutrients, also function as key modulators of immune responses. Vitamin deficiencies can lead to elevated inflammation and weakened immunity, while appropriate supplementation may reduce the impact of pathogenic infections. Vitamin A not only possesses anti-inflammatory properties but also regulates thymus and bone marrow homeostasis, thereby enhancing immune function and fortifying the body's defense against infections [146]. The relationship between Vitamin C and immunity has been a significant area of study. As a potent antioxidant, Vitamin C promotes the proliferation of T cells and NK cells, thereby influencing their immune functions. It has been proposed as a nutritional supplement for various diseases, including cancer [147]. Vitamin D not only promotes the differentiation of monocytes into macrophages but also enhances macrophage chemotaxis and phagocytic function [148], which helps counteract the immune decline associated with aging. In addition to regulating innate immunity, vitamin D modulates adaptive immunity by influencing cytokine and inflammatory factor release from DCs and T cells. Studies have shown a significant inverse relationship between baseline serum vitamin D levels and the development of RA, with early-stage RA patients exhibiting a higher prevalence of severe vitamin D deficiency compared to healthy individuals [149]. Vitamin E, a widely used fat-soluble antioxidant, enhances DC activation and function, promoting the infiltration and activation of tumor antigen-specific T cells. This suggests that vitamin E augments antitumor immune responses by enhancing DC activity [150].

Mineral deficiencies can lead to various diseases, affecting appetite and subsequently digestion, absorption, metabolism, and growth, which indirectly impacts the immune system [151]. For example, calcium deficiency leads to reduced immune cell activity and antibody secretion, making children more susceptible to recurrent infections and the elderly more prone to illness due to weakened immunity. Zinc is crucial for maintaining skin health, controlling the secretion and production of immune regulatory factors, influencing lymphoid organ development, and synthesizing antibodies. Additionally, zinc is involved in nucleic acid and protein metabolism, indirectly affecting immune function [152].

Probiotics and Prebiotics

The gut microbiota continuously transforms dietary components into numerous bioactive metabolites, thereby mediating the regulatory effects of diet on host health. Through complex interactions with immune cells in the small intestine and colon, the gut microbiota modulates T and B cells, influencing immune responses in both the peripheral and central nervous systems. For example, the gut microbiota can induce the transcription factor HNF4γ in small intestinal intraepithelial lymphocytes (IELs) of mice via linoleic acid (LA), regulating IL-18 signaling within cells and promoting the development of $CD4^{+}CD8\alpha\alpha^{+}$ IELs, thus impacting mucosal immunity in the mouse gut [153]. Probiotics and prebiotics are strategies targeting the microbiota to improve host health. In recent years, a variety of therapeutic approaches have emerged in which probiotics also play a large role in anti-inflammatory regulation and immune cell modulation.

Fecal microbiota transplantation from 8-week-old mice into naturally aging mice over 24 months old has been shown to rejuvenate the gut microbiota and significantly alleviate age-related physiological changes [154]. Additionally, supplementation with *Akkermansia*, *Bifidobacterium*, *Lactobacillus*, or acetate can beneficially modulate the gut microbiota, extending the lifespan of Caenorhabditis elegans or mice and improving age-associated diseases [155]. Various bacterial strains, including specific *Lactobacillus* and *Bifidobacterium bifidum*, have been demonstrated to influence the intestinal mucosal barrier, luminal environment, and mucosal immune system. These probiotics can modulate diverse cells within both the innate and acquired immune systems, such as monocytes, macrophages, dendritic cells, NK cells, and lymphocytes [7]. The anti-inflammatory cytokines they release in the gastrointestinal tract play a crucial role in this immune modulation. Among immunological agents, T regulatory cells (Tregs) are pivotal in maintaining immune regulation and tolerance. Strains like *L. reuteri* and *L. casei* enhance anti-inflammatory responses by elevating IL-10 levels and activating Tregs [156]. Furthermore, innovative probiotics, such as *L. rhamnosus* GG and *L. reuteri* DSM 17938, have been shown to inhibit the activation of T cells and NK cells, as well as the release of IFNγ in cultures of PBMCs exposed to Staphylococcus aureus. This indicates that specific probiotic strains can either suppress or stimulate NK cell activity, underscoring their potential in targeted immune therapies [157].

Therefore, various probiotics and prebiotics can improve immune processes by modulating the gut microbiota, thereby reducing the likelihood and severity of age-related diseases.

Caloric Restriction

Caloric restriction has been identified as a pivotal intervention for ameliorating age-related inflammatory changes. Detailed single-cell transcriptomic analyses from studies on aging rats have shown that reducing caloric intake can diminish age-associated inflammation. This reduction in calories notably reverses the pro-inflammatory polarization of M1-type macrophages, increases the proportion of anti-inflammatory macrophages with high phagocytic capabilities, and modifies ligand-receptor interactions between macrophages and stromal cells [59]. In a significant population-based trial in the United States, adults maintaining a 25% calorie-restricted diet over two years experienced a 2% to 3% reduction in biological aging markers, which translates into a 10% to 15% lower mortality risk [158]. Despite challenges in maintaining high levels of calorie restriction due to increased cortisol production and perceived stress, subsequent studies have demonstrated that a 14% calorie reduction does not compromise immune function but rather enhances the immune profile by slowing immune aging and promoting the production of thymic T cells, and PLA2G7 has been identified as a potential anti-aging target [159].

5.1.2. Exercise

The acceleration of aging due to sedentary lifestyles can be significantly countered by replacing inactive periods with 30 min of daily low- or moderate-intensity exercise. This practice is particularly beneficial for individuals over 60, where changes in physical activity levels have a pronounced effect on phenotypic aging [160]. Research indicates that physical activity mitigates senescence and inflammatory markers in both murine models and humans, underscoring exercise's role in countering the adverse effects of poor dietary habits and sedentary behaviors [161–163]. Additionally, these data suggest that exercise can offset the deleterious aspects of senescence brought about by poor diet and sedentary lifestyles.

5.1.3. Sleep

Aging is often accompanied by a decline in sleep quality, which is especially noticeable in middle-aged and older adults. However, the causal relationship between sleep and accelerated aging remains controversial. Elucidating the causal relationship between aging and sleep is greatly limited by the lack of precise measurements of aging. Studies suggest that poor sleep quality may accelerate biological aging, while improved sleep can mitigate the aging acceleration, particularly those accelerations exacerbated by environmental factors like air pollution [164]. Additionally, maintaining a regular sleep routine of approximately 7 h per night can significantly promote both physical and mental health in the aging population [165]. Furthermore, research in older mice highlights the phenomenon of sleep fragmentation, characterized by frequent sleep–wake transitions. Specifically, this condition is linked to a reduction in Hcrt neurons and an increase in their compensatory discharge frequencies, which enhance arousal. In Alzheimer's disease patients, the loss of Hcrt neurons is more pronounced, leading to increased neuronal excitability during disease progression, compromising sleep quality and possibly contributing to the accumulation of β-amyloid, thus exacerbating the aging process [166].

5.2. *Pharmacological Interventions*

In addition to changing lifestyle, medications are a more precise and expedient approach to addressing the effects of aging.

Anti-inflammatory treatments, targeting molecules like IL-1β, TNF-α, IFNAR1, caspase-1, and NLRP3, have demonstrated effectiveness in reducing both normal and accelerated aging in murine models [167–170]. Research indicates that the aging immune system is not inherently weaker; rather, it is restrained by a specific subset of follicular helper T cells (Tfh) that produce interleukin 10 (IL-10). Blockading IL-10 at the time of vaccination in older mice has been shown to restore their antibody responses almost to the levels observed in their younger counterparts [171]. At the cellular level, high-throughput compound screening has identified a TERT activator compound (TAC). This molecule, by activating telomerase reverse transcriptase, promotes TERT transcription in adult somatic cells of humans and mice. It restores youthful expression levels, supports telomere synthesis, and reduces DNA damage signaling at telomeres, effectively reversing signs of cellular aging [172]. Additionally, TPCA-1, a dual inhibitor of NF-κB and Jak-Stat pathways, significantly enhances the expression of the transcription factor Bcl11b, which diminishes from young to early aging stages. This restoration confers youthful characteristics to mammary cells and substantially lowers the incidence of breast cancer, which typically increases with age [173]. Additionally, based on certain clinical outcomes, rapamycin and metformin appear to hold the most promise for practical application. Rapamycin and mTOR inhibitors can regulate cell growth and various cellular processes by reducing the activity of mechanistic target of rapamycin complex 1 (mTORC1). Metformin, an antidiabetic drug, interacts with several known longevity pathways, with effects similar to those of caloric restriction, including enhanced insulin sensitivity [174]. It has also been shown to induce autophagy in $CD4^+$ T cells, shifting them from an inflammatory to a non-inflammatory state, and inhibiting Th17 cell differentiation, thus presenting an approach to the mitigation of aging effects [175]. Currently, however, the clinical application of these drugs is often limited by their toxic side effects, especially in elderly patients with comorbidities, as they may increase the risk of hyperglycemia, hyperlipidemia, nephrotoxicity, impaired wound healing, thrombocytopenia, and immunosuppression. Additionally, these small molecule drugs pose a metabolic burden and challenge to liver function in the elderly [174].

5.3. *Immunomodulatory Therapies*

Healthy lifestyles, such as regular exercise and calorie-restricted diets, have been shown to effectively combat aging and age-related diseases. In recent years, both small molecule pharmacologic interventions and genetic approaches have emerged as promising strategies for age protection. However, these interventions all have limitations. Small molecules often lack specificity, potentially affecting non-target cells and tissues, and may inadvertently inhibit anti-apoptotic pathways, risking off-target effects on healthy tissues. Genetic interventions, while potentially effective, carry the risk of long-term adverse effects, such as tumorigenesis. Immunotherapeutic strategies for cell delivery systems are currently of great interest in the treatment of malignant tumors [176], e.g., the precise delivery of mRNA can target cancer [177]. Thus, there is an urgent need to develop accurate immunotherapy interventions for aging.

Precision immunotherapy presents a promising approach to combating aging and related diseases. Maintaining the number and activity of immune cells is crucial for anti-aging. For example, transferring splenocytes from young mice to ERCC1 knockout mice, which have impaired DNA repair, has been shown to slow aging in the latter. This indicates that senescent immune cells may promote systemic aging, while young immune cell transplantation can decelerate it [178]. Vaccination targeting specific antigens represents a promising strategy to combat aging and age-related diseases by inducing tailored immune responses. For instance, Alzheimer's vaccines reduce Aβ and tau proteins in the brain via antibody production. Targeting molecules like DPP4, IL1-β, and ApoB can slow age-related vascular changes. A CD153 vaccine targets T cells, selectively eliminating pathogenic, highly active, or senescent ones. This targeted clearance of senescent cells may yield more effective outcomes than addressing systemic dysfunction [179]. Immunotherapies for T cell senescence, like CAR-T therapy, which was originally developed for cancer, are gaining traction. CAR-T cells can be engineered to specifically target and eliminate senescent cells, potentially reducing "inflammaging" and enhancing tissue function. Examples include uPAR-targeted CAR-T, which selectively removes senescent cells, rejuvenating aged mice and slowing aging in younger mice [180], and NKG2D-CAR-T, which is capable of targeting

NKG2DL-expressing senescent cells in both humans and mice [181]. Due to T cell memory, a single CAR-T administration could provide long-term treatment and preventive benefits; further research is needed to verify its effectiveness in the elderly. Additionally, mTOR inhibitors targeting downstream pathways of the rapamycin complex show potential for the delaying of age-related diseases by modulating T cell function. They enhance immune responses and reduce infection rates in the elderly by decreasing PD1$^+$ T cell exhaustion and boosting IFN-stimulated gene expression. These inhibitors are generally well tolerated [87]. With the increase in age, the hematopoietic system as the initial source of immune cells will also be affected by aging, which will accelerate the aging of the body [182]. To counteract age-related diseases, manipulating hematopoietic stem cells (HSCs) has been proposed as a potential strategy [183]. One promising approach involves reprogramming HSCs into induced pluripotent stem cells and then re-differentiating them into youthful HSCs [184].

Aging is a modifiable process, potentially influenced by nutritional and pharmacological interventions, and immune aging shares this plasticity. As a major contributor to the onset of chronic diseases, interventions targeting either age-related diseases or the aging process itself hold significant promise for promoting healthy aging (Figure 5). Among these strategies, lifestyle changes, particularly dietary adjustments, are the safest and most controllable and are suitable for the everyday modulation of immune function. Pharmacological and immunomodulatory strategies, while more precise and effective, might carry risks of unknown side effects. Therefore, they are generally considered for urgent situations, such as in the management of age-related diseases, where targeted interventions can address specific dysfunctions.

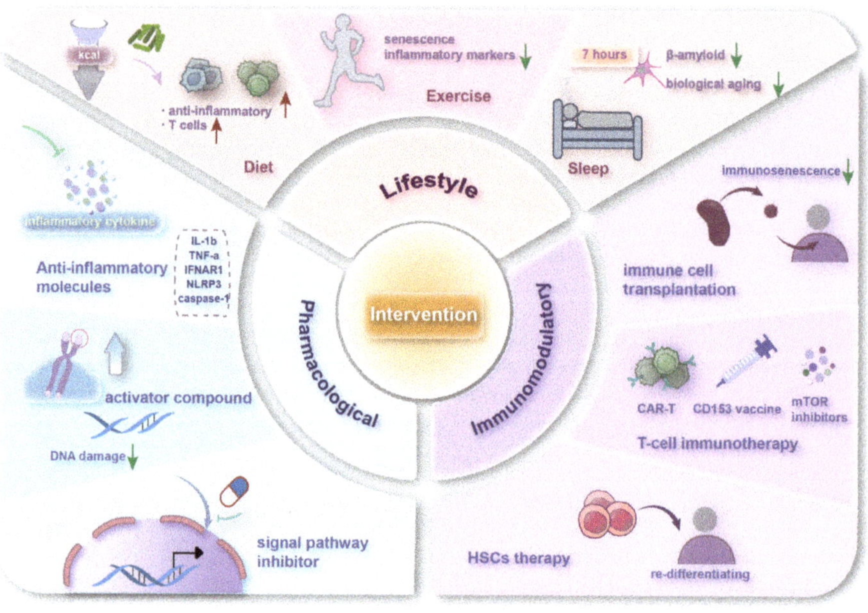

Figure 5. Various potential interventions and strategies that can improve aging immune functions. (1) Lifestyle interventions include adjustments in diet, exercise, and sleep patterns. (2) Pharmacological interventions encompass anti-inflammatory drugs, activators of specific molecules, and inhibitors of certain pro-aging signaling pathways. (3) Immunotherapeutic approaches involve the transplantation of immune cells, CAR-T technologies, and hematopoietic stem cell transplants. The red upward arrow indicates an increase, the green downward arrow indicates a decrease.

6. Discussion and Perspectives

The immune system is influenced by both intrinsic and extrinsic factors and undergoes significant alterations throughout the aging process, leading to immunosenescence. This term refers to the pathological mechanisms underlying various diseases, including infections, autoimmune disorders, and cancers. Immunosenescence is affected by factors such as thymic involution, inflammation, and DNA damage, which activate inflammation-related signaling pathways, ultimately contributing to disease onset and further influencing individual aging. Notably, the aging of the immune system is plastic [185]; both non-genetic environmental factors and pharmacological interventions can modify or accelerate its progression. Exploring these dimensions enhances our understanding of immunosenescence and establishes a foundation for the application of aging intervention strategies.

Although we have discussed the mutual influence between physiological aging and immunosenescence, their relationship remains elusive. From an immunological perspective, a higher physiological age does not necessarily equate to more severe immunosenescence. The degree of immune aging is more directly linked to the incidence and progression of diseases. Elderly individuals with longevity possess powerful immune systems that can rapidly respond to infections and tumors, thus better defending against infections and slowing the aging process [186]. Therefore, it is indispensable to distinguish between physiological aging and immunosenescence. However, due to the immune system's complexity, with hundreds of cell types and thousands of epigenetic markers, establishing biomarkers that accurately reflect a patient's immune status is still an unresolved task [187]. Future research should not only provide a multidimensional assessment of immune capacity, incorporating factors such as cell types, numbers, and functions in both innate and adaptive immunity to inform population immune age, but also conduct longitudinal studies to track changes in immune status over time and validate immune age assessment. The application of multi-omics technologies, which integrate data from genomics, transcriptomics, proteomics, metabolomics, and epigenomics, offers a comprehensive approach to understanding aging at multiple biological levels. These techniques help identify key biomarkers and pathways associated with the aging process, as well as age-related diseases [3,188]. By combining these insights, researchers can uncover how various factors interact to drive aging, enabling more precise interventions, such as personalized medicine, dietary changes, or pharmaceutical treatments, aimed at predicting disease and mortality risk, mitigating the effects of aging, and promoting healthy longevity.

Despite some advances in medical treatment, one major challenge is the considerable differences in immune responses among species. Many current mouse models are adopted in immunological research, but it remains uncertain whether the interventions that are effective in these models can be applied to humans, as findings from mice do not always predict human outcomes. Many important discoveries have yet to be translated to humans; this is likely due to significant biological differences between species. Moreover, due to unknown factors within the body, even the interventions deemed effective still require long-term studies for validation. Considering the risks and effects of interventions, aside from some safe food-derived substances, other intervention strategies require more fundamental and in-depth clinical research. Additionally, individual variability—such as race, genetic background, and life-style—cannot be underestimated, as it complicates the precise personalization of interventions to slow aging. In summary, the conducting of population studies faces challenges such as ethical concerns, safety issues, and questions of efficacy, all of which require urgent resolution.

Furthermore, current pharmacological strategies, including small molecule treatments and broader immunotherapies, lack precision in targeting the mechanisms of aging. These approaches generally produce broad effects rather than addressing specific cellular or molecular targets. A major challenge in this field is achieving selective modulation—either activation or inhibition—of particular proteins or cellular pathways without unavoidably disrupting other physiological processes, such as oncogenic pathways. Therefore, developing targeted therapies that may intervene in the aging process without triggering adverse effects remains a complex and critical area of research, while using food-derived substances for precise nutritional interventions is safe, efficient, and highly acceptable.

Ethical considerations surrounding immune cell transplantation or gene interventions in elderly populations must balance the potential benefits and risks. The benefits may include enhanced immune function and the prevention of age-related diseases. However, concerns arise regarding patient consent, autonomy, safety, and long-term effects, especially considering the potential for unforeseen immune responses or complications in the elderly. Additionally, access to these treatments could exacerbate health disparities. A balanced perspective should emphasize both medical advancements and the necessity for rigorous ethical oversight, patient welfare, and equitable access to these interventions.

In summary, this review delves into the multifaceted transformations within the immune system as it ages and examines the quantitative and qualitative shifts in both innate and adaptive immune cells. It highlights how these changes during aging contribute to an increased vulnerability to infections and other health complications. Lastly, it outlines various potential interventions, primarily focusing on nutritional strategies, to mitigate immunosenescence. This review not only enhances our understanding of aging-related immune alterations but also provides perspectives for targeted strategies to enhance immune resilience in the elderly.

Author Contributions: Conceptualization, J.C.; validation, Y.Y. and S.S.; supervision, C.L. and J.Z., visualization, Y.L., F.B. and R.W.; writing—original draft preparation, W.Y.; writing—review and editing, J.C. All authors have read and agreed to the published version of the manuscript.

Funding: This review was supported by the National Key Research and Development Program of China 2023YFF1103503.

Conflicts of Interest: The authors declare no conflicts of interest.

References

1. Lopez-Otin, C.; Blasco, M.A.; Partridge, L.; Serrano, M.; Kroemer, G. Hallmarks of aging: An expanding universe. *Cell* **2023**, *186*, 243–278. [CrossRef] [PubMed]
2. Dodig, S.; Cepelak, I.; Pavic, I. Hallmarks of senescence and aging. *Biochem. Med.* **2019**, *29*, 030501. [CrossRef] [PubMed]
3. Shen, X.; Wang, C.; Zhou, X.; Zhou, W.; Hornburg, D.; Wu, S.; Snyder, M.P. Nonlinear dynamics of multi-omics profiles during human aging. *Nat. Aging* **2024**. [CrossRef] [PubMed]
4. Liu, Z.; Liang, Q.; Ren, Y.; Guo, C.; Ge, X.; Wang, L.; Cheng, Q.; Luo, P.; Zhang, Y.; Han, X. Immunosenescence: Molecular mechanisms and diseases. *Signal Transduct. Target. Ther.* **2023**, *8*, 200. [CrossRef]
5. Bieri, G.; Schroer, A.B.; Villeda, S.A. Blood-to-brain communication in aging and rejuvenation. *Nat. Neurosci.* **2023**, *26*, 379–393. [CrossRef]
6. Li, X.; Li, C.; Zhang, W.; Wang, Y.; Qian, P.; Huang, H. Inflammation and aging: Signaling pathways and intervention therapies. *Signal Transduct. Target. Ther.* **2023**, *8*, 239. [CrossRef]
7. Mazziotta, C.; Tognon, M.; Martini, F.; Torreggiani, E.; Rotondo, J.C. Probiotics Mechanism of Action on Immune Cells and Beneficial Effects on Human Health. *Cells* **2023**, *12*, 184. [CrossRef]
8. Branzk, N.; Lubojemska, A.; Hardison, S.E.; Wang, Q.; Gutierrez, M.G.; Brown, G.D.; Papayannopoulos, V. Neutrophils sense microbe size and selectively release neutrophil extracellular traps in response to large pathogens. *Nat. Immunol.* **2014**, *15*, 1017–1025. [CrossRef]
9. Yang, F.; Feng, C.; Zhang, X.; Lu, J.; Zhao, Y. The Diverse Biological Functions of Neutrophils, Beyond the Defense Against Infections. *Inflammation* **2017**, *40*, 311–323. [CrossRef]
10. Nikolich-Zugich, J.; Davies, J.S. Homeostatic migration and distribution of innate immune cells in primary and secondary lymphoid organs with ageing. *Clin. Exp. Immunol.* **2017**, *187*, 337–344. [CrossRef]
11. Kolaczkowska, E.; Kubes, P. Neutrophil recruitment and function in health and inflammation. *Nat. Rev. Immunol.* **2013**, *13*, 159–175. [CrossRef] [PubMed]
12. Simell, B.; Vuorela, A.; Ekstrom, N.; Palmu, A.; Reunanen, A.; Meri, S.; Kayhty, H.; Vakevainen, M. Aging reduces the functionality of anti-pneumococcal antibodies and the killing of Streptococcus pneumoniae by neutrophil phagocytosis. *Vaccine* **2011**, *29*, 1929–1934. [CrossRef] [PubMed]
13. Tseng, C.W.; Kyme, P.A.; Arruda, A.; Ramanujan, V.K.; Tawackoli, W.; Liu, G.Y. Innate immune dysfunctions in aged mice facilitate the systemic dissemination of methicillin-resistant S. aureus. *PLoS ONE* **2012**, *7*, e41454. [CrossRef] [PubMed]

14. Lu, R.J.; Taylor, S.; Contrepois, K.; Kim, M.; Bravo, J.I.; Ellenberger, M.; Sampathkumar, N.K.; Benayoun, B.A. Multi-omic profiling of primary mouse neutrophils predicts a pattern of sex and age-related functional regulation. *Nat. Aging* **2021**, *1*, 715–733. [CrossRef] [PubMed]
15. Zhang, D.; Chen, G.; Manwani, D.; Mortha, A.; Xu, C.; Faith, J.J.; Burk, R.D.; Kunisaki, Y.; Jang, J.E.; Scheiermann, C.; et al. Neutrophil ageing is regulated by the microbiome. *Nature* **2015**, *525*, 528–532. [CrossRef]
16. Finger, C.E.; Moreno-Gonzalez, I.; Gutierrez, A.; Moruno-Manchon, J.F.; McCullough, L.D. Age-related immune alterations and cerebrovascular inflammation. *Mol. Psychiatry* **2022**, *27*, 803–818. [CrossRef]
17. Simmons, S.R.; Bhalla, M.; Herring, S.E.; Tchalla, E.Y.I.; Bou Ghanem, E.N. Older but Not Wiser: The Age-Driven Changes in Neutrophil Responses during Pulmonary Infections. *Infect. Immun.* **2021**, *89*, e00653-20. [CrossRef]
18. Barkaway, A.; Rolas, L.; Joulia, R.; Bodkin, J.; Lenn, T.; Owen-Woods, C.; Reglero-Real, N.; Stein, M.; Vazquez-Martinez, L.; Girbl, T.; et al. Age-related changes in the local milieu of inflamed tissues cause aberrant neutrophil trafficking and subsequent remote organ damage. *Immunity* **2021**, *54*, 1494–1510.e7. [CrossRef]
19. Wynn, T.A.; Chawla, A.; Pollard, J.W. Macrophage biology in development, homeostasis and disease. *Nature* **2013**, *496*, 445–455. [CrossRef]
20. Wu, S.; Zhang, R. CD38-expressing macrophages drive age-related NAD(+) decline. *Nat. Metab.* **2020**, *2*, 1186–1187. [CrossRef]
21. Selman, M.; Pardo, A. Fibroageing: An ageing pathological feature driven by dysregulated extracellular matrix-cell mechanobiology. *Ageing Res. Rev.* **2021**, *70*, 101393. [CrossRef] [PubMed]
22. Birjandi, S.Z.; Ippolito, J.A.; Ramadorai, A.K.; Witte, P.L. Alterations in marginal zone macrophages and marginal zone B cells in old mice. *J. Immunol.* **2011**, *186*, 3441–3451. [CrossRef] [PubMed]
23. Zhou, Z.; Yao, J.; Wu, D.; Huang, X.; Wang, Y.; Li, X.; Lu, Q.; Qiu, Y. Type 2 cytokine signaling in macrophages protects from cellular senescence and organismal aging. *Immunity* **2024**, *57*, 513–527.e6. [CrossRef] [PubMed]
24. Ma, S.; Sun, S.; Geng, L.; Song, M.; Wang, W.; Ye, Y.; Ji, Q.; Zou, Z.; Wang, S.; He, X.; et al. Caloric Restriction Reprograms the Single-Cell Transcriptional Landscape of Rattus Norvegicus Aging. *Cell* **2020**, *180*, 984–1001.e22. [CrossRef] [PubMed]
25. Minhas, P.S.; Latif-Hernandez, A.; McReynolds, M.R.; Durairaj, A.S.; Wang, Q.; Rubin, A.; Joshi, A.U.; He, J.Q.; Gauba, E.; Liu, L.; et al. Restoring metabolism of myeloid cells reverses cognitive decline in ageing. *Nature* **2021**, *590*, 122–128. [CrossRef]
26. Shakhar, G.; Lindquist, R.L.; Skokos, D.; Dudziak, D.; Huang, J.H.; Nussenzweig, M.C.; Dustin, M.L. Stable T cell-dendritic cell interactions precede the development of both tolerance and immunity in vivo. *Nat. Immunol.* **2005**, *6*, 707–714. [CrossRef]
27. Mogilenko, D.A.; Shpynov, O.; Andhey, P.S.; Arthur, L.; Swain, A.; Esaulova, E.; Brioschi, S.; Shchukina, I.; Kerndl, M.; Bambouskova, M.; et al. Comprehensive Profiling of an Aging Immune System Reveals Clonal GZMK(+) CD8(+) T Cells as Conserved Hallmark of Inflammaging. *Immunity* **2021**, *54*, 99–115.e12. [CrossRef]
28. Chougnet, C.A.; Thacker, R.I.; Shehata, H.M.; Hennies, C.M.; Lehn, M.A.; Lages, C.S.; Janssen, E.M. Loss of Phagocytic and Antigen Cross-Presenting Capacity in Aging Dendritic Cells Is Associated with Mitochondrial Dysfunction. *J. Immunol.* **2015**, *195*, 2624–2632. [CrossRef]
29. Zhivaki, D.; Kennedy, S.N.; Park, J.; Boriello, F.; Devant, P.; Cao, A.; Bahleda, K.M.; Murphy, S.; McCabe, C.; Evavold, C.L.; et al. Correction of age-associated defects in dendritic cells enables CD4(+) T cells to eradicate tumors. *Cell* **2024**, *187*, 3888–3903.e18. [CrossRef]
30. Qin, L.; Jing, X.; Qiu, Z.; Cao, W.; Jiao, Y.; Routy, J.P.; Li, T. Aging of immune system: Immune signature from peripheral blood lymphocyte subsets in 1068 healthy adults. *Aging* **2016**, *8*, 848–859. [CrossRef]
31. Borzi, R.M.; Mazzetti, I.; Cattini, L.; Uguccioni, M.; Baggiolini, M.; Facchini, A. Human chondrocytes express functional chemokine receptors and release matrix-degrading enzymes in response to C-X-C and C-C chemokines. *Arthritis Rheum.* **2000**, *43*, 1734–1741. [CrossRef] [PubMed]
32. Shaw, A.C.; Goldstein, D.R.; Montgomery, R.R. Age-dependent dysregulation of innate immunity. *Nat. Rev. Immunol.* **2013**, *13*, 875–887. [CrossRef] [PubMed]
33. Gounder, S.S.; Abdullah, B.J.J.; Radzuanb, N.; Zain, F.; Sait, N.B.M.; Chua, C.; Subramani, B. Effect of Aging on NK Cell Population and Their Proliferation at Ex Vivo Culture Condition. *Anal. Cell. Pathol.* **2018**, *2018*, 7871814. [CrossRef] [PubMed]
34. Wlaschek, M.; Maity, P.; Makrantonaki, E.; Scharffetter-Kochanek, K. Connective Tissue and Fibroblast Senescence in Skin Aging. *J. Investig. Dermatol.* **2021**, *141*, 985–992. [CrossRef] [PubMed]
35. Mittelbrunn, M.; Kroemer, G. Hallmarks of T cell aging. *Nat. Immunol.* **2021**, *22*, 687–698. [CrossRef]
36. Lerner, A.; Yamada, T.; Miller, R.A. Pgp-1hi T lymphocytes accumulate with age in mice and respond poorly to concanavalin A. *Eur. J. Immunol.* **1989**, *19*, 977–982. [CrossRef]
37. Linton, P.J.; Haynes, L.; Klinman, N.R.; Swain, S.L. Antigen-independent changes in naive CD4 T cells with aging. *J. Exp. Med.* **1996**, *184*, 1891–1900. [CrossRef]
38. Ernst, D.N.; Hobbs, M.V.; Torbett, B.E.; Glasebrook, A.L.; Rehse, M.A.; Bottomly, K.; Hayakawa, K.; Hardy, R.R.; Weigle, W.O. Differences in the expression profiles of CD45RB, Pgp-1, and 3G11 membrane antigens and in the patterns of lymphokine secretion by splenic CD4+ T cells from young and aged mice. *J. Immunol.* **1990**, *145*, 1295–1302. [CrossRef]
39. Han, G.M.; Zhao, B.; Jeyaseelan, S.; Feng, J.M. Age-associated parallel increase of Foxp3(+)CD4(+) regulatory and CD44(+)CD4(+) memory T cells in SJL/J mice. *Cell Immunol.* **2009**, *258*, 188–196. [CrossRef]
40. Lages, C.S.; Suffia, I.; Velilla, P.A.; Huang, B.; Warshaw, G.; Hildeman, D.A.; Belkaid, Y.; Chougnet, C. Functional regulatory T cells accumulate in aged hosts and promote chronic infectious disease reactivation. *J. Immunol.* **2008**, *181*, 1835–1848. [CrossRef]

41. Jia, Z.; Ren, Z.; Ye, D.; Li, J.; Xu, Y.; Liu, H.; Meng, Z.; Yang, C.; Chen, X.; Mao, X.; et al. Immune-Ageing Evaluation of Peripheral T and NK Lymphocyte Subsets in Chinese Healthy Adults. *Phenomics* **2023**, *3*, 360–374. [CrossRef] [PubMed]
42. Moiseeva, V.; Cisneros, A.; Sica, V.; Deryagin, O.; Lai, Y.; Jung, S.; Andres, E.; An, J.; Segales, J.; Ortet, L.; et al. Senescence atlas reveals an aged-like inflamed niche that blunts muscle regeneration. *Nature* **2023**, *613*, 169–178. [CrossRef] [PubMed]
43. Fulop, T.; Larbi, A.; Pawelec, G.; Khalil, A.; Cohen, A.A.; Hirokawa, K.; Witkowski, J.M.; Franceschi, C. Immunology of Aging: The Birth of Inflammaging. *Clin. Rev. Allergy Immunol.* **2023**, *64*, 109–122. [CrossRef] [PubMed]
44. Carrasco, E.; Gomez de Las Heras, M.M.; Gabande-Rodriguez, E.; Desdin-Mico, G.; Aranda, J.F.; Mittelbrunn, M. The role of T cells in age-related diseases. *Nat. Rev. Immunol.* **2022**, *22*, 97–111. [CrossRef] [PubMed]
45. Rodriguez, I.J.; Lalinde Ruiz, N.; Llano Leon, M.; Martinez Enriquez, L.; Montilla Velasquez, M.D.P.; Ortiz Aguirre, J.P.; Rodriguez Bohorquez, O.M.; Velandia Vargas, E.A.; Hernandez, E.D.; Parra Lopez, C.A. Immunosenescence Study of T Cells: A Systematic Review. *Front. Immunol.* **2020**, *11*, 604591. [CrossRef]
46. Yoshida, S.; Nakagami, H.; Hayashi, H.; Ikeda, Y.; Sun, J.; Tenma, A.; Tomioka, H.; Kawano, T.; Shimamura, M.; Morishita, R.; et al. The CD153 vaccine is a senotherapeutic option for preventing the accumulation of senescent T cells in mice. *Nat. Commun.* **2020**, *11*, 2482. [CrossRef]
47. Hao, Y.; O'Neill, P.; Naradikian, M.S.; Scholz, J.L.; Cancro, M.P. A B-cell subset uniquely responsive to innate stimuli accumulates in aged mice. *Blood* **2011**, *118*, 1294–1304. [CrossRef]
48. Rubtsov, A.V.; Rubtsova, K.; Fischer, A.; Meehan, R.T.; Gillis, J.Z.; Kappler, J.W.; Marrack, P. Toll-like receptor 7 (TLR7)-driven accumulation of a novel CD11c(+) B-cell population is important for the development of autoimmunity. *Blood* **2011**, *118*, 1305–1315. [CrossRef]
49. Rubtsova, K.; Rubtsov, A.V.; Thurman, J.M.; Mennona, J.M.; Kappler, J.W.; Marrack, P. B cells expressing the transcription factor T-bet drive lupus-like autoimmunity. *J. Clin. Investig.* **2017**, *127*, 1392–1404. [CrossRef]
50. Dai, D.; Gu, S.; Han, X.; Ding, H.; Jiang, Y.; Zhang, X.; Yao, C.; Hong, S.; Zhang, J.; Shen, Y.; et al. The transcription factor ZEB2 drives the formation of age-associated B cells. *Science* **2024**, *383*, 413–421. [CrossRef]
51. Lv, J.; Zhang, C.; Liu, X.; Gu, C.; Liu, Y.; Gao, Y.; Huang, Z.; Jiang, Q.; Chen, B.; He, D.; et al. An aging-related immune landscape in the hematopoietic immune system. *Immun. Ageing* **2024**, *21*, 3. [CrossRef] [PubMed]
52. Ma, F.; Cao, Y.; Du, H.; Braikia, F.Z.; Zong, L.; Ollikainen, N.; Bayer, M.; Qiu, X.; Park, B.; Roy, R.; et al. Three-dimensional chromatin reorganization regulates B cell development during ageing. *Nat. Cell. Biol.* **2024**, *26*, 991–1002. [CrossRef] [PubMed]
53. Yu, Y.; Lu, C.; Yu, W.; Lei, Y.; Sun, S.; Liu, P.; Bai, F.; Chen, Y.; Chen, J. B Cells Dynamic in Aging and the Implications of Nutritional Regulation. *Nutrients* **2024**, *16*, 487. [CrossRef] [PubMed]
54. Borodkina, A.V.; Deryabin, P.I.; Giukova, A.A.; Nikolsky, N.N. "Social Life" of Senescent Cells: What Is SASP and Why Study It? *Acta Naturae* **2018**, *10*, 4–14. [CrossRef] [PubMed]
55. Krtolica, A.; Parrinello, S.; Lockett, S.; Desprez, P.Y.; Campisi, J. Senescent fibroblasts promote epithelial cell growth and tumorigenesis: A link between cancer and aging. *Proc. Natl. Acad. Sci. USA* **2001**, *98*, 12072–12077. [CrossRef]
56. Coppe, J.P.; Patil, C.K.; Rodier, F.; Sun, Y.; Munoz, D.P.; Goldstein, J.; Nelson, P.S.; Desprez, P.Y.; Campisi, J. Senescence-associated secretory phenotypes reveal cell-nonautonomous functions of oncogenic RAS and the p53 tumor suppressor. *PLoS Biol.* **2008**, *6*, 2853–2868. [CrossRef]
57. Kuilman, T.; Peeper, D.S. Senescence-messaging secretome: SMS-ing cellular stress. *Nat. Rev. Cancer* **2009**, *9*, 81–94. [CrossRef]
58. Lopes-Paciencia, S.; Saint-Germain, E.; Rowell, M.C.; Ruiz, A.F.; Kalegari, P.; Ferbeyre, G. The senescence-associated secretory phenotype and its regulation. *Cytokine* **2019**, *117*, 15–22. [CrossRef]
59. Mogilenko, D.A.; Shchukina, I.; Artyomov, M.N. Immune ageing at single-cell resolution. *Nat. Rev. Immunol.* **2022**, *22*, 484–498. [CrossRef]
60. Miller, J. The function of the thymus and its impact on modern medicine. *Science* **2020**, *369*, eaba2429. [CrossRef]
61. Elyahu, Y.; Monsonego, A. Thymus involution sets the clock of the aging T-cell landscape: Implications for declined immunity and tissue repair. *Ageing Res. Rev.* **2021**, *65*, 101231. [CrossRef] [PubMed]
62. Wang, H.; Zhang, Y.; Wang, Z.; Zhang, L.; Guo, M.; Cao, C.; Xiao, H. Deciphering Nucleic Acid Binding Proteome of Mouse Immune Organs Reveals Hub Proteins for Aging. *Mol. Cell. Proteom.* **2023**, *22*, 100611. [CrossRef] [PubMed]
63. Oh, J.; Wang, W.; Thomas, R.; Su, D.M. Thymic rejuvenation via FOXN1-reprogrammed embryonic fibroblasts (FREFs) to counteract age-related inflammation. *JCI Insight* **2020**, *5*, e140313. [CrossRef] [PubMed]
64. Ferrucci, L.; Fabbri, E. Inflammageing: Chronic inflammation in ageing, cardiovascular disease, and frailty. *Nat. Rev. Cardiol.* **2018**, *15*, 505–522. [CrossRef] [PubMed]
65. Furman, D.; Campisi, J.; Verdin, E.; Carrera-Bastos, P.; Targ, S.; Franceschi, C.; Ferrucci, L.; Gilroy, D.W.; Fasano, A.; Miller, G.W.; et al. Chronic inflammation in the etiology of disease across the life span. *Nat. Med.* **2019**, *25*, 1822–1832. [CrossRef]
66. Chambers, E.S.; Akbar, A.N. Can blocking inflammation enhance immunity during aging? *J. Allergy Clin. Immunol.* **2020**, *145*, 1323–1331. [CrossRef]
67. Wang, D.; Malo, D.; Hekimi, S. Elevated mitochondrial reactive oxygen species generation affects the immune response via hypoxia-inducible factor-1alpha in long-lived Mclk1+/− mouse mutants. *J. Immunol.* **2010**, *184*, 582–590. [CrossRef]
68. Sanderson, S.L.; Simon, A.K. In aged primary T cells, mitochondrial stress contributes to telomere attrition measured by a novel imaging flow cytometry assay. *Aging Cell* **2017**, *16*, 1234–1243. [CrossRef]

69. Vida, C.; de Toda, I.M.; Cruces, J.; Garrido, A.; Gonzalez-Sanchez, M.; De la Fuente, M. Role of macrophages in age-related oxidative stress and lipofuscin accumulation in mice. *Redox Biol.* **2017**, *12*, 423–437. [CrossRef]
70. Niedernhofer, L.J.; Gurkar, A.U.; Wang, Y.; Vijg, J.; Hoeijmakers, J.H.J.; Robbins, P.D. Nuclear Genomic Instability and Aging. *Annu. Rev. Biochem.* **2018**, *87*, 295–322. [CrossRef]
71. Huang, Z.; Sun, S.; Lee, M.; Maslov, A.Y.; Shi, M.; Waldman, S.; Marsh, A.; Siddiqui, T.; Dong, X.; Peter, Y.; et al. Single-cell analysis of somatic mutations in human bronchial epithelial cells in relation to aging and smoking. *Nat. Genet.* **2022**, *54*, 492–498. [CrossRef] [PubMed]
72. Lee, K.A.; Flores, R.R.; Jang, I.H.; Saathoff, A.; Robbins, P.D. Immune Senescence, Immunosenescence and Aging. *Front. Aging* **2022**, *3*, 900028. [CrossRef] [PubMed]
73. Yousefzadeh, M.J.; Zhao, J.; Bukata, C.; Wade, E.A.; McGowan, S.J.; Angelini, L.A.; Bank, M.P.; Gurkar, A.U.; McGuckian, C.A.; Calubag, M.F.; et al. Tissue specificity of senescent cell accumulation during physiologic and accelerated aging of mice. *Aging Cell* **2020**, *19*, e13094. [CrossRef] [PubMed]
74. Stead, E.R.; Bjedov, I. Balancing DNA repair to prevent ageing and cancer. *Exp. Cell Res.* **2021**, *405*, 112679. [CrossRef] [PubMed]
75. Miller, K.N.; Victorelli, S.G.; Salmonowicz, H.; Dasgupta, N.; Liu, T.; Passos, J.F.; Adams, P.D. Cytoplasmic DNA: Sources, sensing, and role in aging and disease. *Cell* **2021**, *184*, 5506–5526. [CrossRef]
76. Hennekam, R.C.M. Pathophysiology of premature aging characteristics in Mendelian progeroid disorders. *Eur. J. Med. Genet.* **2020**, *63*, 104028. [CrossRef]
77. Blackburn, E.H.; Epel, E.S.; Lin, J. Human telomere biology: A contributory and interactive factor in aging, disease risks, and protection. *Science* **2015**, *350*, 1193–1198. [CrossRef]
78. Desdin-Mico, G.; Soto-Heredero, G.; Mittelbrunn, M. Mitochondrial activity in T cells. *Mitochondrion* **2018**, *41*, 51–57. [CrossRef]
79. Sena, L.A.; Li, S.; Jairaman, A.; Prakriya, M.; Ezponda, T.; Hildeman, D.A.; Wang, C.R.; Schumacker, P.T.; Licht, J.D.; Perlman, H.; et al. Mitochondria are required for antigen-specific T cell activation through reactive oxygen species signaling. *Immunity* **2013**, *38*, 225–236. [CrossRef]
80. Quinn, K.M.; Palchaudhuri, R.; Palmer, C.S.; La Gruta, N.L. The clock is ticking: The impact of ageing on T cell metabolism. *Clin. Transl. Immunol.* **2019**, *8*, e01091. [CrossRef]
81. Smith, A.L.; Whitehall, J.C.; Bradshaw, C.; Gay, D.; Robertson, F.; Blain, A.P.; Hudson, G.; Pyle, A.; Houghton, D.; Hunt, M.; et al. Age-associated mitochondrial DNA mutations cause metabolic remodelling that contributes to accelerated intestinal tumorigenesis. *Nat. Cancer* **2020**, *1*, 976–989. [CrossRef] [PubMed]
82. Singh, B.; Schoeb, T.R.; Bajpai, P.; Slominski, A.; Singh, K.K. Reversing wrinkled skin and hair loss in mice by restoring mitochondrial function. *Cell Death Dis.* **2018**, *9*, 735. [CrossRef] [PubMed]
83. Lujan, S.A.; Longley, M.J.; Humble, M.H.; Lavender, C.A.; Burkholder, A.; Blakely, E.L.; Alston, C.L.; Gorman, G.S.; Turnbull, D.M.; McFarland, R.; et al. Ultrasensitive deletion detection links mitochondrial DNA replication, disease, and aging. *Genome Biol.* **2020**, *21*, 248. [CrossRef] [PubMed]
84. Henson, S.M. CD8+ T-cell senescence: No role for mTOR. *Biochem. Soc. Trans.* **2015**, *43*, 734–739. [CrossRef] [PubMed]
85. Fernandes, S.A.; Demetriades, C. The Multifaceted Role of Nutrient Sensing and mTORC1 Signaling in Physiology and Aging. *Front. Aging* **2021**, *2*, 707372. [CrossRef]
86. Papadopoli, D.; Boulay, K.; Kazak, L.; Pollak, M.; Mallette, F.A.; Topisirovic, I.; Hulea, L. mTOR as a central regulator of lifespan and aging. *F1000Research* **2019**, *8*, 1–21. [CrossRef]
87. Mannick, J.B.; Lamming, D.W. Targeting the biology of aging with mTOR inhibitors. *Nat. Aging* **2023**, *3*, 642–660. [CrossRef]
88. Mannick, J.B.; Del Giudice, G.; Lattanzi, M.; Valiante, N.M.; Praestgaard, J.; Huang, B.; Lonetto, M.A.; Maecker, H.T.; Kovarik, J.; Carson, S.; et al. mTOR inhibition improves immune function in the elderly. *Sci. Transl. Med.* **2014**, *6*, 268ra179. [CrossRef]
89. Xie, X.; Ma, G.; Li, X.; Zhao, J.; Zhao, Z.; Zeng, J. Activation of innate immune cGAS-STING pathway contributes to Alzheimer's pathogenesis in 5xFAD mice. *Nat. Aging* **2023**, *3*, 202–212. [CrossRef]
90. Gulen, M.F.; Samson, N.; Keller, A.; Schwabenland, M.; Liu, C.; Gluck, S.; Thacker, V.V.; Favre, L.; Mangeat, B.; Kroese, L.J.; et al. cGAS-STING drives ageing-related inflammation and neurodegeneration. *Nature* **2023**, *620*, 374–380. [CrossRef]
91. Salminen, A.; Huuskonen, J.; Ojala, J.; Kauppinen, A.; Kaarniranta, K.; Suuronen, T. Activation of innate immunity system during aging: NF-kB signaling is the molecular culprit of inflamm-aging. *Ageing Res. Rev.* **2008**, *7*, 83–105. [CrossRef] [PubMed]
92. Kauppinen, A.; Suuronen, T.; Ojala, J.; Kaarniranta, K.; Salminen, A. Antagonistic crosstalk between NF-kappaB and SIRT1 in the regulation of inflammation and metabolic disorders. *Cell Signal.* **2013**, *25*, 1939–1948. [CrossRef] [PubMed]
93. Wang, Y.; Bi, Y.; Chen, X.; Li, C.; Li, Y.; Zhang, Z.; Wang, J.; Lu, Y.; Yu, Q.; Su, H.; et al. Histone Deacetylase SIRT1 Negatively Regulates the Differentiation of Interleukin-9-Producing CD4(+) T Cells. *Immunity* **2016**, *44*, 1337–1349. [CrossRef] [PubMed]
94. Xue, C.; Yao, Q.; Gu, X.; Shi, Q.; Yuan, X.; Chu, Q.; Bao, Z.; Lu, J.; Li, L. Evolving cognition of the JAK-STAT signaling pathway: Autoimmune disorders and cancer. *Signal Transduct. Target. Ther.* **2023**, *8*, 204. [CrossRef] [PubMed]
95. Wang, B.; Han, J.; Elisseeff, J.H.; Demaria, M. The senescence-associated secretory phenotype and its physiological and pathological implications. *Nat. Rev. Mol. Cell Biol.* **2024**. [CrossRef]
96. Xu, M.; Tchkonia, T.; Ding, H.; Ogrodnik, M.; Lubbers, E.R.; Pirtskhalava, T.; White, T.A.; Johnson, K.O.; Stout, M.B.; Mezera, V.; et al. JAK inhibition alleviates the cellular senescence-associated secretory phenotype and frailty in old age. *Proc. Natl. Acad. Sci. USA* **2015**, *112*, E6301–E6310. [CrossRef]

97. Zheng, R.; Zhang, Y.; Zhang, K.; Yuan, Y.; Jia, S.; Liu, J. The Complement System, Aging, and Aging-Related Diseases. *Int. J. Mol. Sci.* **2022**, *23*, 8689. [CrossRef]
98. Yoshikawa, T.T. Epidemiology and unique aspects of aging and infectious diseases. *Clin. Infect. Dis.* **2000**, *30*, 931–933. [CrossRef]
99. Jiang, N.; He, J.; Weinstein, J.A.; Penland, L.; Sasaki, S.; He, X.S.; Dekker, C.L.; Zheng, N.Y.; Huang, M.; Sullivan, M.; et al. Lineage structure of the human antibody repertoire in response to influenza vaccination. *Sci. Transl. Med.* **2013**, *5*, 171ra119. [CrossRef]
100. Treanor, J.J.; Talbot, H.K.; Ohmit, S.E.; Coleman, L.A.; Thompson, M.G.; Cheng, P.Y.; Petrie, J.G.; Lofthus, G.; Meece, J.K.; Williams, J.V.; et al. Effectiveness of seasonal influenza vaccines in the United States during a season with circulation of all three vaccine strains. *Clin. Infect. Dis.* **2012**, *55*, 951–959. [CrossRef]
101. Valmori, D.; Merlo, A.; Souleimanian, N.E.; Hesdorffer, C.S.; Ayyoub, M. A peripheral circulating compartment of natural naive CD4 Tregs. *J. Clin. Investig.* **2005**, *115*, 1953–1962. [CrossRef] [PubMed]
102. Stahl, E.C.; Brown, B.N. Cell Therapy Strategies to Combat Immunosenescence. *Organogenesis* **2015**, *11*, 159–172. [CrossRef] [PubMed]
103. Simon, A.K.; Hollander, G.A.; McMichael, A. Evolution of the immune system in humans from infancy to old age. *Proc. Biol. Sci.* **2015**, *282*, 20143085. [CrossRef]
104. Solana, R.; Tarazona, R.; Aiello, A.E.; Akbar, A.N.; Appay, V.; Beswick, M.; Bosch, J.A.; Campos, C.; Cantisan, S.; Cicin-Sain, L.; et al. CMV and Immunosenescence: From basics to clinics. *Immun. Ageing* **2012**, *9*, 23. [CrossRef] [PubMed]
105. Nikolich-Zugich, J.; Knox, K.S.; Rios, C.T.; Natt, B.; Bhattacharya, D.; Fain, M.J. SARS-CoV-2 and COVID-19 in older adults: What we may expect regarding pathogenesis, immune responses, and outcomes. *Geroscience* **2020**, *42*, 505–514. [CrossRef] [PubMed]
106. Du, S.Q.; Yuan, W. Mathematical modeling of interaction between innate and adaptive immune responses in COVID-19 and implications for viral pathogenesis. *J. Med. Virol.* **2020**, *92*, 1615–1628. [CrossRef]
107. De Wilde, A.H.; Snijder, E.J.; Kikkert, M.; van Hemert, M.J. Host Factors in Coronavirus Replication. *Curr. Top. Microbiol. Immunol.* **2018**, *419*, 1–42. [CrossRef]
108. Wang, S.; Yao, X.; Ma, S.; Ping, Y.; Fan, Y.; Sun, S.; He, Z.; Shi, Y.; Sun, L.; Xiao, S.; et al. A single-cell transcriptomic landscape of the lungs of patients with COVID-19. *Nat. Cell. Biol.* **2021**, *23*, 1314–1328. [CrossRef]
109. Bassendine, M.F.; Bridge, S.H.; McCaughan, G.W.; Gorrell, M.D. COVID-19 and comorbidities: A role for dipeptidyl peptidase 4 (DPP4) in disease severity? *J. Diabetes* **2020**, *12*, 649–658. [CrossRef]
110. Goronzy, J.J.; Weyand, C.M. Aging, autoimmunity and arthritis: T-cell senescence and contraction of T-cell repertoire diversity - catalysts of autoimmunity and chronic inflammation. *Arthritis Res. Ther.* **2003**, *5*, 225–234. [CrossRef]
111. Collaborators, G.B.D.R.A. Global, regional, and national burden of rheumatoid arthritis, 1990-2020, and projections to 2050: A systematic analysis of the Global Burden of Disease Study 2021. *Lancet Rheumatol.* **2023**, *5*, e594–e610. [CrossRef]
112. Struglics, A.; Okroj, M.; Sward, P.; Frobell, R.; Saxne, T.; Lohmander, L.S.; Blom, A.M. The complement system is activated in synovial fluid from subjects with knee injury and from patients with osteoarthritis. *Arthritis Res. Ther.* **2016**, *18*, 223. [CrossRef] [PubMed]
113. Cao, W.; Zheng, D.; Wang, G.; Zhang, J.; Ge, S.; Singh, M.; Wang, H.; Song, M.; Li, D.; Wang, W.; et al. Modelling biological age based on plasma peptides in Han Chinese adults. *Aging* **2020**, *12*, 10676–10686. [CrossRef] [PubMed]
114. Scarpaci, S.; Frasca, D.; Barattini, P.; Guidi, L.; Doria, G. DNA damage recognition and repair capacities in human naive and memory T cells from peripheral blood of young and elderly subjects. *Mech. Ageing Dev.* **2003**, *124*, 517–524. [CrossRef] [PubMed]
115. Minaga, K.; Watanabe, T.; Hara, A.; Yoshikawa, T.; Kamata, K.; Kudo, M. Plasmacytoid Dendritic Cells as a New Therapeutic Target for Autoimmune Pancreatitis and IgG4-Related Disease. *Front. Immunol.* **2021**, *12*, 713779. [CrossRef] [PubMed]
116. Li, P.; Jiang, M.; Li, K.; Li, H.; Zhou, Y.; Xiao, X.; Xu, Y.; Krishfield, S.; Lipsky, P.E.; Tsokos, G.C.; et al. Glutathione peroxidase 4-regulated neutrophil ferroptosis induces systemic autoimmunity. *Nat. Immunol.* **2021**, *22*, 1107–1117. [CrossRef]
117. Kishimoto, D.; Kirino, Y.; Tamura, M.; Takeno, M.; Kunishita, Y.; Takase-Minegishi, K.; Nakano, H.; Kato, I.; Nagahama, K.; Yoshimi, R.; et al. Dysregulated heme oxygenase-1(low) M2-like macrophages augment lupus nephritis via Bach1 induced by type I interferons. *Arthritis Res. Ther.* **2018**, *20*, 64. [CrossRef]
118. Bolouri, N.; Akhtari, M.; Farhadi, E.; Mansouri, R.; Faezi, S.T.; Jamshidi, A.; Mahmoudi, M. Role of the innate and adaptive immune responses in the pathogenesis of systemic lupus erythematosus. *Inflamm. Res.* **2022**, *71*, 537–554. [CrossRef]
119. Chen, X.; Sun, X.; Yang, W.; Yang, B.; Zhao, X.; Chen, S.; He, L.; Chen, H.; Yang, C.; Xiao, L.; et al. An autoimmune disease variant of IgG1 modulates B cell activation and differentiation. *Science* **2018**, *362*, 700–705. [CrossRef]
120. Crow, M.K. Pathogenesis of systemic lupus erythematosus: Risks, mechanisms and therapeutic targets. *Ann. Rheum. Dis.* **2023**, *82*, 999–1014. [CrossRef]
121. Sullivan, K.E. Complement deficiency and autoimmunity. *Curr. Opin. Pediatr.* **1998**, *10*, 600–606. [CrossRef] [PubMed]
122. Kuchcinski, G.; Rumetshofer, T.; Zervides, K.A.; Lopes, R.; Gautherot, M.; Pruvo, J.P.; Bengtsson, A.A.; Hansson, O.; Jonsen, A.; Sundgren, P.C.M. MRI BrainAGE demonstrates increased brain aging in systemic lupus erythematosus patients. *Front. Aging Neurosci.* **2023**, *15*, 1274061. [CrossRef] [PubMed]
123. Gueye, M.; Preziosa, P.; Ramirez, G.A.; Bozzolo, E.P.; Canti, V.; Margoni, M.; Meani, A.; Moiola, L.; Rovere-Querini, P.; Manfredi, A.A.; et al. Choroid plexus and perivascular space enlargement in neuropsychiatric systemic lupus erythematosus. *Mol. Psychiatry* **2024**, *29*, 359–368. [CrossRef] [PubMed]
124. Torre, L.A.; Bray, F.; Siegel, R.L.; Ferlay, J.; Lortet-Tieulent, J.; Jemal, A. Global cancer statistics, 2012. *CA Cancer J. Clin.* **2015**, *65*, 87–108. [CrossRef] [PubMed]

125. Li, D.; de Glas, N.A.; Hurria, A. Cancer and Aging: General Principles, Biology, and Geriatric Assessment. *Clin. Geriatr. Med.* **2016**, *32*, 1–15. [CrossRef]
126. Kadambi, S.; Loh, K.P.; Dunne, R.; Magnuson, A.; Maggiore, R.; Zittel, J.; Flannery, M.; Inglis, J.; Gilmore, N.; Mohamed, M.; et al. Older adults with cancer and their caregivers - current landscape and future directions for clinical care. *Nat. Rev. Clin. Oncol.* **2020**, *17*, 742–755. [CrossRef]
127. Lopez-Otin, C.; Kroemer, G. Hallmarks of Health. *Cell* **2021**, *184*, 33–63. [CrossRef]
128. Yonekura, S.; Terrisse, S.; Alves Costa Silva, C.; Lafarge, A.; Iebba, V.; Ferrere, G.; Goubet, A.G.; Fahrner, J.E.; Lahmar, I.; Ueda, K.; et al. Cancer Induces a Stress Ileopathy Depending on beta-Adrenergic Receptors and Promoting Dysbiosis that Contributes to Carcinogenesis. *Cancer Discov.* **2022**, *12*, 1128–1151. [CrossRef]
129. Ness, K.K.; Kirkland, J.L.; Gramatges, M.M.; Wang, Z.; Kundu, M.; McCastlain, K.; Li-Harms, X.; Zhang, J.; Tchkonia, T.; Pluijm, S.M.F.; et al. Premature Physiologic Aging as a Paradigm for Understanding Increased Risk of Adverse Health Across the Lifespan of Survivors of Childhood Cancer. *J. Clin. Oncol.* **2018**, *36*, 2206–2215. [CrossRef]
130. Carroll, J.E.; Bower, J.E.; Ganz, P.A. Cancer-related accelerated ageing and biobehavioural modifiers: A framework for research and clinical care. *Nat. Rev. Clin. Oncol.* **2022**, *19*, 173–187. [CrossRef]
131. Lopez-Otin, C.; Pietrocola, F.; Roiz-Valle, D.; Galluzzi, L.; Kroemer, G. Meta-hallmarks of aging and cancer. *Cell Metab.* **2023**, *35*, 12–35. [CrossRef] [PubMed]
132. Geiger, R.; Rieckmann, J.C.; Wolf, T.; Basso, C.; Feng, Y.; Fuhrer, T.; Kogadeeva, M.; Picotti, P.; Meissner, F.; Mann, M.; et al. L-Arginine Modulates T Cell Metabolism and Enhances Survival and Anti-tumor Activity. *Cell* **2016**, *167*, 829–842.e13. [CrossRef] [PubMed]
133. Gnanaprakasam, J.N.R.; Kushwaha, B.; Liu, L.; Chen, X.; Kang, S.; Wang, T.; Cassel, T.A.; Adams, C.M.; Higashi, R.M.; Scott, D.A.; et al. Asparagine restriction enhances CD8(+) T cell metabolic fitness and antitumoral functionality through an NRF2-dependent stress response. *Nat. Metab.* **2023**, *5*, 1423–1439. [CrossRef] [PubMed]
134. Yang, L.; Chu, Z.; Liu, M.; Zou, Q.; Li, J.; Liu, Q.; Wang, Y.; Wang, T.; Xiang, J.; Wang, B. Amino acid metabolism in immune cells: Essential regulators of the effector functions, and promising opportunities to enhance cancer immunotherapy. *J. Hematol. Oncol.* **2023**, *16*, 59. [CrossRef] [PubMed]
135. Sullivan, Z.A.; Khoury-Hanold, W.; Lim, J.; Smillie, C.; Biton, M.; Reis, B.S.; Zwick, R.K.; Pope, S.D.; Israni-Winger, K.; Parsa, R.; et al. $\gamma\delta$ T cells regulate the intestinal response to nutrient sensing. *Science* **2021**, *371*, eaba8310. [CrossRef]
136. Tan, J.; Ni, D.; Wali, J.A.; Cox, D.A.; Pinget, G.V.; Taitz, J.; Daïen, C.I.; Senior, A.; Read, M.N.; Simpson, S.J.; et al. Dietary carbohydrate, particularly glucose, drives B cell lymphopoiesis and function. *Iscience* **2021**, *24*, 102835. [CrossRef]
137. He, C.; Wang, S.; Zhou, C.; He, M.; Wang, J.; Ladds, M.; Lianoudaki, D.; Sedimbi, S.K.; Lane, D.P.; Westerberg, L.S.; et al. CD36 and LC3B initiated autophagy in B cells regulates the humoral immune response. *Autophagy* **2021**, *17*, 3577–3591. [CrossRef]
138. Zhou, X.; Zhu, X.; Li, C.; Li, Y.; Ye, Z.; Shapiro, V.S.; Copland, J.A., 3rd; Hitosugi, T.; Bernlohr, D.A.; Sun, J.; et al. Stearoyl-CoA Desaturase-Mediated Monounsaturated Fatty Acid Availability Supports Humoral Immunity. *Cell Rep.* **2021**, *34*, 108601. [CrossRef]
139. Sanchez, H.N.; Moroney, J.B.; Gan, H.; Shen, T.; Im, J.L.; Li, T.; Taylor, J.R.; Zan, H.; Casali, P. B cell-intrinsic epigenetic modulation of antibody responses by dietary fiber-derived short-chain fatty acids. *Nat. Commun.* **2020**, *11*, 60. [CrossRef]
140. Hao, F.; Tian, M.; Zhang, X.; Jin, X.; Jiang, Y.; Sun, X.; Wang, Y.; Peng, P.; Liu, J.; Xia, C.; et al. Butyrate enhances CPT1A activity to promote fatty acid oxidation and iTreg differentiation. *Proc. Natl. Acad. Sci. USA* **2021**, *118*, e2014681118. [CrossRef]
141. Lai, Y.; Gao, Y.; Lin, J.; Liu, F.; Yang, L.; Zhou, J.; Xue, Y.; Li, Y.; Chang, Z.; Li, J.; et al. Dietary elaidic acid boosts tumoral antigen presentation and cancer immunity via ACSL5. *Cell Metab.* **2024**, *36*, 822–838.e8. [CrossRef] [PubMed]
142. Nava Lauson, C.B.; Tiberti, S.; Corsetto, P.A.; Conte, F.; Tyagi, P.; Machwirth, M.; Ebert, S.; Loffreda, A.; Scheller, L.; Sheta, D.; et al. Linoleic acid potentiates CD8(+) T cell metabolic fitness and antitumor immunity. *Cell Metab.* **2023**, *35*, 633–650.e9. [CrossRef] [PubMed]
143. Jia, M.; Wang, Y.; Wang, J.; Qin, D.; Wang, M.; Chai, L.; Fu, Y.; Zhao, C.; Gao, C.; Jia, J.; et al. Myristic acid as a checkpoint to regulate STING-dependent autophagy and interferon responses by promoting N-myristoylation. *Nat. Commun.* **2023**, *14*, 660. [CrossRef] [PubMed]
144. Bosch, M.; Sanchez-Alvarez, M.; Fajardo, A.; Kapetanovic, R.; Steiner, B.; Dutra, F.; Moreira, L.; Lopez, J.A.; Campo, R.; Mari, M.; et al. Mammalian lipid droplets are innate immune hubs integrating cell metabolism and host defense. *Science* **2020**, *370*, eaay8085. [CrossRef] [PubMed]
145. Schulze, M.B.; Minihane, A.M.; Saleh, R.N.M.; Riserus, U. Intake and metabolism of omega-3 and omega-6 polyunsaturated fatty acids: Nutritional implications for cardiometabolic diseases. *Lancet Diabetes Endocrinol.* **2020**, *8*, 915–930. [CrossRef]
146. Huang, Z.; Liu, Y.; Qi, G.; Brand, D.; Zheng, S.G. Role of Vitamin A in the Immune System. *J. Clin. Med.* **2018**, *7*, 258. [CrossRef]
147. Van Gorkom, G.N.Y.; Klein Wolterink, R.G.J.; Van Elssen, C.; Wieten, L.; Germeraad, W.T.V.; Bos, G.M.J. Influence of Vitamin C on Lymphocytes: An Overview. *Antioxidants* **2018**, *7*, 41. [CrossRef]
148. Autier, P.; Boniol, M.; Pizot, C.; Mullie, P. Vitamin D status and ill health: A systematic review. *Lancet Diabetes Endocrinol.* **2014**, *2*, 76–89. [CrossRef]
149. Jeffery, L.E.; Raza, K.; Hewison, M. Vitamin D in rheumatoid arthritis-towards clinical application. *Nat. Rev. Rheumatol.* **2016**, *12*, 201–210. [CrossRef]

150. Yuan, X.; Duan, Y.; Xiao, Y.; Sun, K.; Qi, Y.; Zhang, Y.; Ahmed, Z.; Moiani, D.; Yao, J.; Li, H.; et al. Vitamin E Enhances Cancer Immunotherapy by Reinvigorating Dendritic Cells via Targeting Checkpoint SHP1. *Cancer Discov.* **2022**, *12*, 1742–1759. [CrossRef]
151. Gombart, A.F.; Pierre, A.; Maggini, S. A Review of Micronutrients and the Immune System-Working in Harmony to Reduce the Risk of Infection. *Nutrients* **2020**, *12*, 236. [CrossRef] [PubMed]
152. Wintergerst, E.S.; Maggini, S.; Hornig, D.H. Immune-enhancing role of vitamin C and zinc and effect on clinical conditions. *Ann. Nutr. Metab.* **2006**, *50*, 85–94. [CrossRef] [PubMed]
153. Song, X.; Zhang, H.; Zhang, Y.; Goh, B.; Bao, B.; Mello, S.S.; Sun, X.; Zheng, W.; Gazzaniga, F.S.; Wu, M.; et al. Gut microbial fatty acid isomerization modulates intraepithelial T cells. *Nature* **2023**, *619*, 837–843. [CrossRef] [PubMed]
154. Ma, J.; Liu, Z.; Gao, X.; Bao, Y.; Hong, Y.; He, X.; Zhu, W.; Li, Y.; Huang, W.; Zheng, N.; et al. Gut microbiota remodeling improves natural aging-related disorders through Akkermansia muciniphila and its derived acetic acid. *Pharmacol. Res.* **2023**, *189*, 106687. [CrossRef]
155. Xiao, Y.; Feng, Y.; Zhao, J.; Chen, W.; Lu, W. Achieving healthy aging through gut microbiota-directed dietary intervention: Focusing on microbial biomarkers and host mechanisms. *J. Adv. Res.* **2024**, *24*, 00092-4. [CrossRef]
156. Cristofori, F.; Dargenio, V.N.; Dargenio, C.; Miniello, V.L.; Barone, M.; Francavilla, R. Anti-Inflammatory and Immunomodulatory Effects of Probiotics in Gut Inflammation: A Door to the Body. *Front. Immunol.* **2021**, *12*, 578386. [CrossRef]
157. Johansson, M.A.; Bjorkander, S.; Mata Forsberg, M.; Qazi, K.R.; Salvany Celades, M.; Bittmann, J.; Eberl, M.; Sverremark-Ekstrom, E. Probiotic Lactobacilli Modulate Staphylococcus aureus-Induced Activation of Conventional and Unconventional T cells and NK Cells. *Front. Immunol.* **2016**, *7*, 273. [CrossRef]
158. Waziry, R.; Ryan, C.P.; Corcoran, D.L.; Huffman, K.M.; Kobor, M.S.; Kothari, M.; Graf, G.H.; Kraus, V.B.; Kraus, W.E.; Lin, D.T.S.; et al. Effect of long-term caloric restriction on DNA methylation measures of biological aging in healthy adults from the CALERIE trial. *Nat. Aging* **2023**, *3*, 248–257. [CrossRef]
159. Spadaro, O.; Youm, Y.; Shchukina, I.; Ryu, S.; Sidorov, S.; Ravussin, A.; Nguyen, K.; Aladyeva, E.; Predeus, A.N.; Smith, S.R.; et al. Caloric restriction in humans reveals immunometabolic regulators of health span. *Science* **2022**, *375*, 671–677. [CrossRef]
160. You, Y.; Chen, Y.; Wang, X.; Wei, M.; Zhang, Q.; Cao, Q. Accelerometer-measured physical activity patterns are associated with phenotypic age: Isotemporal substitution effects. *Heliyon* **2023**, *9*, e19158. [CrossRef]
161. Nilsson, M.I.; Bourgeois, J.M.; Nederveen, J.P.; Leite, M.R.; Hettinga, B.P.; Bujak, A.L.; May, L.; Lin, E.; Crozier, M.; Rusiecki, D.R.; et al. Lifelong aerobic exercise protects against inflammaging and cancer. *PLoS ONE* **2019**, *14*, e0210863. [CrossRef] [PubMed]
162. Rossman, M.J.; Kaplon, R.E.; Hill, S.D.; McNamara, M.N.; Santos-Parker, J.R.; Pierce, G.L.; Seals, D.R.; Donato, A.J. Endothelial cell senescence with aging in healthy humans: Prevention by habitual exercise and relation to vascular endothelial function. *Am. J. Physiol. Heart Circ. Physiol.* **2017**, *313*, H890–H895. [CrossRef] [PubMed]
163. Werner, C.; Furster, T.; Widmann, T.; Poss, J.; Roggia, C.; Hanhoun, M.; Scharhag, J.; Buchner, N.; Meyer, T.; Kindermann, W.; et al. Physical exercise prevents cellular senescence in circulating leukocytes and in the vessel wall. *Circulation* **2009**, *120*, 2438–2447. [CrossRef] [PubMed]
164. Gao, X.; Huang, N.; Guo, X.; Huang, T. Role of sleep quality in the acceleration of biological aging and its potential for preventive interaction on air pollution insults: Findings from the UK Biobank cohort. *Aging Cell* **2022**, *21*, e13610. [CrossRef] [PubMed]
165. Li, Y.; Sahakian, B.J.; Kang, J.; Langley, C.; Zhang, W.; Xie, C.; Xiang, S.; Yu, J.; Cheng, W.; Feng, J. The brain structure and genetic mechanisms underlying the nonlinear association between sleep duration, cognition and mental health. *Nat. Aging* **2022**, *2*, 425–437. [CrossRef]
166. Li, S.B.; Damonte, V.M.; Chen, C.; Wang, G.X.; Kebschull, J.M.; Yamaguchi, H.; Bian, W.J.; Purmann, C.; Pattni, R.; Urban, A.E.; et al. Hyperexcitable arousal circuits drive sleep instability during aging. *Science* **2022**, *375*, eabh3021. [CrossRef]
167. Sciorati, C.; Gamberale, R.; Monno, A.; Citterio, L.; Lanzani, C.; De Lorenzo, R.; Ramirez, G.A.; Esposito, A.; Manunta, P.; Manfredi, A.A.; et al. Pharmacological blockade of TNFalpha prevents sarcopenia and prolongs survival in aging mice. *Aging* **2020**, *12*, 23497–23508. [CrossRef]
168. Gocmez, S.S.; Yazir, Y.; Gacar, G.; Demirtas Sahin, T.; Arkan, S.; Karson, A.; Utkan, T. Etanercept improves aging-induced cognitive deficits by reducing inflammation and vascular dysfunction in rats. *Physiol. Behav.* **2020**, *224*, 113019. [CrossRef]
169. D'Souza, S.S.; Zhang, Y.; Bailey, J.T.; Fung, I.T.H.; Kuentzel, M.L.; Chittur, S.V.; Yang, Q. Type I Interferon signaling controls the accumulation and transcriptomes of monocytes in the aged lung. *Aging Cell* **2021**, *20*, e13470. [CrossRef]
170. Marin-Aguilar, F.; Lechuga-Vieco, A.V.; Alcocer-Gomez, E.; Castejon-Vega, B.; Lucas, J.; Garrido, C.; Peralta-Garcia, A.; Perez-Pulido, A.J.; Varela-Lopez, A.; Quiles, J.L.; et al. NLRP3 inflammasome suppression improves longevity and prevents cardiac aging in male mice. *Aging Cell* **2020**, *19*, e13050. [CrossRef]
171. Almanan, M.; Raynor, J.; Ogunsulire, I.; Malyshkina, A.; Mukherjee, S.; Hummel, S.A.; Ingram, J.T.; Saini, A.; Xie, M.M.; Alenghat, T.; et al. IL-10-producing Tfh cells accumulate with age and link inflammation with age-related immune suppression. *Sci. Adv.* **2020**, *6*, eabb0806. [CrossRef] [PubMed]
172. Shim, H.S.; Iaconelli, J.; Shang, X.; Li, J.; Lan, Z.D.; Jiang, S.; Nutsch, K.; Beyer, B.A.; Lairson, L.L.; Boutin, A.T.; et al. TERT activation targets DNA methylation and multiple aging hallmarks. *Cell* **2024**, *187*, 4030–4042.e13. [CrossRef] [PubMed]
173. Bai, H.; Liu, X.; Lin, M.; Meng, Y.; Tang, R.; Guo, Y.; Li, N.; Clarke, M.F.; Cai, S. Progressive senescence programs induce intrinsic vulnerability to aging-related female breast cancer. *Nat. Commun.* **2024**, *15*, 5154. [CrossRef] [PubMed]
174. Partridge, L.; Fuentealba, M.; Kennedy, B.K. The quest to slow ageing through drug discovery. *Nat. Rev. Drug Discov.* **2020**, *19*, 513–532. [CrossRef] [PubMed]

175. Bharath, L.P.; Agrawal, M.; McCambridge, G.; Nicholas, D.A.; Hasturk, H.; Liu, J.; Jiang, K.; Liu, R.; Guo, Z.; Deeney, J.; et al. Metformin Enhances Autophagy and Normalizes Mitochondrial Function to Alleviate Aging-Associated Inflammation. *Cell Metab.* **2020**, *32*, 44–55.e6. [CrossRef]
176. Ding, Y.; Wang, Y.; Hu, Q. Recent advances in overcoming barriers to cell-based delivery systems for cancer immunotherapy. *Exploration* **2022**, *2*, 20210106. [CrossRef]
177. Zhou, F.; Huang, L.; Li, S.; Yang, W.; Chen, F.; Cai, Z.; Liu, X.; Xu, W.; Lehto, V.P.; Lachelt, U.; et al. From structural design to delivery: mRNA therapeutics for cancer immunotherapy. *Exploration* **2024**, *4*, 20210146. [CrossRef]
178. Yousefzadeh, M.J.; Flores, R.R.; Zhu, Y.; Schmiechen, Z.C.; Brooks, R.W.; Trussoni, C.E.; Cui, Y.; Angelini, L.; Lee, K.A.; McGowan, S.J.; et al. An aged immune system drives senescence and ageing of solid organs. *Nature* **2021**, *594*, 100–105. [CrossRef]
179. Wu, R.; Sun, F.; Zhang, W.; Ren, J.; Liu, G.H. Targeting aging and age-related diseases with vaccines. *Nat. Aging* **2024**, *4*, 464–482. [CrossRef]
180. Amor, C.; Fernandez-Maestre, I.; Chowdhury, S.; Ho, Y.J.; Nadella, S.; Graham, C.; Carrasco, S.E.; Nnuji-John, E.; Feucht, J.; Hinterleitner, C.; et al. Prophylactic and long-lasting efficacy of senolytic CAR T cells against age-related metabolic dysfunction. *Nat. Aging* **2024**, *4*, 336–349. [CrossRef]
181. Yang, D.; Sun, B.; Li, S.; Wei, W.; Liu, X.; Cui, X.; Zhang, X.; Liu, N.; Yan, L.; Deng, Y.; et al. NKG2D-CAR T cells eliminate senescent cells in aged mice and nonhuman primates. *Sci. Transl. Med.* **2023**, *15*, eadd1951. [CrossRef] [PubMed]
182. Kovtonyuk, L.V.; Fritsch, K.; Feng, X.; Manz, M.G.; Takizawa, H. Inflamm-Aging of Hematopoiesis, Hematopoietic Stem Cells, and the Bone Marrow Microenvironment. *Front. Immunol.* **2016**, *7*, 502. [CrossRef] [PubMed]
183. Sun, S.; Han, Y.; Lei, Y.; Yu, Y.; Dong, Y.; Chen, J. Hematopoietic Stem Cell: Regulation and Nutritional Intervention. *Nutrients* **2023**, *15*, 2605. [CrossRef] [PubMed]
184. Wahlestedt, M.; Norddahl, G.L.; Sten, G.; Ugale, A.; Frisk, M.A.; Mattsson, R.; Deierborg, T.; Sigvardsson, M.; Bryder, D. An epigenetic component of hematopoietic stem cell aging amenable to reprogramming into a young state. *Blood* **2013**, *121*, 4257–4264. [CrossRef] [PubMed]
185. Nikolich-Zugich, J. Author Correction: The twilight of immunity: Emerging concepts in aging of the immune system. *Nat. Immunol.* **2018**, *19*, 1146. [CrossRef]
186. Karagiannis, T.T.; Dowrey, T.W.; Villacorta-Martin, C.; Montano, M.; Reed, E.; Belkina, A.C.; Andersen, S.L.; Perls, T.T.; Monti, S.; Murphy, G.J.; et al. Multi-modal profiling of peripheral blood cells across the human lifespan reveals distinct immune cell signatures of aging and longevity. *EBioMedicine* **2023**, *90*, 104514. [CrossRef]
187. Alpert, A.; Pickman, Y.; Leipold, M.; Rosenberg-Hasson, Y.; Ji, X.; Gaujoux, R.; Rabani, H.; Starosvetsky, E.; Kveler, K.; Schaffert, S.; et al. A clinically meaningful metric of immune age derived from high-dimensional longitudinal monitoring. *Nat. Med.* **2019**, *25*, 487–495. [CrossRef]
188. Chu, L.X.; Wang, W.J.; Gu, X.P.; Wu, P.; Gao, C.; Zhang, Q.; Wu, J.; Jiang, D.W.; Huang, J.Q.; Ying, X.W.; et al. Spatiotemporal multi-omics: Exploring molecular landscapes in aging and regenerative medicine. *Mil. Med. Res.* **2024**, *11*, 31. [CrossRef]

nutrients

MDPI

Review

Exploring the Geroprotective Potential of Nutraceuticals

Nadia Alejandra Rivero-Segura [1,†], Emmanuel Alejandro Zepeda-Arzate [1,†], Selma Karime Castillo-Vazquez [1,2], Patrick Fleischmann-delaParra [1], Jessica Hernández-Pineda [3], Edgar Flores-Soto [4], Paola García-delaTorre [5], Edgar Antonio Estrella-Parra [6] and Juan Carlos Gomez-Verjan [1,*]

1 Dirección de Investigación, Instituto Nacional de Geriatría (INGER), Mexico City 10200, Mexico; nrivero@inger.gob.mx (N.A.R.-S.); bm.ezepeda@gmail.com (E.A.Z.-A.); selmakarime@gmail.com (S.K.C.-V.); patrickfp15@gmail.com (P.F.-d.)
2 Posgrado en Ciencias Biológicas, Universidad Nacional Autónoma de México, Mexico City 04510, Mexico
3 Departamento de Infectología e Inmunología, Instituto Nacional de Perinatología, SSA, Mexico City 11000, Mexico; jesspinq@yahoo.com.mx
4 Departamento de Farmacología, Facultad de Medicina, Universidad Nacional Autónoma de México, Avenida Universidad No. 3000, Alcaldía de Coyoacán, Mexico City 04510, Mexico; edgarfloressoto@yahoo.com.mx
5 Unidad de Investigación Epidemiológica y en Servicios de Salud, Área Envejecimiento, Centro Médico Nacional Siglo XXI, Instituto Mexicano del Seguro Social, Mexico City 06720, Mexico; pgarciatorre@gmail.com
6 Laboratorio de Fitoquímica, UBIPRO, FES-Iztacala, Universidad Nacional Autónoma de México, Tlalnepantla de Baz 54090, Mexico; estreparr@iztacala.unam.mx
* Correspondence: jverjan@inger.gob.mx; Tel.:+52-(55)-73809087
† These authors contributed equally to this work.

Abstract: Aging is the result of the accumulation of a wide variety of molecular and cellular damages over time, meaning that "the more damage we accumulate, the higher the possibility to develop age-related diseases". Therefore, to reduce the incidence of such diseases and improve human health, it becomes important to find ways to combat such damage. In this sense, geroprotectors have been suggested as molecules that could slow down or prevent age-related diseases. On the other hand, nutraceuticals are another set of compounds that align with the need to prevent diseases and promote health since they are biologically active molecules (occurring naturally in food) that, apart from having a nutritional role, have preventive properties, such as antioxidant, anti-inflammatory and antitumoral, just to mention a few. Therefore, in the present review using the specialized databases Scopus and PubMed we collected information from articles published from 2010 to 2023 in order to describe the role of nutraceuticals during the aging process and, given their role in targeting the hallmarks of aging, we suggest that they are potential geroprotectors that could be consumed as part of our regular diet or administered additionally as nutritional supplements.

Keywords: nutraceuticals; geroprotectors; aging; hallmarks of aging; bioactive compounds

1. Introduction

Food is the source of energy required for all physiological functions and biological activities, as well as the main source of all essential compounds that cannot be naturally synthesized by our organism [1]. Moreover, the effect of food on our health depends on the balance of our diet and on how efficiently and effectively our organism utilizes nutrients (carbohydrates, lipids, proteins, vitamins and minerals) [2]. Interestingly, a healthy life and, consequently, a long lifespan requires balanced nutrition. In this sense, the availability, amount and quality of food varies depending on factors such as origins, composition and manufacturing process, among others.

Nutrients are traditionally classified according to the function they play in our organism. Therefore, nutrients that contribute energy, such as lipids, proteins and carbohydrates, are known as macronutrients. While trace amounts of vitamins, minerals and other organic compounds, which do not directly contribute to energy metabolism, are known as micronutrients or bioactive compounds [3]. In the past decades, micronutrients have been linked to overall well-being and to health benefits associated with disease prevention and treatment. These compounds were termed "nutraceuticals" by Stephen DeFelice in 1995 by joining the concept of pharmaceutical bioactivity with that of nutrients found in our foods [4]. Interestingly, such a term, consequently, has been cited indifferently in the literature, leading to a varied number of definitions that oftentimes are contradictory. Additionally, nutraceuticals' definition tends to be confused with the term "functional foods", which describes specific types of food that are characterized by a high content of bioactive compounds with relevant effects on well-being and health or in reducing the risk of diseases [5]. Interestingly, to date, there is no consensus nor a unique definition for "nutraceuticals" and/or "functional foods". For the current article, we use the term nutraceutical as biologically active molecules naturally occurring in food that, apart from having a nutritional role, provide health-promoting, disease-curing or prevention properties [6]. A nutraceutical must be understood as a single substance that may be isolated for clinical purposes or consumed as part of a specific food [7].

On the other hand, aging as a process affects different levels of the biological hierarchy and significantly affects molecular pathways that, once altered, are associated with several types of diseases, such as cardiovascular, neurodegenerative, cancer, metabolic disorders and many other syndromes [8]. In this context, current aging research focuses on developing strategies to slow down the detrimental effects of aging. In this sense, over the past years the term "geroprotector" has become significant as potential molecules that target the so-called Hallmarks of Aging [9], delaying the onset of age-related diseases and boosting resilience in older adults [10]. The so-called "geroprotectors" stand out, named as such by Illya Mechnikov, one of the fathers of gerontology, who defined them as an agent that allows protection against the effects of aging, thus increasing life expectancy and healthy life span [11].

Interestingly, some of the current geroprotector molecules have been discovered by repurposing previously approved drugs or already tested compounds (most of which are derived from natural sources consumed in the diet). Recently, the development of different-omics tools has allowed redefining the term by a unifying concept in a more formal way as that intervention that delays, reduces and/or prevents diseases associated with aging and that is characterized by having a simultaneous target of one or several of the pillars of aging [12]. Therefore, in the present review, we summarize the activities of nutraceuticals that target the molecular hallmarks of aging that could be consumed as part of our regular diet or administered additionally as nutritional supplements, suggesting the design of a potential "geroprotective diet" that could be defined as a diet that could reduce or prevent adverse outcomes of aging and that may contain a bioactive compound that targets one or several of the hallmarks of aging.

For the selection of the studies summarized in this review, we performed research by using the boolean operators AND/OR, NOT and Medical Subject Headings (MeSH) words, such as "nutraceutics", "nutraceuticals", "functional food", "aging", "geroprotectors" and each of the hallmarks of aging ("telomere attrition", "epigenetics", "loss of proteostasis", "disabled macroautophagy", "deregulated nutrient sensing", "mitochondrial dysfunction", "cellular senescence", "stem cell exhaustion", "altered intercellular communication", "chronic inflammation", "dysbiosis" and "genomic instability"). We only consider original articles published from 2010 to 2023 in Scopus and Pubmed databases and exclude papers that were in other languages different from English or without data published. A total of 69 articles were included to perform the following review, and the nutraceutical classification was performed according to [13].

2. Nutraceuticals

In essence, from a chemical perspective, classic nutraceutical compounds are considered small chemical structures naturally synthesized in the secondary metabolism of plants. This classification includes characteristic chemical structures, such as anthraquinones, alkaloids, saponins, tannins, essential oils, carotenoids, flavonoids and bitters [6]. However, in recent years, carbohydrates, protein and lipids have also been introduced as nutraceuticals together with other essential micronutrients. Lastly, probiotics, prebiotics and fungal extracts have also been partially considered nutraceuticals due not only to their direct effect as such, but mainly because of their ability to metabolize compounds into more bioactive molecules [14]. For instance, as depicted in Figure 1 most of the common nutraceuticals are found in daily foods.

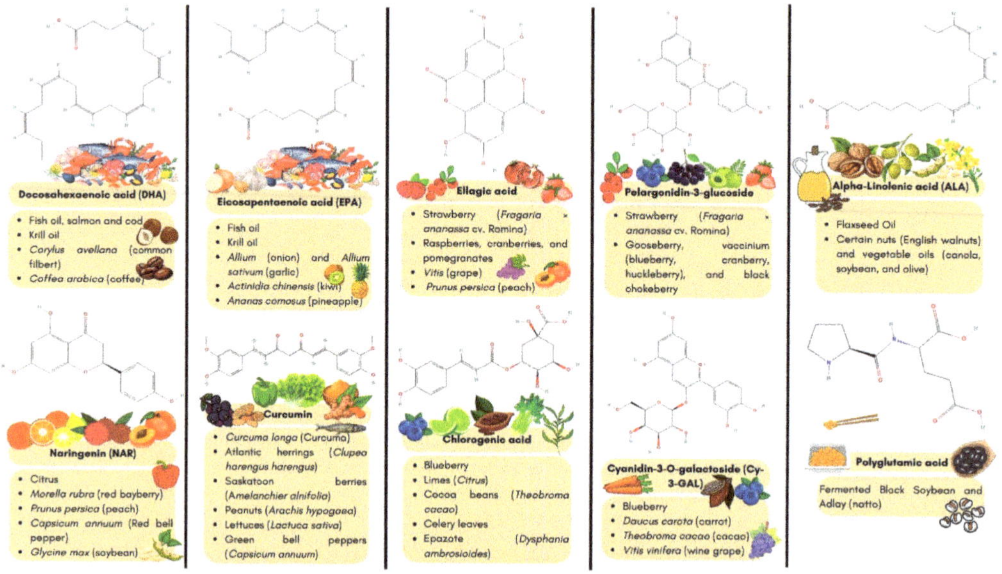

Figure 1. Examples of the enriched sources of the most common nutraceuticals. This figure depicted the most common nutraceuticals found in vegetables, roots, fruits, nuts and seafood. For instance, green bell peppers, lettuce, celery and epazote are enriched in curcumin and chlorogenic acid. However, red bell, peach, soybean, grapes, strawberries and other berries are enriched sources of naringenin, ellagic acid, pelargonidin-3-glucoside and cy-3-gal. Seafood, nuts and vegetable oils are sources enriched in fatty acids, such as DHA, EPA and ALA.

The field of nutraceuticals is still too young compared to those of pharmaceuticals, and there is still a lack of rigorous large-scale clinical trials. Nevertheless, over the last 10 years, there has been significant growth in scientific evidence that suggests that they could be helpful for the treatment of cardiovascular and metabolic diseases, cognitive function, skin physiology, immune function and digestive health [15]. In this sense, the nutraceuticals market is experiencing significant growth due to the increased focus on preventative healthcare, awareness about the benefits of dietary supplements and the growing geriatric population accompanied with rising age-related disease prevalence. It is estimated that the global nutraceutical market (functional food, dietary supplements and functional beverages) is about 554.7 billion USD and will grow to 905.8 billion USD by 2030 [16]. Accordingly, Table 1 provides a summary of selected studies on foods, bioactive compounds or nutraceuticals and their main targets and functions investigated in models of aging or aging-related diseases.

Table 1. General overview of experimental evidence on selected nutraceuticals with potential use in aging or age-related diseases.

Nutraceutical Source	Bioactive Compounds or Organisms	Nutraceutical Classification	Age-Related Target (Pharmacological or Biological Activities)	Refs.
Wheatgrass (*Triticum aestivum*)	Chlorophyll Flavonoids Vitamin C Vitamin E	Phytochemicals Antioxidants Vitamins	Decreases triglycerides in blood. Inhibits growth of leukemia cells. Benefits immunological activity. Decreases oxidative stress.	[17–19]
Aloe vera extract	Quercetin Myricetin Aloin Vanillic acid Palmitic acid Vitamin E Polysaccharides Phenolic compounds	Phytochemicals Antioxidants	Aloe is useful for photoaging since it stimulates fibroblast, which produces collagen and elastin fibers, making the skin more elastic and less wrinkled. Additionally, it inhibits the cyclooxygenase pathway and reduces prostaglandin E2 production from arachidonic acid. Quercetin, which exists in the outer layers of aloe leaf, has a cytoprotective effect on mitochondrial pathways by inhibiting oxidative stress.	[20–22]
Ginseng extract *Panax ginseng*	Ginsenoside C-K Oleanolic acid Ginsenoside Rg1-Rb1, Rd Oleanane Polysaccharides Peptides Phenolic compounds	Phytochemicals	Ginseng exhibited a remarkable antioxidant effect through the enhancement of the cell stress response, mainly by up-regulating heme oxygenase-1. In a rat model of high-fructose diet-induced metabolic disorder, fermented red ginseng reduced hyperlipidemia and hypertension. An aqueous extract of Korean red ginseng rapidly up-regulated endothelial NO synthase (eNOS) via the phosphoinositide 3-kinase (PI3K)/Akt-pathway in human umbilical vein endothelial cells (HUVEC).	[23–25]
Seaweed species *Hypnea musiformis, Ochtodes secundiramea, Padina gymnospora, Codium tomentosum, and Pterocladiella capillacea*	Fucoidan Fucoxanthin Phycoerythin Alginic acid Polysaccharides Carotenoids Taurine	Phytochemicals Amino acid	Seaweed is reported to ameliorate or prevent $A\beta_{25-35}$ aggregation and inhibit AChE and BuChE levels in vitro. MeOH extracts of seaweed *S. muticum* and *S. polyschides* exhibited the highest neuroprotective effects against dopamine-induced neurotoxicity in SH-SY5Y cells.	[26,27]

Table 1. *Cont.*

Nutraceutical Source	Bioactive Compounds or Organisms	Nutraceutical Classification	Age-Related Target (Pharmacological or Biological Activities)	Refs.
Echinacea purpurea extracts	Caffeic acid β-sitosterol Phenolic compounds	Phytochemicals	After 8 weeks of *Echinacea consumption*, a significant increase in NK cell cytotoxic activity was observed. Serum cytokine levels of IL-2, IFN-γ, and TNF-α also significantly increased. In vitro gastrointestinal digestion on the phenolic composition of Echinacea extracts showed significant reductions in IL-6, IL-8, and PGE2 levels in vitro.	[28,29]
Goji berry (*Lycium barbarum)* extract	L. barbarum polysaccharides (LBPs) Pectic polysaccharides Lycopene Beta-carotene Lutein Zeaxanthin Phenolic compounds Rutin	Phytochemicals Antioxidants	Improve mitochondrial function and decrease oxidative stress via Nrf2-Maf and NOS signaling pathways. Improve cognitive performance in aged rats by decreased astrogliosis.	[30]
Chiang-Da (*Gymnema inodorum*) leaf extracts	(3β, 16β)-16,28-dihydroxyolean-12-en-3-yl-*O*-β-D-glucopyranosyl-β-D-glucopyranosiduronic acid (GIA1)	Phytochemicals	Induces anti-hyperglycemic mechanisms by reducing α-glucosidase activity and glucose transport of SGLT.	[31]
Strawberry *(Fragaria x ananassa cv. Romina)* extracts	Ellagic acid Pelargonidin-3-glucoside (Phenolic compounds) K^+, Mg^+, P^+ and Ca^{2+} (Minerals)	Phytochemicals Essential trace elements	Induces DAF-16/FOXO and SKN-1/NRF2 pathways. Delay β-amyloid induced paralysis Reduced β-amyloid aggregation Prevents oxidative stress in *C. elegans*.	[32]
Fish hydrolysate	Eicosapentaenoic acid (EPA) Docosahexaenoic acid (DHA)	Fatty acids	Improved memory performance in aged mice. Regulates gut microbiota. Regulates corticosterone levels. Increased the expression of the mitochondrial respiratory chain (ND1, ND2, ND5, and ND6). Improving total skeletal muscle mass, muscle strength and physical performance in older adults.	[33,34]
Blueberry (*Vaccinium uliginosum* L.) extracts	Polyphenolic compounds Cyaniding-3-O-galactoside (Antocyanin) Pyruvic acid Chlorogenic acid	Phytochemicals	Promotes recovery from cell injury and improves survival of hippocampal pyramidal neurons. Increases antioxidant defenses via ERK signaling pathway in the hippocampus of a senescence-accelerated mouse model.	[35,36]

Table 1. *Cont.*

Nutraceutical Source	Bioactive Compounds or Organisms	Nutraceutical Classification	Age-Related Target (Pharmacological or Biological Activities)	Refs.
Tempeh (soybean fermentation)	Daidzein Genistein Polyphenols Low-molecular-weight soluble dietary fiber Tempeh isoflavone Peptides: Ala-Val, Gly-Leu, Gly-Phe, Pro-Leu, Ala-Phe, Asp-Met, Asp-Tyr, Pro-Ala-Pro, Ile-Ala-Lys, Arg-Ile-Tyr and Val-Ile-Lys-Pro.	Phytochemicals Dietary fiber Antioxidants Proteins and amino acids	Induces Anti-inflammatory and immunomodulatory components. Improve antioxidative activity and increase both SOD and CAT gene expression. Induces anti-hypertensive activity via ACE inhibitor peptide Induces neuroprotection and GABA synthesis in six-month-old senescence-accelerated mice.	[37–40]
Curcumin C3 complex	Polyphenolic orange-yellow pigments: curcumin, demethoxycurcuminbis-demethoxycurcumin	Phytochemicals	Decreased IL-6 concentration and gene expression. Prevents senescent cell accumulation. Improve antioxidant capacity. Upregulate TERT gene expression Increased telomere length in aged rats. Upregulate TERT gene expression Increased telomere length in aged rats (17 months old).	[41]
Blueberry (*Vaccinium uliginosum* L.) extract	Flavonoids (anthocyanidins) Polyphenols (procyanidin) Phenolic acid Pyruvic acid Chlorogenic acid			
Astragalus membranaceus	Astragaloside IV Kaempferol Quercetin Isorhamnetin Triterpene saponins			
Amelanchier ovalis berries ethanolic extract	Gallic acid p-hydroxybenzoic acid Protocatechinic acid	Phytochemicals	Promotes proliferation, lifespan and survival rate of *Saccharomyces cerevisiae* Y-564 exposed to oxidative stress.	[42]
Krill oil	Astaxanthin Choline Omega-3 DHA EPA	Phytochemicals Vitamin precursors Fatty acids	mTOR-p70s6k/Muscular strength and cognitive function The administration of krill oil to a mixed-sex aged C57BL/6 mouse model increased force production (increased grip strength, increased contraction and tetanic strength in the extensor digitorum longus muscle) without altering Ca^{2+} homeostasis in the excitation-contraction coupling mechanism or mitochondrial Ca^{2+} uptake processes.	[43]

Table 1. *Cont.*

Nutraceutical Source	Bioactive Compounds or Organisms	Nutraceutical Classification	Age-Related Target (Pharmacological or Biological Activities)	Refs.
Lycium ruthenicum Murr ethanolic extract	Anthocyanins Lycibarbar spermidine B N1-Dihydrocaffeoyl N10-trans-caffeoyl-spermidine)	Phytochemicals	Prevents oxidative damage by increasing SOD and glutathione peroxidase concentration in a murine model of accelerated aging induced by D-galactose.	[44]
Fermented Black Soybean and Adlay (FBA)	Nattokinase Polyglutamic acid Isoflavones	Proteins and amino acids Phytochemicals	Improves body composition in aged mice (increased gastrocnemius muscle and decreased fat accumulation). Interestingly, it reduced the expression of GLB1 and $p16^{INK4A}$ genes involved in senescence. Counteracts oxidative stress. Decrease inflammation markers MCP-1, IL-6 and IL-10 in aged mice. Improves aging-related gut microbial dysbiosis promoting the growth of beneficial microbes (Alistipes, Anaeroplasma, Coriobacteriaceae UCG002, and Parvibacter).	[45]
Soybean	Daidzein Genistein Glycitein Acetyldaidzin Acetylgenistin Acetylglycitin	Phytochemicals	Induces anti-photoaging in murine models exposed to UVB radiation.	[46]
Fermented milk	*Lactobacillus paracasei* *Lactobacillus plantarum*	Prebiotics or Probiotics	Improve symptoms associated with allergic rhinitis. Reduces airway hyperresponsiveness, asthma and systemic proinflammatory factors (IL-4, IL-5, and IL-3).	[47,48]

3. Nutraceuticals as Geroprotectors

According to Illya Mechnikov, geroprotectors are agents that protect against the adverse effects of aging, thus increasing life expectancy and healthy lifespan [11]. Moreover, the current advances in gerosciences help us to understand biological mechanisms that are closely associated with aging, also known as the Hallmarks of Aging (genomic instability, telomere attrition, epigenetic alterations, loss of proteostasis, deregulated nutrient sensing, mitochondrial dysfunction, cellular senescence, stem cell exhaustion and altered intercellular communication, disabled macroautophagy, chronic inflammation and dysbiosis) [9]. Each hallmark contributes to the aging process and together determines the aging phenotype while meeting a number of criteria: (i) it should manifest during normal aging; (ii) its experimental aggravation should accelerate aging; and (iii) its experimental amelioration should retard the normal aging process and, hence, increase healthy lifespan [49]. Altogether, with this information, we can redefine geroprotector in a more formal way as an intervention targeting simultaneously one or more of the hallmarks of aging delaying, reducing and/or preventing age-related diseases [12]. Hence, in the following section, we aim to suggest the potential properties of nutraceuticals as geroprotectors by summarizing the most outstanding studies where nutraceuticals target the hallmarks of aging.

3.1. Telomere Attrition

Telomeres are the caps at the ends of chromosomes that protect them from fraying and fusing with other chromosomes or becoming damaged by an external agent. In this sense, telomere attrition is an interesting hallmark of aging since telomeres become shorter in each cell division till it is so short that it generates a DNA damage signaling response. In this context, telomerase (a reverse transcriptase enzyme) adds bases to the ends of telomeres and is responsible for telomere length. In most adult cells, telomerase activity is low, leading to progressive telomere shortening with each division till telomeres grow shorter and DNA becomes damaged, conditioning the fate of cells to senescence or death [50]. Several studies show a correlation between shorter telomere length and increased risk of age-related diseases. Hence, it has been proposed that certain compounds, including polyphenols, triterpenes, sesquiterpenes, xanthones and alkaloids derived from natural products have the potential to modulate telomerase activity, suggesting their possible use as anti-aging agents [51].

It has been reported that a three-month administration of vitamin C, zinc and vitamin D3 results in a reduction in the rate of telomere shortening and an increase in telomere length, suggesting that these vitamins may promote telomerase activity [52]. It is noteworthy that the acid form at a low concentration of vitamin A (retinoic acid) reduces $p16^{INK4A}$ expression and induces telomerase activity, increasing the lifespan of human oral keratinocytes [53]. Moreover, both in vitro and in vivo studies, as well as epidemiological studies, demonstrate that the use of vitamin C, vitamin D and vitamin E are associated with individuals with longer telomere lengths than their counterparts [54].

Interestingly, a one-year randomized, double-blind, placebo-controlled study performed in older adults infected with cytomegalovirus and administered with TA-65 (a compound derived from the herb *Astragalus membranaceus* [55]) demonstrated a significant increase in telomere length [56]. Other studies demonstrated that polyphenols, such as curcumin and resveratrol, obtained from grapes, cherries and blueberries, among others, are being studied for their potential anti-aging effects [51]. For instance, the administration of resveratrol to aged C57Bl/6 mice over a one-year period resulted in a decrease in telomere attrition. Notably, this treatment has been observed to have a differential effect according to sex, restoring the length of telomeres significantly in brain cells of aged female mice but not in males [57]. Omega-3 fatty acids (EPA, DHA and polyunsaturated fatty acids (PUFAs)) have been demonstrated to possess anti-inflammatory properties, which are associated with the maintenance of telomere length [55]. A randomized pilot study of older adults with mild cognitive impairment who received supplementation with linolenic acid (LA), EPA or DHA for six months demonstrated that telomere shortening was attenuated. Furthermore, elevated levels of DHA in erythrocytes were associated with decreased telomere shortening [58].

Although telomerase is a promising therapeutic target, overexpression of the enzyme occurs in approximately 90% of cancers, which makes telomerase-related anti-aging therapies somewhat controversial due to potential adverse effects [59]. Therefore, the results of compounds interacting with telomerase should be taken with caution, and research overcoming such limitations are currently needed.

3.2. Epigenetics

Epigenetic changes influence the aging process in a number of ways, including reduced levels of core histone marks, changes in the patterns of histone post-translational modifications, DNA methylation and altered expression of non-coding RNAs (ncRNAs). These changes can trigger aberrant gene expression and genomic instability [60]. Nevertheless, epigenetic changes can also act transgenerationally, influencing the lifespan of the offspring, as demonstrated in studies performed in *C. elegans*, where deficiencies of any of the chromatin modifier components result in lifespan extension for up to three generations [61].

Given that epigenetic modifications are reversible, the use of nutraceuticals represents an attractive strategy for regulating epigenetically active enzymes or the epigenome itself [62]. Indeed, it has been demonstrated that polyphenols, flavonoids and organosulfur compounds in foods (vitamin A, vitamin C, vitamin E, curcumin and resveratrol) exert epi-nutraceutical effects, modifying DNA methylation patterns, histone modifications and regulating miRNA expression [63]. Although the exact effect mechanisms by which vitamins interact with the DNA methylation process is yet to be investigated, there are some experiments that suggest that vitamin A administered in vitro to human embryonic stem cells (hESCs), induced hypermethylation of most genes and a decrease in the H3K27me3 repressive mark [64]. Additionally, it has been reported that vitamin E enhances the expression of DNMT1 and LINE-1. On the other hand, vitamin C seems to activate TET hydroxylase enzymes. Similarly, curcumin has been shown to inhibit DNMT1 and DNMT3B, as well as to inhibit HATs, HDAC2 and HDAC8 [65]. Moreover, curcumin upregulates the expression of miR-15, -16, -9 and -181 and downregulates expression of miR-125b-5p, -19a, -19b, -27a and -130a [66]. Finally, resveratrol was also shown in vitro on HCC1806 breast cancer cells to inhibit DNMT enzymatic activity and to modulate HDAC activity in different models both in vivo and in vitro [67].

Retinoic acid is a micronutrient that modifies one-carbon metabolism; consequently, deficiencies in such micronutrients result in decreased DNA methylation due to the availability of methyl groups [68]. A study involving older adults reported that multivitamin supplementation with vitamins B3, C and D, omega-3 fish oils (EPA and DHA), resveratrol, olive phenols and astaxanthin for 12 weeks demonstrates that older adults with an initial epigenetic age acceleration of $\geq$2 years at baseline showed a significant reduction in both epigenetic age and its acceleration after the supplementation [69]. In addition, curcumin and resveratrol, along with other polyphenols, have been reported to activate sirtuin 1 (SIRT1), which deacetylates histones, leading to the regulation of various biological functions, including metabolism, cellular senescence, inflammatory processes and stress, as it deacetylates p53, forkhead transcription factor, NF-κB and the LX receptors [70,71].

3.3. *Loss of Proteostasis*

The coordinated action of proteostasis networks, including chaperones, the ubiquitin-proteasome system (UPS) and the lysosome-autophagy system, enables proteostasis by affecting the key processes of protein synthesis, translation regulation, protein folding and protein clearance [72,73]. These networks detect and rectify alterations in the proteome; with aging, the maintenance of proteostasis is compromised in cells and organs under resting and stress conditions [72].

Chaperone proteins, such as heat shock proteins (Hsp), can be modulated by certain nutraceuticals, including curcumin and proanthocyanidins present in cranberry extract [74]. An in vitro model of rat glioma C6 cells treated with curcumin (3–10 μM) and exposed to arsenite, cadmium chloride or heat (42 °C for 30 min) showed the synthesis of Hsp27 and Hsp70 proteins [75]. Interestingly, Hsp27 is responsible for protein degradation and controls apoptosis by regulating Akt activation, while Hsp70 has functions in protein folding and unfolding [76,77]. This mechanism of action of Hsp has been studied for the removal of peptide plaques associated with neurodegenerative diseases that occur in the aging population, such as Alzheimer's disease (AD) [76]. It is noteworthy that the proanthocyanidins present in cranberry extract promote proteostasis, in addition to improving the lifespan of *C. elegans* in an AD model. In this model, an IIS-dependent increase in HSF-1 activity has been observed, resulting in the reduction in β-amyloid (Aβ) peptide species and increased protein solubility in old worms [78].

The UPS can be activated through nutraceuticals, such as linolenic acid present in dairy products, and olein, which can be consumed through palm oil; these compounds can exert conformational changes that favor the entry of substrates through the proteolytic chamber [74,79]. Oleuropein, a compound derived from olives and olive oil, has been observed to stimulate proteasome activity in vitro on human embryonic fibroblasts; as a result, the lifespan of these cells was extended by the 20S proteasome alpha channels [80]. Similarly, resveratrol has been examined in AD transgenic mice (3xTg-AD) supplemented from two to ten months old with resveratrol, resulting in an increase in neprilysin, an enzyme responsible for amyloid degradation. Additionally, there was an increase in the levels of the central subunits of the 20S proteasome, as well as an increase in the concentration of Hsp70 and ubiquitinated proteins (improving proteostasis activity) [81]. The study of nutraceuticals that maintain or promote proteostasis is not only beneficial for the study of aging but also for the prevention and treatment of diseases related to age-related protein aggregation [74].

3.4. *Disabled Macroautophagy*

A key protective mechanism of aging is the cellular recycling process called autophagy. During autophagy, damaged cellular components are delivered to acidic vesicles called lysosomes, which secure the degradation and recycling of the components [82]. As we age, autophagy becomes less efficient and contributes to the aging process and the concomitant development of age-related diseases. Particularly, macroautophagy is induced in response to different stressors, such as nutrient or growth factor deprivation, hypoxia, damaged proteins and organelles and genotoxic stress, among others. This response is tightly regulated by a variety of signaling pathways, including mTOR and AMPK, to secure appropriate fine-tuning [83]. While research on manipulating autophagy for longevity benefits in humans is still ongoing, the potential of autophagy-based interventions for promoting healthy aging and preventing diseases is a very promising area of research.

In this sense, the use of quercetin and curcumin induces autophagy via downregulated mTOR, P53 and P21 protein expression and decreased the phosphorylation of AKT, mTOR and P70S6K, respectively, on a model of atherosclerosis and melanoma [84,85]. Studies showed that puerarin contained in *Pueraria lobata* (Kudzu) restored autophagy through activation of AMPK and significantly restored LC3B-II and decreased p62 protein content [86]. Moreover, berberine (located on red berries) increased cell viability and autophagy via up-regulation phosphorylation of the insulin receptor and its downstream signaling molecules AMPK, Akt and eNOS in a diabetes rat model [87]. On the other hand, curcumin exerts neuroprotective effects by attenuating autophagic activities through mediating the PI3K/Akt/mTOR pathway, while also suppressing an inflammatory reaction by regulating the TLR4/p38/MAPK pathway on Sprague Dawley male rats [88].

Additionally, retinoic acid has been shown to protect the liver from injury by promoting autophagy, which is dependent on Foxo3/p-Akt/Foxo1 signaling [89]. In this sense, FOXO1 and FOXO3 increase autophagic flux through core ATG gene expression [83]. The relationship between autophagy and apoptosis is represented by the interaction between the anti-apoptotic protein Bcl-2 and the autophagy protein Beclin 1. Interestingly, the use of nutraceutics, such as xanthohumol, baicalin, apigenin, tetrahydropalmatine and emodin, can regulate this complex Bcl-2/Beclin 1 in age-related diseases [90–95].

3.5. *Deregulated Nutrient Sensing*

The ability to sense and respond to fluctuations in environmental nutrient levels is fundamental for life. Different pathways that detect intracellular and extracellular levels of sugars, amino acids, lipids and surrogate metabolites are integrated and coordinated at the organismal level through hormonal signals [96]. Nutrient-sensing pathways are commonly deregulated in age-related pathologies, such as stroke, Alzheimer's disease, breast cancer and liver cancer. According to several authors, the major nutrient-sensing pathways are insulin and insulin/insulin-like growth factor-1 (IGF-1) signaling IIS, mechanistic target of rapamycin (mTOr), AMP-activated protein kinase (AMPK) and sirtuins (SIRTs).

It is noteworthy that in a *C. elegans* model treated with various concentrations of myoinositol (vitamin B8 found in legumes, beans, nuts and other seeds), lifespan increased with attenuation of the IIS pathway in an AKT- and DAF-16-dependent manner. The same study tested the effect of myoinositol on Hs68 cells, resulting in the inhibition of phosphoinositide 3-kinase (PI3K), down-regulation of PI3K expression and inhibition of AKT phosphorylation, which promoted DAF-16 activation; such a mechanism is associated with longevity [97].

In humans, it has been demonstrated that centenarians exhibit specific energetic and metabolic characteristics that contribute to their longevity. For instance, centenarians maintain normal glucose levels and insulin sensitivity [98]. This observation has been associated with a higher plasma proportion of IGF-I/IGFBP-3, indicating greater bioavailability of IGF-1, which means a more effective insulin action response in centenarians [99,100].

Another interesting target is mTOR, an evolutionarily conserved nutrient-sensing protein kinase that regulates growth and metabolism in all eukaryotic cells. In AD, the use of a selenium derivative, quercetin and curcumin downregulates Akt/mTOR signaling, induces autophagy and inhibits Aβ generation, respectively [101–103].

On the other side, AMPK has been identified as a longevity kinase, which can be activated by nutraceuticals, such as resveratrol, genistein, gallic acid and betaine [104]. Genistein (principally found in soybean products) has been tested with vascular smooth muscle cells, where an increase in the phosphorylation of LKB1 and AMPK induces autophagy through the negative regulation of mTOR. Furthermore, genistein exhibits a multimodal mechanism of action, exerting antioxidant, anti-inflammatory, autophagy and senescence-promoting effects; these properties render it a promising nutraceutical candidate in the field of geroscience [105].

The activation of SIRTs appears to be a key factor in increasing life expectancy. It has been demonstrated that certain compounds derived from foods, including resveratrol, fisetin and quercetin, can activate SIRT1 [106]. Indeed, in a murine model in which resveratrol was administered, it was observed that SIRT1 activation was associated with AMPK activation and increased levels of NAD(+) in skeletal muscle [107]. AMPK, in turn, contributes to the prolongation of longevity of IIS signaling, thus indicating that these pathways interact in an intricate manner during the process of aging [108].

3.6. Mitochondrial Dysfunction

Mitochondrial dysfunction during aging is characterized by a loss of efficiency in the electron transport chain and reductions in the synthesis of high-energy molecules, such as ATP [109]. Mitochondrial dysfunction leads to an increased release of reactive oxygen species (ROS) produced during oxidative phosphorylation, causing oxidative damage [110]. In this sense, several efforts to counteract oxidative stress using nutraceuticals have been performed. For instance, isoquercetin and emodin reduce ROS and MDA production while increasing SOD and CAT activity [111,112], while apigenin increased SOD and GSH-Px activities and decreased ROS and MDA levels in a macular degeneration model [113].

Moreover, in a recent systematic review, sodium nitrite, PUFA, nicotinamide riboside, urolithin A and whey protein powder improve mitochondrial function in terms of oxidative capacity, antioxidant capacity, mitochondrial volume, bioenergetic capacity and mitochondrial activity, including biogenesis and function [114]. Moreover, another interesting bioactive compound is the α-lipoic acid found in red meat, carrots, beets, spinach, broccoli and potatoes; it is a highly effective antioxidant that inhibits ROS production and improves mitochondrial ATP production [115]. Similarly, it has been reported in 20-month-old male Wistar rats that pea protein combined with inulin (a type of dietary fiber found in various plants) improves mitochondrial activity and biogenesis [116]. Another well-studied polyphenol found in wine is resveratrol, which improves mitochondrial energy metabolism via PGC1α and SIRT1 activation [117]. As well, quercetin, a flavonoid found in fruits and vegetables, has been reported to reduce the prevalence of age-related macular degeneration [118]. This effect could be associated with preventing oxidative stress by ROS scavenging and ameliorates mitochondrial dysfunction via the AMPK/SIRT1 signaling pathways, as has been shown with quercetin [118,119].

3.7. Cellular Senescence

Cellular senescence (CS) is an antagonist hallmark of aging defined as a state of irreversible growth arrest and exit from the cell cycle in response to different types of damage (DNA damage, mutations, telomere attrition, ROS and epigenetic disturbances, among others). Senescent cells can secrete a pathogenic senescence-associated secretory phenotype (SASP) that disrupts tissue homeostasis, resulting in loss of tissue repair and the induction of a characteristic inflammatory phenotype associated with aging called *inflammaging*; additionally, under normal conditions and development, SASP could lead to the regeneration signals of tissue [120].

Senescent cells accumulated during aging have been associated with different age-related diseases [121]. CS can be induced by several factors, such as multiple tissue dysfunction, cancer and other pathological conditions related to inflammation. The use of nutraceutics belonging to the flavonoid family (quercetin, fisetin and procyanidin C1) and to the sesquiterpene family (Cis-Nerolidol) have been shown to reduce the expression of genes associated with SASP factors and CCND1, CCNE1, CDK1 and CDK2, respectively [122–125]. It has been proven that the use of nutraceutical products, such as fisetin and procyanidin C1, have the ability to reduce SASP markers in multiple tissues [123,124]. In recent years, there has been an enormous interest in the development of the so-called senolytics (which is a class of molecules that selectively induce the death of senescent cells) [126]. Interestingly, among the most studied senolytic drugs are dasatinib, quercetin, fisetin and navitoclax. In particular, quercetin and fisetin have been isolated from onions, apples, grapes, berries, broccoli, citrus fruits, cherries, green tea, coffee, red wine, capers, strawberries, apples, grapes and cucumbers. Even so, recently, luteolin contained in peppers, parsley, celery and broccoli was implicated in the suppression of the SASP factor [127]. Moreover, the combination of dasatinib and quercetin has been shown to act as a potent senolytic that ameliorates numerous age-related disorders related to intestinal senescence and inflammation through SASP markers reduction [125].

3.8. Stem Cell Exhaustion

Stem cells have the potential to give rise to all tissue types, serving as a source of cells for repairing tissues in the organisms; however, as we become older, tissues experience a progressive decline in homeostatic and regenerative capacities that have been associated with degenerative deleterious alterations in systemic cues and niches that are involved in both the stem cell activity and quantity [128]. In this context, stem cells have a unique metabolism and nutrient needs as compared with other differentiated cell types.

Interestingly, it has been reported that both stem cell quantity and quality are influenced by diets and dietary patterns [129]. For instance, the intestinal stem cells (Lgr5+ and 4+) are stimulated by ketogenic diets leading to the generation of ketones in mitochondria by the 3-hydroxy-3-methylglutaryl-CoA synthase 2. However, the inhibition of this impairs stemness in Lgr5+ stem cells [130]. Interestingly, it has been reported that an oral nutritional supplement containing resveratrol, vitamin D3 and inositol hexaphosphate (Longevinex®, Las Vegas, NV, USA) maintains stem cell survival of retinal pigment epithelium in older adults. Additionally, authors report beneficial effects on structure and visual function in living human eyes [131].

3.9. Altered Intercellular Communication

Aging has been associated with progressive alterations in cellular communication that comprise cellular signaling, leading to cellular pathological states. There are several examples of such in different diseases, for instance, carcinogenesis, neurodegeneration and cardiovascular hypertrophy. In this regard, baicalin has a significant neuroprotective effect in stroke by increasing GABA(A)R $\alpha1$, GABA(A)R $\gamma2$ and KCC2 mRNA and protein levels, facilitating neurological function and suppressing ischemia-induced neuronal damage [132]. On the other hand, vitamin D has shown to be an interesting immunomodulatory molecule that could interact with immune cells [133]. Another interesting compound is epicatechin present in cocoa powder, which induces SOD activity, IGF-1 pathways and mitochondrial biogenesis [133,134]. Additionally, curcumin and ginsenosides induce neuroprotective effects by increasing the expression of both *Creb* and *Bdnf* genes and concomitantly leading to reduced apoptosis [135–137].

3.10. Chronic Inflammation (Inflammaging)

Most older individuals develop *"inflammaging"*, a condition characterized by elevated levels of blood inflammatory markers that carries high susceptibility to chronic morbidity, disability, frailty and premature death [138]. In recent years, increasing evidence suggests that nutrient interventions (natural or synthetic) have an influence in delaying and preventing *inflammaging*. For instance, resveratrol present in several nutraceutics, grapes or fruits, has been shown to inhibit NF-κB-regulated cytokines; data suggest that resveratrol possesses a significant effect against inflammation through the SIRT1 pathway [139]. Quercetin is a pentahydroxyflavone found on several fruits and vegetables, including onions, capers, apples, berries, tea, tomatoes, grapes, brassica vegetables and shallots. It is also found in various nuts, seeds, barks, flowers and leaves [140,141]. Quercetin has been shown to be an inhibitory compound of cyclooxygenase and lipoxygenase enzymes, both quite crucial in the mediation of prostaglandins and leukotrienes in inflammation [141–144]. Additionally, it was reported that quercetin could reduce the secretion of anti-inflammatory cytokines, such as IL-10, and pro-inflammatory cytokines, such as TNF-α, IL-1B and IL-6 [145]. Furthermore, we recently found that quercetin blocks airway smooth muscle contraction by inhibiting L-VDCC and SOCC [146]. Another interesting compound is epigallocatechin-3-gallate, which is present in several nutritional supplements and is the main component of green tea; such compounds have shown to exert anti-inflammatory effects since it can inhibit (PI3K)/Akt/mTOR pathway quite related to aging [147].

Other nutraceuticals from the families of bioactive compounds, such as flavonoids (baicalin and carthamin yellow), flavonols (icariin), flavones (apigenin and scutellarin), alkaloids (berberine), stilbenoids (resveratrol) and adenosine analogs (cordycepin), have been associated with a reduction in blood inflammatory markers (TNF-α, IL-1β, IL-6, IL-18, TGF-β1 and TGF-β2) in in vivo and in vitro models [148–154]. Moreover, the use of γ-tocotrienol repressed inflammasome activation, caspase-1 cleavage and interleukin (IL) 1β secretion in murine macrophages, implicating NLRP3 inflammasome inhibition, thereby delaying the progression of type 2 diabetes [155].

3.11. Dysbiosis

In recent years, many of the modern multifactorial diseases, such as neurodegenerative and metabolic disorders, have shown increasing evidence of an abnormal microbiome structure (dysbiosis), which affects the taxonomic composition, as well as the metagenomic function of the microbial community [156]. Moreover, studies on centenarians showed a youth-associated gut microbiome characterized by an over-representation of a Bacteroides-dominated enterotype [157] and a quite diverse gut virome, including a virus associated with Clostridia [158]. The gut microbiota can metabolize compounds from nutraceuticals to produce new absorbable small molecules, which have active effects. Also, nutraceuticals can regulate the composition of gut microbiota and its secretions, and such secretions may play a therapeutic role [159]. For instance, several flavonoids, polysaccharides and saponins can promote the growth of beneficial gut microbiota; a good example of such is the metabolism of ginsenosides, which are susceptible to Lactobacillus, Bacteroides, Bifidobacterium and their metabolic enzymes, which give rise to secondary ginsenosides with different pharmacological properties (antiangiogenic and anticancer) and are more easily absorbable in the circulation [159].

Another interesting result is the use of a mixture of quercetin (highly present in nutraceuticals and known to have senolytic potential) with dasatinib, which was shown to induce changes in the intestinal microbiota, specifically in the ileum of treated mice versus controls. These changes are associated with (1) a slightly lower Firmicutes-Bacteroidetes ratio, which correlated negatively with inflammatory and senescence markers in all the intestinal sections and (2) increased Akkermansia abundance, which has been linked with reduced intestinal permeability and gut-to-blood leakage of endotoxins, thereby alleviating diet-induced obesity and insulin resistance [125].

3.12. Genomic Instability

Genomic instability is a strong contender as a major cause of aging. The accumulation of DNA damage and mutations within a cell's genome over time causes a remarkable instability in the DNA sequence, leading to indels, chromosomal abnormalities, substitutions, breaks, gaps and aberrant 3-D structures, among others. Interestingly, genomic instability is associated with all the hallmarks of aging, since in the end all lead to DNA damage, such as ROS or RNS. On the other side, genomic instability has been associated with the development of several age-related diseases, such as cancer or neurodegeneration, as well as with mortality [160]; therefore, such a process is quite important in geroprotectors development. In this context, it has been suggested that exogenous antioxidants derived from food, such as fruits, vegetables, cereals, mushrooms, nuts and spices, aid in ameliorating the DNA damage caused by oxidative stress triggered by aging per se or by the exposure to xenobiotics, such as of cigarette smoking, alcohol, radiation or environmental toxins [161]. For instance, anthocyanins from grape juice decrease the oxidative damage in cells [162]. Also, polyphenols and melanoidins obtained from coffee brews (caffeoylquinic acids, feruloylquinic acids, dicaffeoylquinic acids and p-coumaroylquinic acids) augmented active oxygen-scavenging activity [163]. Moreover, flavonoids are a set of compounds contained in several fruits with quite powerful antioxidant properties useful against ROS. On the other hand, resveratrol isolated from *Polygonum cuspidatum* but also contained in several nutraceuticals stimulates SIRT2 activity, increasing DNA stability in yeast of *S. cerevisiae* [164]. Another interesting set of compounds contained in nutraceuticals are curcumins derived from *Curcuma longa.* They decrease lipid peroxidation and β-carotene inducing quenching activities associated with protecting biological systems from ROS-mediated damage [165].

4. State of the Art on Nutraceuticals Research

As we describe in Table 2, it has been reported that nutraceuticals target at least one of the twelve hallmarks of aging. In many cases, such analyses were performed experimentally; however, this represents high costs, and depending on the experimental model, the results are quite heterogeneous, representing a limitation for translational medicine. In this sense, computational techniques (chemoinformatics), such as network pharmacology, molecular modeling, quantitative or qualitative "structure-activity" relationships at different quantic levels (QSAR 2-D, 3-D, 4-D) and artificial intelligence methods, have been widely used in the discovery of potential nutraceuticals that may be applied as geroprotectors [166]. Such approaches lead to overpassing the limitations of experimental strategies and concomitantly accelerating and optimizing drug discovery. Moreover, in recent years the field of Artificial Intelligence technologies, including machine-learning methods, such as deep-learning, SVM and cognitive computing, have stimulated the process of chemical data processing obtained from omics data and clinical trials, leading to obtaining novel solutions to achieve healthy aging or to repurpose compounds to be used against age-related diseases. Although such a topic is out of the scope of the present review, we encourage our readers to refer to [167,168].

Table 2. Nutraceuticals and the hallmarks of aging. Of particular interest are bioactive compounds contained in nutraceuticals, such as vitamins A, D, and E, as well as curcumin, resveratrol, quercetin, apigenin, genistein, and fatty acids such as EPA and DHA, which have been shown to act on at least three hallmarks of aging, indicating potential as geroprotectors. Telomere Attrition (T.A.), Epigenetic Mechanisms (EP.), Loss of Proteostasis (L.P.), Disabled Macroautophagy (D.M.), Deregulated Nutrient-Sensing (D.N.S.), Mitochondrial dysfunction (M.D.), Cellular Senescence (C.S.), Stem Cell Exhaustion (S.C.E.), Altered Intercellular Communication (A.I.C.), Chronic Inflammation (C.I.), Dysbiosis (Dys.) and Genomic Instability (G.I.). (✔) Targeted hallmarks by nutraceuticals.

Nutraceutical Source	Bioactive Compounds	Hallmarks of Aging												Refs.
		T.A.	EP.	L.P.	D.M.	D.N.S.	M.D.	C.S.	S.C.E.	A.I.C.	C.I.	Dys.	G.I.	
Dairy products, Cod liver oil, Fish oil, Beef liver, Carrot seed oil, Palm fruit oil	Vitamin A (retinoic acid)	✔	✔		✔									[53,68,89]
Wheatgrass, Acerola cherry extract, Rosehip extract, Camu camu extract, Sea buckthorn oil	Vitamin C	✔	✔											[52,54,63,69]
Cod liver oil, Fish oil, Lanolin	Vitamin D	✔	✔						✔	✔				[52,54,69,131,133]
Wheat seed oil, Sunflower oil, Almond oil, Soybean oil, Acai berry extract	Vitamin E	✔	✔								✔			[54,63,155]
Turmeric extract, Curcumin C3 complex,	Curcumin	✔	✔	✔	✔	✔				✔			✔	[51,63,70,74,75,85, 88,102,135]
Grape seed extract Red wine extract Blueberry extract Cranberry extract Peanut extract	Resveratrol	✔	✔	✔		✔	✔		✔		✔		✔	[51,57,63,69,70,81, 104,106,107,117, 131,139,164]
Aloe vera extract Quercetin supplements Multi-antioxidant formulas Onion extract Apple extract Broccoli extract	Quercetin				✔	✔	✔	✔			✔	✔		[84,103,106,118, 125,127,145,169]
Fish oil Krill oil Seaweed oil	EPA	✔	✔				✔							[55,58,69,114]

Table 2. *Cont.*

Nutraceutical Source	Bioactive Compounds	Hallmarks of Aging												Refs.
		T.A.	EP.	L.P.	D.M.	D.N.S.	M.D.	C.S.	S.C.E.	A.I.C.	C.I.	Dys.	G.I.	
Fish oil Krill oil Seaweed oil	DHA	✔	✔				✔							[55,58,69,114]
Sunflower oil Corn oil Soybean oil Grape seed oil Hemp seed oil	Linoleic Acid	✔		✔										[58,74]
Chamomile extract Parsley extract Celery seed extract Citrus bioflavonoid complex	Apigenin				✔		✔				✔			[93,113]
Tempeh Red clover extract Soybean Soy isoflavone supplements	Genistein				✔	✔			✔		✔			[104,105]
Amelanchier ovalis berries ethanolic extract Pomegranate extract Green tea extract Grape seed extract Acai berry extract	Gallic acid					✔								[104]
Wheat germ Quinoa	Betaine					✔								[104]
Grape seed extract Pine bark extract Cocoa extracct	Procyanidin C1							✔						[123]
Carrots, peppers, thyme, broccoli, onion leaves, cabbages, apple skins, rosemary, parsley, and spinach	Luteolin							✔						[127]

Table 2. *Cont.*

Nutraceutical Source	Bioactive Compounds	Hallmarks of Aging												Refs.
		T.A.	EP.	L.P.	D.M.	D.N.S.	M.D.	C.S.	S.C.E.	A.I.C.	C.I.	Dys.	G.I.	
Krill oil Seaweed-based supplemnts	Astaxanthin		✔											[69]
Olive leaf extract Olive oil	Oleuropein			✔										[80]
Scutellaria baicalensis root extract	Baicalin				✔					✔	✔			[94,132,149]
Multivitamin/mineral supplements α-lipoic acid supplements	α-lipoic acid						✔							[115]
Inulin supplements Prebiotic supplements Probiotic and prebiotic combination Chicory root extract	Inulin						✔							[116]
Cocoa extract Dark Chocolate Green tea extract	Epicatechin									✔				[134]
Soy isoflavone supplements Soy-based products	Daidzein	✔	✔				✔			✔	✔			[170–173]
Seaweed oil Tomato extract Palm fruit oil Carrot seed oil Mixed carotenoid supplements	Carotenoids	✔			✔		✔	✔			✔		✔	[174–177]
Brown algae extract Seaweed-based supplements	Fucoxanthin				✔						✔			[178–180]
Brown algae extract Seaweed-based supplements	Fucoidan				✔			✔		✔	✔			[181–185]

Table 2. *Cont.*

Nutraceutical Source	Bioactive Compounds	Hallmarks of Aging												Refs.
		T.A.	EP.	L.P.	D.M.	D.N.S.	M.D.	C.S.	S.C.E.	A.I.C.	C.I.	Dys.	G.I.	
Ginseng root extract Ginseng containing supplements	Ginsenosides C-K				✔						✔			[186–188]
Korean Red Ginseng extract American Ginseng extract	Ginsenosides Rg1-Rb1, Rd	✔					✔	✔	✔		✔			[189–193]
Aloe vera gel *Aloe vera* extract	Aloin		✔	✔	✔				✔	✔	✔			[194–199]

5. Concluding Remarks State of the Art on Nutraceuticals Research

The concept of food as medicine has been around for centuries, with Hippocrates famously stating, "*Let food be thy medicine and medicine be thy food*". As we age, our bodies become more susceptible to disease and decline in function. This has led to increased interest in strategies like consuming nutraceuticals to promote healthy aging. This review explores the potential of nutraceuticals as geroprotectors, substances that may delay or prevent age-related diseases. We highlight that not all nutraceuticals qualify as geroprotectors. Only those that target multiple hallmarks of aging, as defined by criteria established by researchers like Moskalev et al., have this potential. Vitamin D, curcumin, resveratrol, quercetin, genistein, gallic acid, baicalin, lipoic acid and epicatechin are already reported as geroprotectors at the geroproctector.org [200]. In this context, several nutraceuticals, including vitamins A, C and E, EPA, DHA, daidzein and aloin, among others, may be considered promising geroprotectors that require further research, including clinical trials and studies investigating the molecular mechanisms of these nutraceuticals. Moreover, understanding their optimal dosage and long-term safety in humans is also crucial. Future studies should also explore the potential benefits of combining different nutraceuticals and how they interact with lifestyle factors like exercise. By continuing research in this field, we may develop strategies to improve health span and promote longevity.

Author Contributions: N.A.R.-S.: Conceptualization, Investigation, Writing—Original Draft, Writing Review and Editing; Supervision and Project administration. E.A.Z.-A.: Investigation, Data Curation, Writing—Original Draft and Editing and Visualization. S.K.C.-V.: Investigation, Writing—Original Draft, Writing Review and Editing; P.F.-d.: Investigation, Writing—Original Draft. J.H.-P.: Conceptualization, Investigation, Writing—Original Draft. E.F.-S.: Conceptualization, Investigation, Writing—Original Draft. P.G.-d.: Conceptualization. E.A.E.-P.: Conceptualization J.C.G.-V.: Validation; Formal analysis, Writing—Review and Editing and Visualization. All authors have read and agreed to the published version of the manuscript.

Funding: This study was supported by the Consejo Nacional de Humanidades, Ciencias y Tecnologías, México; Project 319706. The publication of this paper was conducted by the Instituto Nacional de Geriatría, México, and financially supported by the Consejo Nacional de Humanidades, Ciencias y Tecnologías, México; Project 319706 "¿Se pueden revertir el envejecimiento con fármacos?".

Acknowledgments: This study was part of a registered project at the Instituto Nacional de Geriatría DI-PI-008-2021 (JCGV). This paper serves as a fulfillment of CVU: 1337238 for obtaining an M.Sc. degree in the Posgrado en Ciencias Biológicas UNAM (SKCV). We thank the Consejo Nacional de Humanidades, Ciencias y Tecnologías (CONAHCYT) for funding and for the support of this research through the project 319706 "¿Se pueden revertir el envejecimiento con fármacos?".

Conflicts of Interest: The authors declare no conflicts of interest.

References

1. Lean, M.E.J. Principles of Human Nutrition. *Medicine* **2015**, *43*, 61–65. [CrossRef]
2. Longo, V.D.; Anderson, R.M. Nutrition, Longevity and Disease: From Molecular Mechanisms to Interventions. *Cell* **2022**, *185*, 1455–1470. [CrossRef] [PubMed]
3. Combet, E.; Buckton, C. Micronutrient Deficiencies, Vitamin Pills and Nutritional Supplements. *Medicine* **2015**, *43*, 66–72. [CrossRef]
4. DeFelice, S.L. The Nutraceutical Revolution: Its Impact on Food Industry R&D. *Trends Food Sci. Technol.* **1995**, *6*, 59–61.
5. Aronson, J.K. Defining "Nutraceuticals": Neither Nutritious nor Pharmaceutical. *Br. J. Clin. Pharmacol.* **2017**, *83*, 8–19. [CrossRef]
6. Sachdeva, V.; Roy, A.; Bharadvaja, N. Current Prospects of Nutraceuticals: A Review. *Curr. Pharm. Biotechnol.* **2020**, *21*, 884–896. [CrossRef]
7. Nicoletti, M. Nutraceuticals and Botanicals: Overview and Perspectives. *Int. J. Food Sci. Nutr.* **2012**, *63* (Suppl. S1), 2–6. [CrossRef]
8. da Costa, J.P.; Vitorino, R.; Silva, G.M.; Vogel, C.; Duarte, A.C.; Rocha-Santos, T. A Synopsis on Aging-Theories, Mechanisms and Future Prospects. *Ageing Res. Rev.* **2016**, *29*, 90–112. [CrossRef]
9. López-Otín, C.; Blasco, M.A.; Partridge, L.; Serrano, M.; Kroemer, G. Hallmarks of Aging: An Expanding Universe. *Cell* **2023**, *186*, 243–278. [CrossRef]
10. Trendelenburg, A.U.; Scheuren, A.C.; Potter, P.; Müller, R.; Bellantuono, I. Geroprotectors: A Role in the Treatment of Frailty. *Mech. Ageing Dev.* **2019**, *180*, 11–20. [CrossRef]

11. Metchnikoff, E. *The Prolongation of Life: Optimistic Studies*; William Heinemann Ltd.: London, UK, 1910.
12. Moskalev, A.; Chernyagina, E.; Kudryavtseva, A.; Shaposhnikov, M. Geroprotectors: A Unified Concept and Screening Approaches. *Aging Dis.* **2017**, *8*, 354–363. [CrossRef] [PubMed]
13. Hodjat, M.; Khalid, M.; Asghari, M.; Atri, S.; Rahimifard, M.; Nejad, S.M.; Baeeri, M. Nutrients and Nutraceuticals in Aging. In *Nutrients and Nutraceuticals for Active & Healthy Ageing*; Springer: Singapore, 2020; pp. 63–109, ISBN 9789811535512.
14. Shokryazdan, P.; Faseleh Jahromi, M.; Liang, J.B.; Ho, Y.W. Probiotics: From Isolation to Application. *J. Am. Coll. Nutr.* **2017**, *36*, 666–676. [CrossRef] [PubMed]
15. Puri, V.; Nagpal, M.; Singh, I.; Singh, M.; Dhingra, G.A.; Huanbutta, K.; Dheer, D.; Sharma, A.; Sangnim, T. A Comprehensive Review on Nutraceuticals: Therapy Support and Formulation Challenges. *Nutrients* **2022**, *14*, 4637. [CrossRef]
16. Nutraceutical Industry 2024. Available online: https://www.reportlinker.com/market-report/Nutraceutical/11559/Nutraceutical?term=nutraceutical%20industry&matchtype=b&loc_interest=&loc_physical=9047091&utm_group=standard&utm_term=nutraceutical%20industry&utm_campaign=ppc&utm_source=google_ads&utm_medium=paid_ads&utm_content=transactionnel-4&hsa_acc=7082072004&hsa_cam=15072279998&hsa_grp=129918785895&hsa_ad=559945523670&hsa_src=g&hsa_tgt=kwd-1461105442248&hsa_kw=nutraceutical%20industry&hsa_mt=b&hsa_net=adwords&hsa_ver=3&gad_source=1&gclid=Cj0KCQjwh7K1BhCZARIsAKOrVqECD74ko4F-SUwa-XYfwmAbyFK14yPv6EIMRXEIuA377ovdzlPiEtkaAgNtEALw_wcB (accessed on 2 August 2024).
17. Afroz, R.D.; Nurunabbi, A.S.M.; Khan, M.I. Effects of Wheatgrass (*Triticum aestivum*) Juice on Serum Cholesterol of Experimentally Induced Hypercholesterolaemic Male Long Evans Rat. *Banglad. J. Physiol. Pharmacol.* **2014**, *27*, 21–27. [CrossRef]
18. Bagwe, S.M.; Kale, P.P.; Bhatt, L.K.; Prabhavalkar, K.S. Herbal Approach in the Treatment of Pancytopenia. *J. Complement. Integr. Med.* **2017**, *14*, 20160053. [CrossRef] [PubMed]
19. Bar-Sela, G.; Cohen, M.; Ben-Arye, E.; Epelbaum, R. The Medical Use of Wheatgrass: Review of the Gap Between Basic and Clinical Applications. *Mini Rev. Med. Chem.* **2015**, *15*, 1002–1010. [CrossRef]
20. Surjushe, A.; Vasani, R.; Saple, D.G. Aloe Vera: A Short Review. *Indian J. Dermatol.* **2008**, *53*, 163–166. [CrossRef]
21. Cho, S.; Lee, S.; Lee, M.-J.; Lee, D.H.; Won, C.-H.; Kim, S.M.; Chung, J.H. Dietary Aloe Vera Supplementation Improves Facial Wrinkles and Elasticity and It Increases the Type I Procollagen Gene Expression in Human Skin in Vivo. *Ann. Dermatol.* **2009**, *21*, 6–11. [CrossRef]
22. Lee, H.; Choi, W.; Ro, H.; Kim, G.; Lee, H. Skin Antiaging Effects of the Fermented Outer Layers of Leaf Skin of Aloe Barbadensis Miller Associated with the Enhancement of Mitochondrial Activities of UVb-Irradiated Human Skin Fibroblasts. *Appl. Sci.* **2021**, *11*, 5660. [CrossRef]
23. Mancuso, C.; Santangelo, R. Panax Ginseng and Panax Quinquefolius: From Pharmacology to Toxicology. *Food Chem. Toxicol.* **2017**, *107*, 362–372. [CrossRef]
24. Kim, Y.-M.; Namkoong, S.; Yun, Y.-G.; Hong, H.-D.; Lee, Y.-C.; Ha, K.-S.; Lee, H.; Kwon, H.J.; Kwon, Y.-G.; Kim, Y.-M. Water Extract of Korean Red Ginseng Stimulates Angiogenesis by Activating the PI3K/Akt-Dependent ERK1/2 and eNOS Pathways in Human Umbilical Vein Endothelial Cells. *Biol. Pharm. Bull.* **2007**, *30*, 1674–1679. [CrossRef]
25. Dattilo, S.; Mancuso, C.; Koverech, G.; Di Mauro, P.; Ontario, M.L.; Petralia, C.C.; Petralia, A.; Maiolino, L.; Serra, A.; Calabrese, E.J.; et al. Heat Shock Proteins and Hormesis in the Diagnosis and Treatment of Neurodegenerative Diseases. *Immun. Ageing* **2015**, *12*, 20. [CrossRef] [PubMed]
26. Jannat, K.; Balakrishnan, R.; Han, J.-H.; Yu, Y.-J.; Kim, G.-W.; Choi, D.-K. The Neuropharmacological Evaluation of Seaweed: A Potential Therapeutic Source. *Cells* **2023**, *12*, 2652. [CrossRef] [PubMed]
27. Silva, J.; Alves, C.; Pinteus, S.; Mendes, S.; Pedrosa, R. Seaweeds' Neuroprotective Potential Set in Vitro on a Human Cellular Stress Model. *Mol. Cell. Biochem.* **2020**, *473*, 229–238. [CrossRef] [PubMed]
28. Ávila-Gálvez, M.Á.; Giménez-Bastida, J.A.; Karadeniz, B.; Romero-Reyes, S.; Espín, J.C.; Pelvan, E.; González-Sarrías, A. Polyphenolic Characterization and Anti-Inflammatory Effect of In Vitro Digested Extracts of L. Plant Parts in an Inflammatory Model of Human Colon Cells. *Int. J. Mol. Sci.* **2024**, *25*, 1744. [CrossRef]
29. Lee, S.-K.; Lee, D.-R.; Kim, H.-L.; Choi, B.-K.; Kwon, K.-B. A Randomized, Double-Blind, Placebo-Controlled Study on Immune Improvement Effects of Ethanolic Extract of *Echinacea purpurea* (L.) Moench in Korean Adults. *Phytother. Res.* **2024**, *38*, 3645–3659. [CrossRef]
30. Ruíz-Salinas, A.K.; Vázquez-Roque, R.A.; Díaz, A.; Pulido, G.; Treviño, S.; Floran, B.; Flores, G. The Treatment of Goji Berry (*Lycium barbarum*) Improves the Neuroplasticity of the Prefrontal Cortex and Hippocampus in Aged Rats. *J. Nutr. Biochem.* **2020**, *83*. [CrossRef]
31. Srinuanchai, W.; Nooin, R.; Pitchakarn, P.; Karinchai, J.; Suttisansanee, U.; Chansriniyom, C.; Jarussophon, S.; Temviriyanukul, P.; Nuchuchua, O. Inhibitory Effects of *Gymnema inodorum* (Lour.) Decne Leaf Extracts and Its Triterpene Saponin on Carbohydrate Digestion and Intestinal Glucose Absorption. *J. Ethnopharmacol.* **2021**, *266*, 113398. [CrossRef]
32. Navarro-Hortal, M.D.; Romero-Márquez, J.M.; Esteban-Muñoz, A.; Sánchez-González, C.; Rivas-García, L.; Llopis, J.; Cianciosi, D.; Giampieri, F.; Sumalla-Cano, S.; Battino, M.; et al. Strawberry (Fragaria × Ananassa Cv. Romina) Methanolic Extract Attenuates Alzheimer's Beta Amyloid Production and Oxidative Stress by SKN-1/NRF and DAF-16/FOXO Mediated Mechanisms in C. Elegans. *Food Chem.* **2022**, *372*, 131272. [CrossRef]

33. Mougin, C.; Chataigner, M.; Lucas, C.; Leyrolle, Q.; Pallet, V.; Layé, S.; Bouvret, E.; Dinel, A.-L.; Joffre, C. Dietary Marine Hydrolysate Improves Memory Performance and Social Behavior through Gut Microbiota Remodeling during Aging. *Foods* **2023**, *12*, 4199. [CrossRef]
34. Xu, D.; Lu, Y.; Yang, X.; Pan, D.; Wang, Y.; Yin, S.; Wang, S.; Sun, G. Effects of Fish Oil-Derived N-3 Polyunsaturated Fatty Acid on Body Composition, Muscle Strength and Physical Performance in Older People: A Secondary Analysis of a Randomised, Double-Blind, Placebo-Controlled Trial. *Age Ageing* **2022**, *51*, afac274. [CrossRef] [PubMed]
35. Tan, L.; Yang, H.P.; Pang, W.; Lu, H.; Hu, Y.D.; Li, J.; Lu, S.J.; Zhang, W.Q.; Jiang, Y.G. Cyanidin-3-O-Galactoside and Blueberry Extracts Supplementation Improves Spatial Memory and Regulates Hippocampal ERK Expression in Senescence-Accelerated Mice. *Biomed. Environ. Sci.* **2014**, *27*, 186–196.
36. Boespflug, E.L.; Eliassen, J.C.; Dudley, J.A.; Shidler; Kalt, W.; Summer, S.S.; Stein, A.L.; Stover, A.N.; Krikorian, R. Enhanced Neural Activation with Blueberry Supplementation in Mild Cognitive Impairment. *Nutr. Neurosci.* **2018**, *21*, 297–305. [CrossRef] [PubMed]
37. Chan, Y.-C.; Lee, I.-T.; Wang, M.-F.; Yeh, W.-C.; Liang, B.-C. Tempeh Attenuates Cognitive Deficit, Antioxidant Imbalance, and Amyloid β of Senescence-Accelerated Mice by Modulating Nrf2 Expression via MAPK Pathway. *J. Funct. Foods* **2018**, *50*, 112–119. [CrossRef]
38. Qiao, Y.; Zhang, K.; Zhang, Z.; Zhang, C.; Sun, Y.; Feng, Z. Fermented Soybean Foods: A Review of Their Functional Components, Mechanism of Action and Factors Influencing Their Health Benefits. *Food Res. Int.* **2022**, *158*, 111575. [CrossRef]
39. Das, G.; Paramithiotis, S.; Sundaram Sivamaruthi, B.; Wijaya, C.H.; Suharta, S.; Sanlier, N.; Shin, H.-S.; Patra, J.K. Traditional Fermented Foods with Anti-Aging Effect: A Concentric Review. *Food Res. Int.* **2020**, *134*, 109269. [CrossRef] [PubMed]
40. Tamam, B.; Syah, D.; Suhartono, M.T.; Kusuma, W.A.; Tachibana, S.; Lioe, H.N. Proteomic Study of Bioactive Peptides from Tempe. *J. Biosci. Bioeng.* **2019**, *128*, 241–248. [CrossRef]
41. Selim, A.M.; Nooh, M.M.; El-Sawalhi, M.M.; Ismail, N.A. Amelioration of Age-Related Alterations in Rat Liver: Effects of Curcumin C3 Complex, Astragalus Membranaceus and Blueberry. *Exp. Gerontol.* **2020**, *137*, 110982. [CrossRef]
42. Asyakina, L.; Atuchin, V.; Drozdova, M.; Kozlova, O.; Prosekov, A. Ex Vivo and In Vitro Antiaging and Antioxidant Extract Activity of the Amelanchier Ovalis from Siberia. *Int. J. Mol. Sci.* **2022**, *23*, 15156. [CrossRef]
43. Singlár, Z.; Szentesi, P.; Fodor, J.; Angyal, Á.; Csernoch, L.; Sztretye, M. Assessing the Potential of Nutraceuticals as Geroprotectors on Muscle Performance and Cognition in Aging Mice. *Antioxidants* **2021**, *10*, 1415. [CrossRef]
44. Cui, B.; Liu, L.; Shi, T.; Yin, M.; Feng, X.; Shan, Y. The Ethanolic Extract of Lycium Ruthenicum Ameliorates Age-Related Physiological Damage in Mice. *Molecules* **2023**, *28*, 7615. [CrossRef]
45. Koh, Y.C.; Kuo, L.H.; Chang, Y.Y.; Tung, Y.C.; Lo, Y.C.; Pan, M.H. Modulatory Effect of Fermented Black Soybean and Adlay on Gut Microbiota Contributes to Healthy Aging. *Mol. Nutr. Food Res.* **2023**, *67*, 2200700. [CrossRef] [PubMed]
46. Huang, C.-C.; Hsu, B.-Y.; Wu, N.-L.; Tsui, W.-H.; Lin, T.-J.; Su, C.-C.; Hung, C.-F. Anti-Photoaging Effects of Soy Isoflavone Extract (aglycone and Acetylglucoside Form) from Soybean Cake. *Int. J. Mol. Sci.* **2010**, *11*, 4782–4795. [CrossRef] [PubMed]
47. Ishida, Y.; Nakamura, F.; Kanzato, H.; Sawada, D.; Hirata, H.; Nishimura, A.; Kajimoto, O.; Fujiwara, S. Clinical Effects of Lactobacillus Acidophilus Strain L-92 on Perennial Allergic Rhinitis: A Double-Blind, Placebo-Controlled Study. *J. Dairy Sci.* **2005**, *88*, 527–533. [CrossRef] [PubMed]
48. Liu, Y.-W.; Liao, T.-W.; Chen, Y.-H.; Chiang, Y.-C.; Tsai, Y.-C. Oral Administration of Heat-Inactivated Lactobacillus Plantarum K37 Modulated Airway Hyperresponsiveness in Ovalbumin-Sensitized BALB/c Mice. *PLoS ONE* **2014**, *9*, e100105. [CrossRef]
49. López-Otín, C.; Blasco, M.A.; Partridge, L.; Serrano, M.; Kroemer, G. The Hallmarks of Aging. *Cell* **2013**, *153*, 1194–1217. [CrossRef]
50. Ganesan, K.; Xu, B. Telomerase Inhibitors from Natural Products and Their Anticancer Potential. *Int. J. Mol. Sci.* **2017**, *19*, 13. [CrossRef]
51. Jacczak, B.; Rubiś, B.; Totoń, E. Potential of Naturally Derived Compounds in Telomerase and Telomere Modulation in Skin Senescence and Aging. *Int. J. Mol. Sci.* **2021**, *22*, 6381. [CrossRef]
52. Tsatsakis, A.; Renieri, E.; Tsoukalas, D.; Buga, A.M.; Sarandi, E.; Vakonaki, E.; Fragkiadaki, P.; Alegakis, A.; Nikitovic, D.; Calina, D.; et al. A Novel Nutraceutical Formulation Increases Telomere Length and Activates Telomerase Activity in Middle-aged Rats. *Mol. Med. Rep.* **2023**, *28*, 232. [CrossRef]
53. You, Y.O.; Lee, G.; Min, B.M. Retinoic Acid Extends the in Vitro Life Span of Normal Human Oral Keratinocytes by Decreasing p16(INK4A) Expression and Maintaining Telomerase Activity. *Biochem. Biophys. Res. Commun.* **2000**, *268*, 268–274. [CrossRef]
54. Galiè, S.; Canudas, S.; Muralidharan, J.; García-Gavilán, J.; Bulló, M.; Salas-Salvadó, J. Impact of Nutrition on Telomere Health: Systematic Review of Observational Cohort Studies and Randomized Clinical Trials. *Adv. Nutr.* **2020**, *11*, 576. [CrossRef] [PubMed]
55. Schellnegger, M.; Hofmann, E.; Carnieletto, M.; Kamolz, L.-P. Unlocking Longevity: The Role of Telomeres and Its Targeting Interventions. *Front. Aging* **2024**, *5*, 1339317. [CrossRef] [PubMed]
56. Salvador, L.; Singaravelu, G.; Harley, C.B.; Flom, P.; Suram, A.; Raffaele, J.M. A Natural Product Telomerase Activator Lengthens Telomeres in Humans: A Randomized, Double Blind, and Placebo Controlled Study. *Rejuvenation Res.* **2016**, *19*, 478. [CrossRef] [PubMed]

57. Pal, K.; Raghuram, G.V.; Dsouza, J.; Shinde, S.; Jadhav, V.; Shaikh, A.; Rane, B.; Tandel, H.; Kondhalkar, D.; Chaudhary, S.; et al. A pro-Oxidant Combination of Resveratrol and Copper down-Regulates Multiple Biological Hallmarks of Ageing and Neurodegeneration in Mice. *Sci. Rep.* **2022**, *12*, 17209. [CrossRef] [PubMed]
58. O'Callaghan, N.; Parletta, N.; Milte, C.M.; Benassi-Evans, B.; Fenech, M.; Howe, P.R. Telomere Shortening in Elderly Individuals with Mild Cognitive Impairment May Be Attenuated with ω-3 Fatty Acid Supplementation: A Randomized Controlled Pilot Study. *Nutrition* **2014**, *30*, 489–491. [CrossRef]
59. Guo, J.; Huang, X.; Dou, L.; Yan, M.; Shen, T.; Tang, W.; Li, J. Aging and Aging-Related Diseases: From Molecular Mechanisms to Interventions and Treatments. *Signal Transduct. Target. Ther.* **2022**, *7*, 391. [CrossRef]
60. Pal, S.; Tyler, J.K. Epigenetics and Aging. *Sci. Adv.* **2016**, *2*, e1600584. [CrossRef]
61. Greer, E.L.; Maures, T.J.; Ucar, D.; Hauswirth, A.G.; Mancini, E.; Lim, J.P.; Benayoun, B.A.; Shi, Y.; Brunet, A. Transgenerational Epigenetic Inheritance of Longevity in C. Elegans. *Nature* **2011**, *479*, 365. [CrossRef]
62. Martínez-Iglesias, O.; Naidoo, V.; Carrera, I.; Corzo, L.; Cacabelos, R. Natural Bioactive Products as Epigenetic Modulators for Treating Neurodegenerative Disorders. *Pharmaceuticals* **2023**, *16*, 216. [CrossRef]
63. Huang, D.; Cui, L.; Ahmed, S.; Zainab, F.; Wu, Q.; Wang, X.; Yuan, Z. An Overview of Epigenetic Agents and Natural Nutrition Products Targeting DNA Methyltransferase, Histone Deacetylases and microRNAs. *Food Chem. Toxicol.* **2019**, *123*, 574–594. [CrossRef]
64. Bar-El Dadon, S.; Reifen, R. Vitamin A and the Epigenome. *Crit. Rev. Food Sci. Nutr.* **2017**, *57*, 2404–2411. [CrossRef] [PubMed]
65. Khajebishak, Y.; Alivand, M.; Faghfouri, A.H.; Moludi, J.; Payahoo, L. The Effects of Vitamins and Dietary Pattern on Epigenetic Modification of Non-Communicable Diseases. *Int. J. Vitam. Nutr. Res.* **2023**, *93*, 362–377. [CrossRef] [PubMed]
66. Hassan, F.-U.; Rehman, M.S.-U.; Khan, M.S.; Ali, M.A.; Javed, A.; Nawaz, A.; Yang, C. Curcumin as an Alternative Epigenetic Modulator: Mechanism of Action and Potential Effects. *Front. Genet.* **2019**, *10*, 514. [CrossRef] [PubMed]
67. Fernandes, G.F.S.; Silva, G.D.B.; Pavan, A.R.; Chiba, D.E.; Chin, C.M.; Dos Santos, J.L. Epigenetic Regulatory Mechanisms Induced by Resveratrol. *Nutrients* **2017**, *9*, 1201. [CrossRef]
68. Joven, J.; Micol, V.; Segura-Carretero, A.; Alonso-Villaverde, C.; Menéndez, J.A. Polyphenols and the Modulation of Gene Expression Pathways: Can We Eat Our Way out of the Danger of Chronic Disease? *Crit. Rev. Food Sci. Nutr.* **2014**, *54*, 985–1001. [CrossRef]
69. McGee, K.C.; Sullivan, J.; Hazeldine, J.; Schmunk, L.J.; Martin-Herranz, D.E.; Jackson, T.; Lord, J.M. A Combination Nutritional Supplement Reduces DNA Methylation Age Only in Older Adults with a Raised Epigenetic Age. *GeroScience* **2024**, *46*, 4333–4347. [CrossRef]
70. Park, L.K.; Friso, S.; Choi, S.W. Nutritional Influences on Epigenetics and Age-Related Disease. *Proc. Nutr. Soc.* **2012**, *71*, 75–83. [CrossRef]
71. Julien, C.; Tremblay, C.; Émond, V.; Lebbadi, M.; Salem, N., Jr.; Bennett, D.A.; Calon, F. SIRT1 Decrease Parallels the Accumulation of Tau in Alzheimer Disease. *J. Neuropathol. Exp. Neurol.* **2009**, *68*, 48. [CrossRef]
72. Kaushik, S.; Cuervo, A.M. Proteostasis and Aging. *Nat. Med.* **2015**, *21*, 1406–1415. [CrossRef]
73. Brehme, M.; Sverchkova, A.; Voisine, C. Proteostasis Network Deregulation Signatures as Biomarkers for Pharmacological Disease Intervention. *Curr. Opin. Syst. Biol.* **2019**, *15*, 74–81. [CrossRef]
74. Cuanalo-Contreras, K.; Moreno-Gonzalez, I. Natural Products as Modulators of the Proteostasis Machinery: Implications in Neurodegenerative Diseases. *Int. J. Mol. Sci.* **2019**, *20*, 4666. [CrossRef]
75. Kato, K.; Ito, H.; Kamei, K.; Iwamoto, I. Stimulation of the Stress-Induced Expression of Stress Proteins by Curcumin in Cultured Cells and in Rat Tissues in Vivo. *Cell Stress Chaperones* **1998**, *3*, 152. [CrossRef]
76. Maiti, P.; Manna, J.; Veleri, S.; Frautschy, S. Molecular Chaperone Dysfunction in Neurodegenerative Diseases and Effects of Curcumin. *BioMed Res. Int.* **2014**, *2014*, 495091. [CrossRef] [PubMed]
77. Rane, M.J.; Pan, Y.; Singh, S.; Powell, D.W.; Wu, R.; Cummins, T.; Chen, Q.; McLeish, K.R.; Klein, J.B. Heat Shock Protein 27 Controls Apoptosis by Regulating Akt Activation. *J. Biol. Chem.* **2003**, *278*, 27828–27835. [CrossRef]
78. Guo, H.; Cao, M.; Zou, S.; Ye, B.; Dong, Y. Cranberry Extract Standardized for Proanthocyanidins Alleviates β-Amyloid Peptide Toxicity by Improving Proteostasis Through HSF-1 in Caenorhabditis Elegans Model of Alzheimer's Disease. *J. Gerontol. A Biol. Sci. Med. Sci.* **2016**, *71*, 1564. [CrossRef]
79. Chondrogianni, N.; Gonos, E.S. Proteasome Activation as a Novel Antiaging Strategy. *IUBMB Life* **2008**, *60*, 651–655. [CrossRef] [PubMed]
80. Katsiki, M.; Chondrogianni, N.; Chinou, I.; Rivett, A.J.; Gonos, E.S. The Olive Constituent Oleuropein Exhibits Proteasome Stimulatory Properties in Vitro and Confers Life Span Extension of Human Embryonic Fibroblasts. *Rejuvenation Res.* **2007**, *10*, 157–172. [CrossRef] [PubMed]
81. Corpas, R.; Griñán-Ferré, C.; Rodríguez-Farré, E.; Pallàs, M.; Sanfeliu, C. Resveratrol Induces Brain Resilience Against Alzheimer Neurodegeneration Through Proteostasis Enhancement. *Mol. Neurobiol.* **2019**, *56*, 1502–1516. [CrossRef]
82. Nieto-Torres, J.L.; Hansen, M. Macroautophagy and Aging: The Impact of Cellular Recycling on Health and Longevity. *Mol. Aspects Med.* **2021**, *82*, 101020. [CrossRef]
83. Wong, S.Q.; Kumar, A.V.; Mills, J.; Lapierre, L.R. Autophagy in Aging and Longevity. *Hum. Genet.* **2020**, *139*, 277. [CrossRef]
84. Cao, H.; Jia, Q.; Shen, D.; Yan, L.; Chen, C.; Xing, S. Quercetin Has a Protective Effect on Atherosclerosis via Enhancement of Autophagy in ApoE Mice. *Exp. Ther. Med.* **2019**, *18*, 2451–2458. [CrossRef]

85. Zhao, G.; Han, X.; Zheng, S.; Li, Z.; Sha, Y.; Ni, J.; Sun, Z.; Qiao, S.; Song, Z. Curcumin Induces Autophagy, Inhibits Proliferation and Invasion by Downregulating AKT/mTOR Signaling Pathway in Human Melanoma Cells. *Oncol. Rep.* **2016**, *35*, 1065–1074. [CrossRef] [PubMed]
86. Liu, B.; Wu, Z.; Li, Y.; Ou, C.; Huang, Z.; Zhang, J.; Liu, P.; Luo, C.; Chen, M. Puerarin Prevents Cardiac Hypertrophy Induced by Pressure Overload through Activation of Autophagy. *Biochem. Biophys. Res. Commun.* **2015**, *464*, 908–915. [CrossRef] [PubMed]
87. Geng, F.-H.; Li, G.-H.; Zhang, X.; Zhang, P.; Dong, M.-Q.; Zhao, Z.-J.; Zhang, Y.; Dong, L.; Gao, F. Berberine Improves Mesenteric Artery Insulin Sensitivity through up-Regulating Insulin Receptor-Mediated Signalling in Diabetic Rats. *Br. J. Pharmacol.* **2016**, *173*, 1569–1579. [CrossRef] [PubMed]
88. Huang, L.; Chen, C.; Zhang, X.; Li, X.; Chen, Z.; Yang, C.; Liang, X.; Zhu, G.; Xu, Z. Neuroprotective Effect of Curcumin Against Cerebral Ischemia-Reperfusion Via Mediating Autophagy and Inflammation. *J. Mol. Neurosci.* **2018**, *64*, 129–139. [CrossRef]
89. Zhong, C.; Pu, L.-Y.; Fang, M.-M.; Gu, Z.; Rao, J.-H.; Wang, X.-H. Retinoic Acid Receptor α Promotes Autophagy to Alleviate Liver Ischemia and Reperfusion Injury. *World J. Gastroenterol.* **2015**, *21*, 12381–12391. [CrossRef]
90. Kim, S.-H.; Chung, D.-K.; Lee, Y.J.; Song, C.-H.; Ku, S.-K. Neuroprotective Effects of Danggui-Jakyak-San on Rat Stroke Model through Antioxidant/antiapoptotic Pathway. *J. Ethnopharmacol.* **2016**, *188*, 123–133. [CrossRef]
91. Ahn, S.M.; Kim, H.N.; Kim, Y.R.; Choi, Y.W.; Kim, C.M.; Shin, H.K.; Choi, B.T. Emodin from Polygonum Multiflorum Ameliorates Oxidative Toxicity in HT22 Cells and Deficits in Photothrombotic Ischemia. *J. Ethnopharmacol.* **2016**, *188*, 13–20. [CrossRef]
92. Sun, R.; Song, Y.; Li, S.; Ma, Z.; Deng, X.; Fu, Q.; Qu, R.; Ma, S. Levo-Tetrahydropalmatine Attenuates Neuron Apoptosis Induced by Cerebral Ischemia-Reperfusion Injury: Involvement of c-Abl Activation. *J. Mol. Neurosci.* **2018**, *65*, 391–399. [CrossRef]
93. Pang, Q.; Zhao, Y.; Chen, X.; Zhao, K.; Zhai, Q.; Tu, F. Apigenin Protects the Brain against Ischemia/Reperfusion Injury via Caveolin-1/VEGF In Vitro and In Vivo. *Oxidative Med. Cell. Longev.* **2018**, *2018*, 7017204. [CrossRef]
94. Zheng, W.-X.; Cao, X.-L.; Wang, F.; Wang, J.; Ying, T.-Z.; Xiao, W.; Zhang, Y.; Xing, H.; Dong, W.; Xu, S.-Q.; et al. Baicalin Inhibiting Cerebral Ischemia/hypoxia-Induced Neuronal Apoptosis via MRTF-A-Mediated Transactivity. *Eur. J. Pharmacol.* **2015**, *767*, 201–210. [CrossRef] [PubMed]
95. Wang, C.C.; Ho, Y.H.; Hung, C.F.; Kuo, J.R.; Wang, S.J. Xanthohumol, an Active Constituent from Hope, Affords Protection against Kainic Acid-Induced Excitotoxicity in Rats. *Neurochem. Int.* **2020**, *133*, 104629. [CrossRef] [PubMed]
96. Efeyan, A.; Comb, W.C.; Sabatini, D.M. Nutrient-Sensing Mechanisms and Pathways. *Nature* **2015**, *517*, 302–310. [CrossRef] [PubMed]
97. Yang, N.-C.; Chin, C.-Y.; Zheng, Y.-X.; Lee, I. The Attenuation of Insulin/IGF-1 Signaling Pathway Plays a Crucial Role in the Myo-Inositol-Alleviated Aging in Caenorhabditis Elegans. *Int. J. Mol. Sci.* **2023**, *24*, 6194. [CrossRef]
98. Franceschi, C.; Passarino, G.; Mari, D.; Monti, D. Centenarians as a 21st Century Healthy Aging Model: A Legacy of Humanity and the Need for a World-Wide Consortium (WWC100+). *Mech. Ageing Dev.* **2017**, *165*, 55–58. [CrossRef]
99. Paolisso, G.; Ammendola, S.; Del Buono, A.; Gambardella, A.; Riondino, M.; Tagliamonte, M.R.; Rizzo, M.R.; Carella, C.; Varricchio, M. Serum Levels of Insulin-Like Growth Factor-I (IGF-I) and IGF-Binding Protein-3 in Healthy Centenarians: Relationship with Plasma Leptin and Lipid Concentrations, Insulin Action, and Cognitive Function. *J. Clin. Endocrinol. Metab.* **1997**, *82*, 2204–2209. [CrossRef]
100. Vitale, G.; Pellegrino, G.; Vollery, M.; Hofland, L.J. ROLE of IGF-1 System in the Modulation of Longevity: Controversies and New Insights From a Centenarians' Perspective. *Front. Endocrinol.* **2019**, *10*, 431907. [CrossRef]
101. Chang, C.-H.; Bijian, K.; Wernic, D.; Su, J.; da Silva, S.D.; Yu, H.; Qiu, D.; Asslan, M.; Alaoui-Jamali, M.A. A Novel Orally Available Seleno-Purine Molecule Suppresses Triple-Negative Breast Cancer Cell Proliferation and Progression to Metastasis by Inducing Cytostatic Autophagy. *Autophagy* **2019**, *15*, 1376–1390. [CrossRef]
102. Wang, C.; Zhang, X.; Teng, Z.; Zhang, T.; Li, Y. Downregulation of PI3K/Akt/mTOR Signaling Pathway in Curcumin-Induced Autophagy in APP/PS1 Double Transgenic Mice. *Eur. J. Pharmacol.* **2014**, *740*, 312–320. [CrossRef]
103. Ji, Y.; Li, L.; Ma, Y.-X.; Li, W.-T.; Li, L.; Zhu, H.-Z.; Wu, M.-H.; Zhou, J.-R. Quercetin Inhibits Growth of Hepatocellular Carcinoma by Apoptosis Induction in Part via Autophagy Stimulation in Mice. *J. Nutr. Biochem.* **2019**, *69*, 108–119. [CrossRef]
104. Pignatti, C.; D'Adamo, S.; Stefanelli, C.; Flamigni, F.; Cetrullo, S. Nutrients and Pathways That Regulate Health Span and Life Span. *Geriatrics* **2020**, *5*, 95. [CrossRef]
105. Mas-Bargues, C.; Borrás, C.; Viña, J. Genistein, a Tool for Geroscience. *Mech. Ageing Dev.* **2022**, *204*, 111665. [CrossRef] [PubMed]
106. Bai, X.; Yao, L.; Ma, X.; Xu, X. Small Molecules as SIRT Modulators. *Mini-Rev. Med. Chem.* **2018**, *18*, 1151–1157. [CrossRef]
107. Price, N.L.; Gomes, A.P.; Ling, A.J.; Duarte, F.V.; Martin-Montalvo, A.; North, B.J.; Agarwal, B.; Ye, L.; Ramadori, G.; Teodoro, J.S.; et al. SIRT1 Is Required for AMPK Activation and the Beneficial Effects of Resveratrol on Mitochondrial Function. *Cell Metab.* **2012**, *15*, 675–690. [CrossRef] [PubMed]
108. Zhao, L.; Cao, J.; Hu, K.; He, X.; Yun, D.; Tong, T.; Han, L. Sirtuins and Their Biological Relevance in Aging and Age-Related Diseases. *Aging Dis.* **2020**, *11*, 927. [CrossRef] [PubMed]
109. Nicolson, G.L. Mitochondrial Dysfunction and Chronic Disease: Treatment With Natural Supplements. *Integr. Med. A Clin. J.* **2014**, *13*, 35.
110. van der Rijt, S.; Molenaars, M.; McIntyre, R.L.; Janssens, G.E.; Houtkooper, R.H. Integrating the Hallmarks of Aging Throughout the Tree of Life: A Focus on Mitochondrial Dysfunction. *Front. Cell Dev. Biol.* **2020**, *8*, 594416. [CrossRef]
111. Dai, Y.; Zhang, H.; Zhang, J.; Yan, M. Isoquercetin Attenuates Oxidative Stress and Neuronal Apoptosis after Ischemia/reperfusion Injury via Nrf2-Mediated Inhibition of the NOX4/ROS/NF-κB Pathway. *Chem. Biol. Interact.* **2018**, *284*, 32–40. [CrossRef]

112. Leung, S.W.; Lai, J.H.; Wu, J.C.-C.; Tsai, Y.-R.; Chen, Y.-H.; Kang, S.-J.; Chiang, Y.-H.; Chang, C.-F.; Chen, K.-Y. Neuroprotective Effects of Emodin against Ischemia/Reperfusion Injury through Activating ERK-1/2 Signaling Pathway. *Int. J. Mol. Sci.* **2020**, *21*, 2899. [CrossRef]
113. Zhang, Y.; Yang, Y.; Yu, H.; Li, M.; Hang, L.; Xu, X. Apigenin Protects Mouse Retina against Oxidative Damage by Regulating the Nrf2 Pathway and Autophagy. *Oxidative Med. Cell. Longev.* **2020**, *2020*, 9420704. [CrossRef]
114. Lippi, L.; Uberti, F.; Folli, A.; Turco, A.; Curci, C.; d'Abrosca, F.; de Sire, A.; Invernizzi, M. Impact of Nutraceuticals and Dietary Supplements on Mitochondria Modifications in Healthy Aging: A Systematic Review of Randomized Controlled Trials. *Aging Clin. Exp. Res.* **2022**, *34*, 2659–2674. [CrossRef] [PubMed]
115. Shanaida, M.; Lysiuk, R.; Mykhailenko, O.; Hudz, N.; Abdulsalam, A.; Gontova, T.; Oleshchuk, O.; Ivankiv, Y.; Shanaida, V.; Lytkin, D.; et al. Alpha-Lipoic Acid: An Antioxidant with Anti-Aging Properties for Disease Therapy. *Curr. Med. Chem.* **2024**, *31*, e190424229159. [CrossRef] [PubMed]
116. Lewis Luján, L.M.; McCarty, M.F.; Di Nicolantonio, J.J.; Gálvez Ruiz, J.C.; Rosas-Burgos, E.C.; Plascencia-Jatomea, M.; Iloki Assanga, S.B. Nutraceuticals/Drugs Promoting Mitophagy and Mitochondrial Biogenesis May Combat the Mitochondrial Dysfunction Driving Progression of Dry Age-Related Macular Degeneration. *Nutrients* **2022**, *14*, 1985. [CrossRef] [PubMed]
117. Lagouge, M.; Argmann, C.; Gerhart-Hines, Z.; Meziane, H.; Lerin, C.; Daussin, F.; Messadeq, N.; Milne, J.; Lambert, P.; Elliott, P.; et al. Resveratrol Improves Mitochondrial Function and Protects against Metabolic Disease by Activating SIRT1 and PGC-1alpha. *Cell* **2006**, *127*, 1109–1122. [CrossRef]
118. Gopinath, B.; Liew, G.; Kifley, A.; Flood, V.M.; Joachim, N.; Lewis, J.R.; Hodgson, J.M.; Mitchell, P. Dietary Flavonoids and the Prevalence and 15-Y Incidence of Age-Related Macular Degeneration. *Am. J. Clin. Nutr.* **2018**, *108*, 381–387. [CrossRef]
119. Chang, Y.-Y.; Lee, Y.-J.; Hsu, M.-Y.; Wang, M.; Tsou, S.-C.; Chen, C.-C.; Lin, J.-A.; Hsiao, Y.-P.; Lin, H.-W. Protective Effect of Quercetin on Sodium Iodate-Induced Retinal Apoptosis through the Reactive Oxygen Species-Mediated Mitochondrion-Dependent Pathway. *Int. J. Mol. Sci.* **2021**, *22*, 4056. [CrossRef]
120. Zhang, L.; Pitcher, L.E.; Yousefzadeh, M.J.; Niedernhofer, L.J.; Robbins, P.D.; Zhu, Y. Cellular Senescence: A Key Therapeutic Target in Aging and Diseases. *J. Clin. Investig.* **2022**, *132*, e158450. [CrossRef]
121. Regulski, M.J. Cellular Senescence: What, Why, and How. *Wounds A Compend. Clin. Res. Pract.* **2017**, *29*, 168–174.
122. Biazi, B.I.; Zanetti, T.A.; Baranoski, A.; Corveloni, A.C.; Mantovani, M.S. Cis-Nerolidol Induces Endoplasmic Reticulum Stress and Cell Death in Human Hepatocellular Carcinoma Cells through Extensive CYP2C19 and CYP1A2 Oxidation. *Basic Clin. Pharmacol. Toxicol.* **2017**, *121*, 334–341. [CrossRef]
123. Xu, Q.; Fu, Q.; Li, Z.; Liu, H.; Wang, Y.; Lin, X.; He, R.; Zhang, X.; Ju, Z.; Campisi, J.; et al. The Flavonoid Procyanidin C1 Has Senotherapeutic Activity and Increases Lifespan in Mice. *Nat. Metab.* **2021**, *3*, 1706–1726. [CrossRef]
124. Yousefzadeh, M.J.; Zhu, Y.; McGowan, S.J.; Angelini, L.; Fuhrmann-Stroissnigg, H.; Xu, M.; Ling, Y.Y.; Melos, K.I.; Pirtskhalava, T.; Inman, C.L.; et al. Fisetin Is a Senotherapeutic That Extends Health and Lifespan. *EBioMedicine* **2018**, *36*, 18–28. [CrossRef]
125. Saccon, T.D.; Nagpal, R.; Yadav, H.; Cavalcante, M.B.; Nunes, A.D.D.C.; Schneider, A.; Gesing, A.; Hughes, B.; Yousefzadeh, M.; Tchkonia, T.; et al. Senolytic Combination of Dasatinib and Quercetin Alleviates Intestinal Senescence and Inflammation and Modulates the Gut Microbiome in Aged Mice. *J. Gerontol. A Biol. Sci. Med. Sci.* **2021**, *76*, 1895–1905. [CrossRef] [PubMed]
126. Kirkland, J.L.; Tchkonia, T. Senolytic Drugs: From Discovery to Translation. *J. Intern. Med.* **2020**, *288*, 518–536. [CrossRef] [PubMed]
127. Mbara, K.C.; Devnarain, N.; Owira, P.M.O. Potential Role of Polyphenolic Flavonoids as Senotherapeutic Agents in Degenerative Diseases and Geroprotection. *Pharmaceut. Med.* **2022**, *36*, 331–352. [CrossRef] [PubMed]
128. Ahmed, A.S.I.; Sheng, M.H.; Wasnik, S.; Baylink, D.J.; Lau, K.-H.W. Effect of Aging on Stem Cells. *World J. Exp. Med.* **2017**, *7*, 1–10. [CrossRef]
129. Stover, P.J.; Field, M.S.; Brawley, H.N.; Angelin, B.; Iversen, P.O.; Frühbeck, G. Nutrition and Stem Cell Integrity in Aging. *J. Intern. Med.* **2022**, *292*, 587–603. [CrossRef]
130. Cheng, C.-W.; Biton, M.; Haber, A.L.; Gunduz, N.; Eng, G.; Gaynor, L.T.; Tripathi, S.; Calibasi-Kocal, G.; Rickelt, S.; Butty, V.L.; et al. Ketone Body Signaling Mediates Intestinal Stem Cell Homeostasis and Adaptation to Diet. *Cell* **2019**, *178*, 1115–1131.e15. [CrossRef]
131. Richer, S.; Patel, S.; Sockanathan, S.; Ulanski, L.J., 2nd; Miller, L.; Podella, C. Resveratrol Based Oral Nutritional Supplement Produces Long-Term Beneficial Effects on Structure and Visual Function in Human Patients. *Nutrients* **2014**, *6*, 4404–4420. [CrossRef]
132. Dai, J.; Chen, L.; Qiu, Y.-M.; Li, S.-Q.; Xiong, W.-H.; Yin, Y.-H.; Jia, F.; Jiang, J.-Y. Activations of GABAergic Signaling, HSP70 and MAPK Cascades Are Involved in Baicalin's Neuroprotection against Gerbil Global Ischemia/reperfusion Injury. *Brain Res. Bull.* **2013**, *90*, 1–9. [CrossRef]
133. Ruggiero, C.; Tafaro, L.; Cianferotti, L.; Tramontana, F.; Macchione, I.G.; Caffarelli, C.; Virdis, A.; Ferracci, M.; Rinonapoli, G.; Mecocci, P.; et al. Targeting the Hallmarks of Aging with Vitamin D: Starting to Decode the Myth. *Nutrients* **2024**, *16*, 906. [CrossRef]
134. McDonald, C.M.; Ramirez-Sanchez, I.; Oskarsson, B.; Joyce, N.; Aguilar, C.; Nicorici, A.; Dayan, J.; Goude, E.; Abresch, R.T.; Villarreal, F.; et al. (-)-Epicatechin Induces Mitochondrial Biogenesis and Markers of Muscle Regeneration in Adults with Becker Muscular Dystrophy. *Muscle Nerve* **2021**, *63*, 239–249. [CrossRef] [PubMed]

135. Kunnumakkara, A.B.; Bordoloi, D.; Padmavathi, G.; Monisha, J.; Roy, N.K.; Prasad, S.; Aggarwal, B.B. Curcumin, the Golden Nutraceutical: Multitargeting for Multiple Chronic Diseases. *Br. J. Pharmacol.* **2017**, *174*, 1325–1348. [CrossRef]
136. Long, H.-Z.; Cheng, Y.; Zhou, Z.-W.; Luo, H.-Y.; Wen, D.-D.; Gao, L.-C. PI3K/AKT Signal Pathway: A Target of Natural Products in the Prevention and Treatment of Alzheimer's Disease and Parkinson's Disease. *Front. Pharmacol.* **2021**, *12*, 648636. [CrossRef] [PubMed]
137. Nie, L.; Xia, J.; Li, H.; Zhang, Z.; Yang, Y.; Huang, X.; He, Z.; Liu, J.; Yang, X. Ginsenoside Rg1 Ameliorates Behavioral Abnormalities and Modulates the Hippocampal Proteomic Change in Triple Transgenic Mice of Alzheimer's Disease. *Oxidative Med. Cell. Longev.* **2017**, *2017*, 6473506. [CrossRef] [PubMed]
138. Ferrucci, L.; Fabbri, E. Inflammageing: Chronic Inflammation in Ageing, Cardiovascular Disease, and Frailty. *Nat. Rev. Cardiol.* **2018**, *15*, 505–522. [CrossRef]
139. Haigis, M.C.; Guarente, L.P. Mammalian Sirtuins--Emerging Roles in Physiology, Aging, and Calorie Restriction. *Genes Dev.* **2006**, *20*, 2913–2921. [CrossRef]
140. Hollman, P.C.H.; Arts, I.C.W. Flavonols, Flavones and Flavanols—Nature, Occurrence and Dietary Burden. *J. Sci. Food Agric.* **2000**, *80*, 1081–1093. [CrossRef]
141. Mlcek, J.; Jurikova, T.; Skrovankova, S.; Sochor, J. Quercetin and Its Anti-Allergic Immune Response. *Molecules* **2016**, *21*, 623. [CrossRef]
142. Xiao, X.; Shi, D.; Liu, L.; Wang, J.; Xie, X.; Kang, T.; Deng, W. Quercetin Suppresses Cyclooxygenase-2 Expression and Angiogenesis through Inactivation of P300 Signaling. *PLoS ONE* **2011**, *6*, e22934. [CrossRef]
143. Kim, H.P.; Mani, I.; Iversen, L.; Ziboh, V.A. Effects of Naturally-Occurring Flavonoids and Biflavonoids on Epidermal Cyclooxygenase and Lipoxygenase from Guinea-Pigs. *Prostaglandins Leukot. Essent. Fat. Acids* **1998**, *58*, 17–24. [CrossRef]
144. Lee, K.M.; Hwang, M.K.; Lee, D.E.; Lee, K.W.; Lee, H.J. Protective Effect of Quercetin against Arsenite-Induced COX-2 Expression by Targeting PI3K in Rat Liver Epithelial Cells. *J. Agric. Food Chem.* **2010**, *58*, 5815–5820. [CrossRef]
145. Seo, M.-J.; Lee, Y.-J.; Hwang, J.-H.; Kim, K.-J.; Lee, B.-Y. The Inhibitory Effects of Quercetin on Obesity and Obesity-Induced Inflammation by Regulation of MAPK Signaling. *J. Nutr. Biochem.* **2015**, *26*, 1308–1316. [CrossRef]
146. Flores-Soto, E.; Romero-Martínez, B.S.; Solís-Chagoyán, H.; Estrella-Parra, E.A.; Avila-Acevedo, J.G.; Gomez-Verjan, J.C.; Reyes-García, J.; Casas-Hernández, M.F.; Sommer, B.; Montaño, L.M. Chamaecyparis Lawsoniana and Its Active Compound Quercetin as Ca2+ Inhibitors in the Contraction of Airway Smooth Muscle. *Molecules* **2024**, *29*, 2284. [CrossRef] [PubMed]
147. Peairs, A.; Dai, R.; Gan, L.; Shimp, S.; Rylander, M.N.; Li, L.; Reilly, C.M. Epigallocatechin-3-Gallate (EGCG) Attenuates Inflammation in MRL/lpr Mouse Mesangial Cells. *Cell. Mol. Immunol.* **2010**, *7*, 123–132. [CrossRef] [PubMed]
148. Guo, H.; Zhu, L.; Tang, P.; Chen, D.; Li, Y.; Li, J.; Bao, C. Carthamin Yellow Improves Cerebral Ischemia-reperfusion Injury by Attenuating Inflammation and Ferroptosis in Rats. *Int. J. Mol. Med.* **2021**, *47*, 52. [CrossRef] [PubMed]
149. Yang, S.; Wang, H.; Yang, Y.; Wang, R.; Wang, Y.; Wu, C.; Du, G. Baicalein Administered in the Subacute Phase Ameliorates Ischemia-Reperfusion-Induced Brain Injury by Reducing Neuroinflammation and Neuronal Damage. *Biomed. Pharmacother.* **2019**, *117*, 109102. [CrossRef]
150. Chen, H.-L.; Jia, W.-J.; Li, H.-E.; Han, H.; Li, F.; Zhang, X.-L.-N.; Li, J.-J.; Yuan, Y.; Wu, C.-Y. Scutellarin Exerts Anti-Inflammatory Effects in Activated Microglia/Brain Macrophage in Cerebral Ischemia and in Activated BV-2 Microglia Through Regulation of MAPKs Signaling Pathway. *Neuromol. Med.* **2020**, *22*, 264–277. [CrossRef]
151. Mo, Z.-T.; Liao, Y.-L.; Zheng, J.; Li, W.-N. Icariin Protects Neurons from Endoplasmic Reticulum Stress-Induced Apoptosis after OGD/R Injury via Suppressing IRE1α-XBP1 Signaling Pathway. *Life Sci.* **2020**, *255*, 117847. [CrossRef]
152. Dai, M.; Chen, B.; Wang, X.; Gao, C.; Yu, H. Icariin Enhance Mild Hypothermia-Induced Neuroprotection via Inhibiting the Activation of NF-κB in Experimental Ischemic Stroke. *Metab. Brain Dis.* **2021**, *36*, 1779–1790. [CrossRef]
153. Wei, P.; Wang, K.; Luo, C.; Huang, Y.; Misilimu, D.; Wen, H.; Jin, P.; Li, C.; Gong, Y.; Gao, Y. Cordycepin Confers Long-Term Neuroprotection via Inhibiting Neutrophil Infiltration and Neuroinflammation after Traumatic Brain Injury. *J. Neuroinflamm.* **2021**, *18*, 137. [CrossRef]
154. Li, Z.; Geng, Y.-N.; Jiang, J.-D.; Kong, W.-J. Antioxidant and Anti-Inflammatory Activities of Berberine in the Treatment of Diabetes Mellitus. *Evid.-Based Complement. Altern. Med.* **2014**, *2014*, 289264. [CrossRef] [PubMed]
155. Kim, Y.; Wang, W.; Okla, M.; Kang, I.; Moreau, R.; Chung, S. Suppression of NLRP3 Inflammasome by γ-Tocotrienol Ameliorates Type 2 Diabetes. *J. Lipid Res.* **2016**, *57*, 66–76. [CrossRef] [PubMed]
156. Levy, M.; Kolodziejczyk, A.A.; Thaiss, C.A.; Elinav, E. Dysbiosis and the Immune System. *Nat. Rev. Immunol.* **2017**, *17*, 219–232. [CrossRef] [PubMed]
157. Pang, S.; Chen, X.; Lu, Z.; Meng, L.; Huang, Y.; Yu, X.; Huang, L.; Ye, P.; Chen, X.; Liang, J.; et al. Longevity of Centenarians Is Reflected by the Gut Microbiome with Youth-Associated Signatures. *Nat. Aging* **2023**, *3*, 436–449. [CrossRef]
158. Johansen, J.; Atarashi, K.; Arai, Y.; Hirose, N.; Sørensen, S.J.; Vatanen, T.; Knip, M.; Honda, K.; Xavier, R.J.; Rasmussen, S.; et al. Centenarians Have a Diverse Gut Virome with the Potential to Modulate Metabolism and Promote Healthy Lifespan. *Nat. Microbiol.* **2023**, *8*, 1064–1078. [CrossRef]
159. Milshteyn, A.; Colosimo, D.A.; Brady, S.F. Accessing Bioactive Natural Products from the Human Microbiome. *Cell Host Microbe* **2018**, *23*, 725–736. [CrossRef]
160. López-Gil, L.; Pascual-Ahuir, A.; Proft, M. Genomic Instability and Epigenetic Changes during Aging. *Int. J. Mol. Sci.* **2023**, *24*, 14279. [CrossRef]

161. Xu, D.-P.; Li, Y.; Meng, X.; Zhou, T.; Zhou, Y.; Zheng, J.; Zhang, J.-J.; Li, H.-B. Natural Antioxidants in Foods and Medicinal Plants: Extraction, Assessment and Resources. *Int. J. Mol. Sci.* **2017**, *18*, 96. [CrossRef]
162. Muñoz-Espada, A.C.; Wood, K.V.; Bordelon, B.; Watkins, B.A. Anthocyanin Quantification and Radical Scavenging Capacity of Concord, Norton, and Marechal Foch Grapes and Wines. *J. Agric. Food Chem.* **2004**, *52*, 6779–6786. [CrossRef]
163. Yashin, A.; Yashin, Y.; Wang, J.Y.; Nemzer, B. Antioxidant and Antiradical Activity of Coffee. *Antioxidants* **2013**, *2*, 230–245. [CrossRef]
164. Howitz, K.T.; Bitterman, K.J.; Cohen, H.Y.; Lamming, D.W.; Lavu, S.; Wood, J.G.; Zipkin, R.E.; Chung, P.; Kisielewski, A.; Zhang, L.-L.; et al. Small Molecule Activators of Sirtuins Extend Saccharomyces Cerevisiae Lifespan. *Nature* **2003**, *425*, 191–196. [CrossRef]
165. Rahmah, N.N.; Tama, N.D.; Wulandari, N.R.; Nurcholis, W. Potential of β-Carotene as Anti-Aging Serum: A Narrative Review. *Int. J. Life Sci. Pharma Res.* **2021**, *12*, 1999–2004. [CrossRef]
166. Huang, A.; Huo, Y.; Zhong, Y.; Yang, W. AI Technology for Anti-Aging: An Overview. In Proceedings of the 2023 International Conference on Intelligent Supercomputing and BioPharma (ISBP), Zhuhai, China, 6–8 January 2023; IEEE: New York, NY, USA.
167. Carpio, L.E.; Sanz, Y.; Gozalbes, R.; Barigye, S.J. Computational Strategies for the Discovery of Biological Functions of Health Foods, Nutraceuticals and Cosmeceuticals: A Review. *Mol. Divers.* **2021**, *25*, 1425–1438. [CrossRef] [PubMed]
168. Zhavoronkov, A.; Bischof, E.; Lee, K.-F. Artificial Intelligence in Longevity Medicine. *Nat. Aging* **2021**, *1*, 5–7. [CrossRef] [PubMed]
169. Dower, J.I.; Geleijnse, J.M.; Gijsbers, L.; Schalkwijk, C.; Kromhout, D.; Hollman, P.C. Supplementation of the Pure Flavonoids Epicatechin and Quercetin Affects Some Biomarkers of Endothelial Dysfunction and Inflammation in (Pre)Hypertensive Adults: A Randomized Double-Blind, Placebo-Controlled, Crossover Trial. *J. Nutr.* **2015**, *145*, 1459–1463. [CrossRef] [PubMed]
170. Alshehri, M.M.; Sharifi-Rad, J.; Herrera-Bravo, J.; Jara, E.L.; Salazar, L.A.; Kregiel, D.; Uprety, Y.; Akram, M.; Iqbal, M.; Martorell, M.; et al. Therapeutic Potential of Isoflavones with an Emphasis on Daidzein. *Oxidative Med. Cell. Longev.* **2021**, *2021*, 6331630. [CrossRef] [PubMed]
171. Kim, M.-S.; Hong, C.Y.; Lee, S.H. The Phytoestrogenic Effect of Daidzein in Human Dermal Fibroblasts. *J. Soc. Cosmet. Sci. Korea* **2014**, *40*, 279–287.
172. Guo, J.M.; Kang, G.Z.; Xiao, B.X.; Liu, D.H.; Zhang, S. Effect of Daidzein on Cell Growth, Cell Cycle, and Telomerase Activity of Human Cervical Cancer in Vitro. *Int. J. Gynecol. Cancer* **2004**, *14*, 882–888. [CrossRef]
173. Yoshino, M.; Naka, A.; Sakamoto, Y.; Shibasaki, A.; Toh, M.; Tsukamoto, S.; Kondo, K.; Iida, K. Dietary Isoflavone Daidzein Promotes Tfam Expression That Increases Mitochondrial Biogenesis in C2C12 Muscle Cells. *J. Nutr. Biochem.* **2015**, *26*, 1193–1199. [CrossRef]
174. Russo, M.; Moccia, S.; Bilotto, S.; Spagnuolo, C.; Durante, M.; Lenucci, M.S.; Mita, G.; Volpe, M.G.; Aquino, R.P.; Russo, G.L. A Carotenoid Extract from a Southern Italian Cultivar of Pumpkin Triggers Nonprotective Autophagy in Malignant Cells. *Oxidative Med. Cell. Longev.* **2017**, *2017*, 7468538. [CrossRef]
175. Sztretye, M.; Dienes, B.; Gönczi, M.; Czirják, T.; Csernoch, L.; Dux, L.; Szentesi, P.; Keller-Pintér, A. Astaxanthin: A Potential Mitochondrial-Targeted Antioxidant Treatment in Diseases and with Aging. *Oxidative Med. Cell. Longev.* **2019**, *2019*, 3849692. [CrossRef]
176. Metibemu, D.S.; Ogungbe, I.V. Carotenoids in Drug Discovery and Medicine: Pathways and Molecular Targets Implicated in Human Diseases. *Molecules* **2022**, *27*, 6005. [CrossRef] [PubMed]
177. Bartali, B.; Semba, R.D. Carotenoids and Healthy Aging: The Fascination Continues. *Am. J. Clin. Nutr.* **2021**, *113*, 259–260. [CrossRef] [PubMed]
178. Zhu, Y.; Cheng, J.; Min, Z.; Yin, T.; Zhang, R.; Zhang, W.; Hu, L.; Cui, Z.; Gao, C.; Xu, S.; et al. Effects of Fucoxanthin on Autophagy and Apoptosis in SGC-7901cells and the Mechanism. *J. Cell. Biochem.* **2018**, *119*, 7274–7284. [CrossRef] [PubMed]
179. Zhang, L.; Wang, H.; Fan, Y.; Gao, Y.; Li, X.; Hu, Z.; Ding, K.; Wang, Y.; Wang, X. Fucoxanthin Provides Neuroprotection in Models of Traumatic Brain Injury via the Nrf2-ARE and Nrf2-Autophagy Pathways. *Sci. Rep.* **2017**, *7*, 46763. [CrossRef]
180. Lee, A.-H.; Shin, H.-Y.; Park, J.-H.; Koo, S.Y.; Kim, S.M.; Yang, S.-H. Fucoxanthin from Microalgae Phaeodactylum Tricornutum Inhibits pro-Inflammatory Cytokines by Regulating Both NF-κB and NLRP3 Inflammasome Activation. *Sci. Rep.* **2021**, *11*, 543. [CrossRef]
181. Cheng, Y.; Pan, X.; Wang, J.; Li, X.; Yang, S.; Yin, R.; Ma, A.; Zhu, X. Fucoidan Inhibits NLRP3 Inflammasome Activation by Enhancing p62/SQSTM1-Dependent Selective Autophagy to Alleviate Atherosclerosis. *Oxidative Med. Cell. Longev.* **2020**, *2020*, 3186306. [CrossRef]
182. Lee, J.H.; Lee, S.H.; Choi, S.H.; Asahara, T.; Kwon, S.M. The Sulfated Polysaccharide Fucoidan Rescues Senescence of Endothelial Colony-Forming Cells for Ischemic Repair. *Stem Cells* **2015**, *33*, 1939–1951. [CrossRef]
183. Yu, W.-C.; Chen, Y.-L.; Hwang, P.-A.; Chen, T.-H.; Chou, T.-C. Fucoidan Ameliorates Pancreatic β-Cell Death and Impaired Insulin Synthesis in Streptozotocin-Treated β Cells and Mice via a Sirt-1-Dependent Manner. *Mol. Nutr. Food Res.* **2017**, *61*, 1700136. [CrossRef]
184. Wang, T.; Zhu, M.; He, Z.Z. Low-Molecular-Weight Fucoidan Attenuates Mitochondrial Dysfunction and Improves Neurological Outcome After Traumatic Brain Injury in Aged Mice: Involvement of Sirt3. *Cell. Mol. Neurobiol.* **2016**, *36*, 1257–1268. [CrossRef]
185. Li, J.; Chen, K.; Li, S.; Feng, J.; Liu, T.; Wang, F.; Zhang, R.; Xu, S.; Zhou, Y.; Zhou, S.; et al. Protective Effect of Fucoidan from Fucus Vesiculosus on Liver Fibrosis via the TGF-β1/Smad Pathway-Mediated Inhibition of Extracellular Matrix and Autophagy. *Drug Des. Devel. Ther.* **2016**, *10*, 619–630.

186. Oh, J.M.; Kim, E.; Chun, S. Ginsenoside Compound K Induces Ros-Mediated Apoptosis and Autophagic Inhibition in Human Neuroblastoma Cells In Vitro and In Vivo. *Int. J. Mol. Sci.* **2019**, *20*, 4279. [CrossRef] [PubMed]
187. Arafa, E.-S.A.; Refaey, M.S.; Abd El-Ghafar, O.A.; Hassanein, E.H.M.; Sayed, A.M. The Promising Therapeutic Potentials of Ginsenosides Mediated through p38 MAPK Signaling Inhibition. *Heliyon* **2021**, *7*, e08354. [CrossRef] [PubMed]
188. Baik, I.-H.; Kim, K.-H.; Lee, K.-A. Antioxidant, Anti-Inflammatory and Antithrombotic Effects of Ginsenoside Compound K Enriched Extract Derived from Ginseng Sprouts. *Molecules* **2021**, *26*, 4102. [CrossRef]
189. Wang, S.; Qiao, J.; Jiang, C.; Pan, D.; Yu, S.; Chen, J.; Liu, S.; Zhang, P.; Zhao, D.; Liu, M. Ginsenoside Rg1 Delays Chronological Aging in a Yeast Model via CDC19- and SDH2-Mediated Cellular Metabolism. *AntiOxidative Redox Signal.* **2023**, *12*, 296. [CrossRef] [PubMed]
190. Zhou, Y.; Liu, J.; Cai, S.; Liu, D.; Jiang, R.; Wang, Y. Protective Effects of Ginsenoside Rg1 on Aging Sca-1+ Hematopoietic Cells. *Mol. Med. Rep.* **2015**, *12*, 3621–3628. [CrossRef]
191. Sun, M.; Ji, Y.; Zhou, S.; Chen, R.; Yao, H.; Du, M. Ginsenoside Rb3 Inhibits Osteoclastogenesis via ERK/NF-κB Signaling Pathway in Vitro and in Vivo. *Oral Dis.* **2023**, *29*, 3460–3471. [CrossRef]
192. Zhou, S.; Ji, Y.; Yao, H.; Guo, H.; Zhang, Z.; Wang, Z.; Du, M. Application of Ginsenoside Rd in Periodontitis With Inhibitory Effects on Pathogenicity, Inflammation, and Bone Resorption. *Front. Cell. Infect. Microbiol.* **2022**, *12*, 813953. [CrossRef]
193. Yin, L.H.; Cheng, W.X.; Qin, Z.S.; Sun, K.M.; Zhong, M.; Wang, J.K.; Gao, W.Y.; Yu, Z.H. Effects of Ginsenoside Rg-1 on the Proliferation and Osteogenic Differentiation of Human Periodontal Ligament Stem Cells. *Chin. J. Integr. Med.* **2015**, *21*, 676–681. [CrossRef]
194. Park, M.Y.; Kwon, H.J.; Sung, M.K. Evaluation of Aloin and Aloe-Emodin as Anti-Inflammatory Agents in Aloe by Using Murine Macrophages. *Biosci. Biotechnol. Biochem.* **2009**, *73*, 828–832. [CrossRef]
195. Zimbone, S.; Romanucci, V.; Zarrelli, A.; Giuffrida, M.L.; Sciacca, M.F.M.; Lanza, V.; Campagna, T.; Maugeri, L.; Petralia, S.; Consoli, G.M.L.; et al. Exploring the Therapeutic Potential of Aloin: Unraveling Neuroprotective and Anticancer Mechanisms, and Strategies for Enhanced Stability and Delivery. *Sci. Rep.* **2024**, *14*, 16731. [CrossRef] [PubMed]
196. He, J.; Zhang, W.; Zhou, X.; Yan, W.; Wang, Z. Aloin Induced Apoptosis by Enhancing Autophagic Flux through the PI3K/AKT Axis in Osteosarcoma. *Chin. Med.* **2021**, *16*, 1–21. [CrossRef]
197. Vidakovic, M.; Marinello, J.; Lahtela-Kakkonen, M.; Matulis, D.; Linkuvienė, V.; Michel, B.Y.; Navakauskiene, R.; Christodoulou, M.S.; Passarella, D.; Klimasauskas, S.; et al. New Insights into the Epigenetic Activities of Natural Compounds. *OBM Genet.* **2018**, *2*, 029. [CrossRef]
198. Li, P.; Kong, J.; Chen, Z.; Huang, S.; Lv, G.; Wei, B.; Wei, J.; Jing, K.; Quan, J.; Chu, J. Aloin Promotes Osteogenesis of Bone-Marrow-Derived Mesenchymal Stem Cells via the ERK1/2-Dependent Runx2 Signaling Pathway. *J. Nat. Med.* **2019**, *73*, 104–113. [CrossRef] [PubMed]
199. Chen, T.; Li, P.; Qiu, J.; Hu, W.; Li, S.; Shi, H.; Qiu, X.; Huang, D.; Gao, W.; Liang, A. Aloin Regulates Matrix Metabolism and Apoptosis in Human Nucleus Pulposus Cells via the TAK1/NF-κB/NLRP3 Signaling Pathway. *Stem Cells Int.* **2022**, *2022*, 5865011. [CrossRef]
200. Moskalev, A.; Chernyagina, E.; de Magalhães, J.P.; Barardo, D.; Thoppil, H.; Shaposhnikov, M.; Budovsky, A.; Fraifeld, V.E.; Garazha, A.; Tsvetkov, V.; et al. Geroprotectors.org: A New, Structured and Curated Database of Current Therapeutic Interventions in Aging and Age-Related Disease. *Aging* **2015**, *7*, 616–628. [CrossRef] [PubMed]

nutrients

MDPI

Review

Possible Extracellular Signals to Ameliorate Sarcopenia in Response to Medium-Chain Triglycerides (8:0 and 10:0) in Frail Older Adults

Osamu Ezaki

Institute of Women's Health Science, Showa Women's University, Tokyo 154-8533, Japan; ezaki1952@yahoo.co.jp

Abstract: In frail older adults (mean age 85 years old), a 3-month supplementation with a low dose (6 g/day) of medium-chain triglycerides (MCTs; C8:0 and C10:0) given at a meal increased muscle mass and function, relative to supplementation with long-chain triglycerides (LCTs), but it decreased fat mass. The reduction in fat mass was partly due to increased postprandial energy expenditure by stimulation of the sympathetic nervous system (SNS). However, the extracellular signals to ameliorate sarcopenia are unclear. The following three potential extracellular signals to increase muscle mass and function after MCT supplementation are discussed: (1) Activating SNS—the hypothesis for this is based on evidence that a beta2-adrenergic receptor agonist acutely (1–24 h) markedly upregulates isoforms of peroxisomal proliferator-activated receptor gamma coactivator-1alpha (PGC-1alpha) mRNAs, promotes mitochondrial biogenesis, and chronically (~1 month) induces muscle hypertrophy. (2) An increased concentration of plasma acyl-ghrelin stimulates growth hormone secretion. (3) A nitrogen-sparing effect of ketone bodies, which fuel skeletal muscle, may promote muscle protein synthesis and prevent muscle protein breakdown. This review will help guide clinical trials of using MCTs to treat primary (age-related) sarcopenia.

Keywords: acyl-ghrelin; beta2-adrenergic receptor; PGC-1alpha isoform; sympathetic nervous system; sarcopenia; muscle mass; medium chain fatty acid; medium chain triglyceride; ketone body; protein-sparing effect

1. Introduction

Medium-chain triglycerides (MCTs; C8:0 and C10:0) are promising nutrients for the treatment of sarcopenia [1]. Recently, a combined data analysis of two clinical trials with frail older adults (mean age 85 years old) showed that relative to supplementation with 6 g long-chain triglycerides (LCTs)/day, supplementation with 6 g MCTs/day for 3 months statistically significantly increased body weight, body mass index (BMI), muscle mass (right arm muscle area, left calf circumference), and strength/function (right-hand grip strength, right and left knee extension times, walking speed, and number of iterations in leg open and close test) and decreased fat mass (right triceps skinfold thickness) [2]. These increases in favored measures of muscle mass and function were observed with and without supplementation with L-leucine (1.2 g/day) and vitamin D-enriched essential amino acids [2]. The significant increase in body weight in the MCTs-supplemented group is somewhat surprising because dietary MCTs have been considered to reduce fat mass and body weight (while not affecting muscle mass) relative to dietary LCTs [3–7]. The reduction in fat mass was partly due to increased postprandial energy expenditure by stimulation of the sympathetic nervous system (SNS) (Section 4.1). However, the mechanisms of increased muscle mass and function after MCT supplementation in frail older adults are unclear.

The increases in muscle mass and function after MCT supplementation might be due to the following characteristics of participants and protocols in the clinical trials with frail older adults [1,2,8] when compared with other MCT trials with healthy young adults [3–7]: (1) Participants in a nursing home were very old adults (mean age was 85 years old). The response to MCTs may differ in young and older muscle cells. (2) The chronic effects of MCT supplementation for 3 months with a meal based on dietary reference intake in Japan were examined and, therefore, they were unlikely to suffer from severe malnutrition. (3) An MCT dose of 6 g/day is much less than that which aims to increase ketone bodies in the blood (usually 20~40 g/day of MCTs) [9]. A 6 g/day intake of MCTs (75% C8:0 and 25% C10:0) corresponds to about 4% of habitual energy intake (overall mean habitual energy intake was 1430 kcal/day), 15% of habitual fat intake (overall mean fat intake was 39 g/day), and 0.14 g/kg body weight (overall mean body weight was 42.6 kg) in this population.

Marked progress has been made in understanding the etiology of sarcopenia in the past two decades, which may lead to new treatments for this condition. Peroxisomal proliferator-activated receptor gamma coactivator-1alpha (PGC-1alpha) is an inducible transcriptional coactivator of mitochondrial biogenesis and cellular energy metabolism in skeletal muscle that may ameliorate muscle atrophy [10]. A reduction in both RNA and protein expression of PGC-1alpha and downstream signaling targets in slow- and fast-twitch muscle fibers (i.e., multinuclear muscle cells containing fibers of actin and myosin) have been considered major causes of mitochondrial dysfunction during aging [11]. This mitochondrial dysfunction leads to the depletion of adenosine triphosphate (ATP) in muscle fibers, which may result in muscle atrophy [11,12].

The expression of the PGC-1alpha gene is regulated by two different promoters—proximal (canonical) and alternative [13–15]. A beta2-adrenergic receptor (AR) agonist activates the alternative promoter, not the proximal promoter [13]. Via a beta2-AR/cAMP/PKA/cAMP-response element-binding protein (CREB) signaling cascade [16] (Figure 1), an activation of beta2-AR led to increases in the isoforms of PGC-1alpha mRNAs (named PGC-1alpha-b and PGC-1alpha-c) in mouse [17] and human [18] skeletal muscles and PGC-1alpha protein in mouse skeletal muscle [15], and to an increase in mitochondrial function in C2C12 cells [19]. Chronically, beta2-AR agonist (formoterol) administration for 4 weeks in old rats attenuated age-related muscle wasting and weakness [20]. Therefore, MCT supplementation may ameliorate muscle atrophy by way of SNS activation.

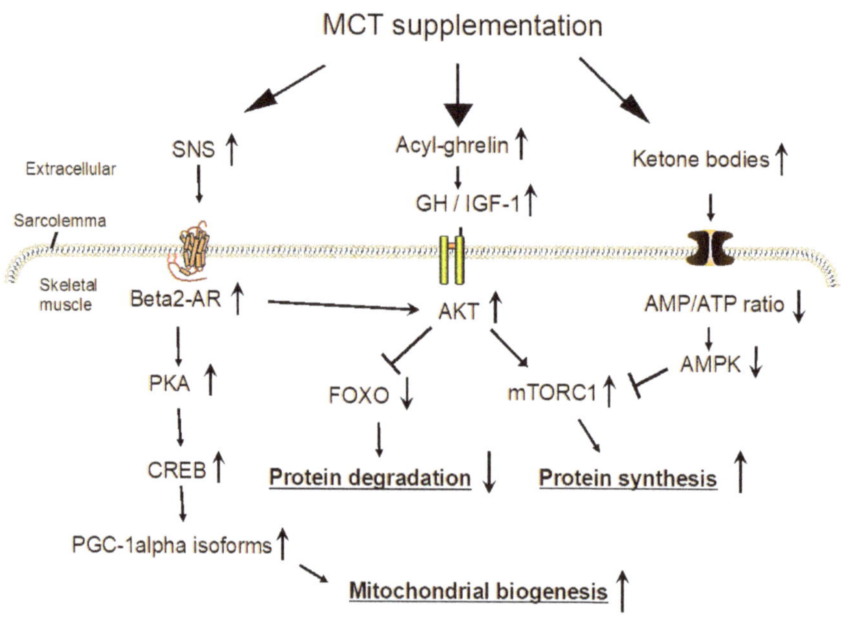

Figure 1. Three possible signaling pathways causing increased muscle mass and function in response to supplementation with medium-chain triglycerides (C8:0 and C10:0) in frail older adults. They are,

from left to right: (1) SNS activation, (2) acyl-ghrelin, and (3) a nitrogen-sparing effect of ketone bodies. Note that each of their main signaling pathways is shown here. See the text for detail. The lines with arrows indicate activation, whereas lines with flat ends indicate inhibition between two molecules. The upward line after the word represents the activated state and the downward line indicates the inhibited state after MCT supplementation relative to controls. AMP, adenosine monophosphate; AMPK, AMP-activated protein kinase; AR, adrenergic receptor; CREB, cAMP-response element-binding protein; FOXO, forkhead box protein O; GH, growth hormone; IGF, insulin-like growth factor; MCT, medium-chain triglyceride; mTORC, mechanistic (mammalian) target of rapamycin complex; PKA, protein kinase A; PGC, peroxisomal proliferator-activated receptor gamma coactivator; SNS, sympathetic nervous system.

This review outlines dietary MCTs' fate (Section 2) and summarizes previous reports describing the mechanisms of MCTs' effects on the amelioration of muscle atrophy (Section 3). Then, taking the characteristics of the clinical trials' participants and protocols into consideration, the following three possible extracellular signals causing ameliorating effects of MCTs on primary sarcopenia are discussed: (1) SNS activation (Section 4), (2) acyl-ghrelin (Section 5), and (3) the nitrogen-sparing effect of ketone bodies (Section 6).

2. Outline of Metabolisms in Dietary MCTs

Dietary MCTs can hydrolyze to medium-chain fatty acids (MCFAs) by lingual lipase in the oral cavity and stomach [21]. MCFAs in the stomach are used to form acyl-ghrelin, which can bind and activate the growth hormone secretagogues receptor (GHSR) [22]. Discovered in the stomach in 1999, ghrelin requires C8:0 to bind and activate GHSR [23]. Ghrelin O-acyltransferase (GOAT) acylates proghrelin with C8:0 to produce acyl-proghrelin, which is converted to acyl-ghrelin by the prohormone convertase [22,24]. Other MCFAs (C6:0 or C10:0) may substitute for C8:0 [25]. MCFAs in the intestinal lumen reduce the concentration of plasma gastric inhibitory peptide (GIP), which leads to decreased fat mass [26,27]. Note that increased concentrations of GIP promote obesity [28]. MCFAs absorbed in the intestine are transferred to the liver via the portal vein and metabolized to acetyl-CoA in the mitochondria [29]. Surplus production of acetyl-CoA that neither enters the TCA cycle nor participates in fatty acid synthesis leads to the formation of ketone bodies (beta-hydroxybutyrate [BHB] and acetoacetate). MCFAs are metabolized more rapidly and completely than LCFAs [30]. This rapid oxidation leads to greater ketone body formation. MCFAs that escape metabolism in the liver and ketone bodies produced by the liver (ketone bodies are not metabolized in the liver) are distributed in the blood via the circulatory system [31]. Peripheral tissues such as the brain, skeletal muscle, and heart can use both MCFAs [32,33] and ketone bodies [34,35] as energy sources and signaling molecules.

3. Previous Studies Elucidated Cellular Mechanisms for Amelioration of Muscle Atrophy in Response to MCT Supplementation

The mechanisms of MCTs' ameliorating effects on muscle atrophy have been elucidated using animal models of energy restriction [36], immobilization [37], and cachexia [38,39] with their appropriate controls.

Three-week-old male Wistar rats were malnourished under a 40% energy-restricted diet containing either MCTs (65% C8:0 and 19% C10:0; MCTs corresponding to 16% of total energy intake and 86% of total fat intake) or LCTs for 15 days and were then sacrificed [36]. The total skeletal muscle weight of the right leg tended to be 15% higher in the MCT-fed group than in the LCT-fed group ($p = 0.10$) [36]. Plasma albumin and insulin concentrations were higher in the MCT-fed mice than in the LCT-fed mice. A two-fold increase in plasma insulin concentration in the MCT-fed group may have contributed to increased protein synthesis in the liver and muscles.

Nine-week-old male Wistar rats fed a diet containing either MCTs (60% C8:0 and 22% C10:0; MCTs corresponding to 14% of total energy intake and 86% of total fat intake) or LCTs for 3 days underwent unilateral hindlimb immobilization [37]. Rats were sacrificed after immobilization for 3, 7, and 14 days. The immobilized soleus muscle mass in the LCT-fed rats decreased markedly compared to that in the contralateral muscle; however, these losses were partially suppressed in the MCT-fed mice with an inhibition of increased muscle ring-finger protein (MuRF-1) content in immobilized muscle. These data suggest that dietary MCT partially alleviated immobilization-induced muscle atrophy by inhibiting the ubiquitin-proteasome pathway.

Seven-week-old male ICR mice were fed a regular chow diet containing either MCT oil (MCFAs not specified) or LCT oil for 1 week [38]. Then, lipopolysaccharide (LPS) was injected into the mice, and 24 h later, they were killed and plasma and tissues were obtained. MCT supplementation suppressed the LPS-induced decrease in plasma ketone bodies and citrate synthase activity (a marker of mitochondrial enzyme activity) in tibialis anterior muscles but did not affect the LPS-induced reduction in PGC-1alpha protein in tibialis anterior muscles relative to LCTs. The authors suggested that the amount of PGC-1alpha protein might not change for 1 week of supplementation.

Six-week-old male Sprague–Dawley rats received a gastric tube, and 48 h later, LPS was injected [39]. Then, rats received enteral nutrition (EN, 100 kcal/kg body weight) containing different doses of C8:0 (0, 0.25, 0.5, and 1 g/kg body weight, respectively) via the gastric tube for 3 days. After subcutaneous injection of L-^{13}C-valine to measure the fractional synthetic rate of muscle proteins, rats were sacrificed. LPS led to decreased body weight and insulin resistance and increased serum inflammatory cytokines, inhibited AKT, inhibited mechanistic target of rapamycin complex (mTORC)1, inhibited forkhead box protein O (FOXO)-1 phosphorylation in muscle (extensor digitorum longus), increased the expression of atrogin-1 and muscle ring-finger protein (MuRF)-1 in muscle, and decreased the fractional synthetic rate of muscle proteins. C8:0-rich EN and simple EN (a dose of 0 g/kg body weight of C8:0) significantly ameliorated these LPS-induced cachexia phenotypes. Importantly, C8:0-rich EN showed a pronounced ameliorating effect compared with simple EN and also increased acyl-ghrelin in circulation, both in a dose-dependent manner.

In a cell experiment in C2C12 myotubes, MCFA (C10:0 or C12:0)-treated cells displayed decreased lipid accumulation, increased mitochondrial oxidative capacity, and increased amounts of PGC-1alpha protein compared with LCFA (C16:0, C18:1, or C18:2)-treated cells [40].

These results indicated that MCT supplementation under several pathophysiological conditions could increase muscle mass and function relative to appropriate controls, with promoted mitochondrial biogenesis, increased protein synthesis, and inhibition of protein degradation. However, aged model mice and their upstream signals (extracellular signals) have yet to be examined.

4. Sympathetic Nervous System Activation

Acute SNS activation promotes mitochondrial biogenesis via increased SNS-sensitive PGC-1alpha isoform [13,41]. Note that muscle mass did not increase in transgenic mice with skeletal muscle-specific overexpression of either PGC-1alpha-a (canonical form), PGC-1alpha-b (isoform), or PGC-1alpha-c (isoform), although they showed mitochondrial biogenesis [41]. Instead, 25-week-old PGC-1alpha overexpressing transgenic mice showed muscle atrophy with depletion of ATP by way of the uncoupling of oxidative phosphorylation [42]. Promoting mitochondrial biogenesis may prevent or treat muscle atrophy but does not increase muscle mass under normal conditions.

4.1. MCT Supplementation Increases Energy Expenditure during the Postprandial Period by Activating the SNS

In healthy mostly overweight adults, MCTs combined with energy-restricted diets resulted in greater fat mass loss and body weight loss than did LCTs combined with energy-restricted diets, and their mechanisms have been elucidated [43–47]. However, the energy restriction per se markedly alters the metabolism in the whole body (e.g., decreased plasma insulin and glucose concentrations, increased plasma ketone bodies and free fatty acids concentrations, and others), which may affect the impact of MCTs on energy metabolism [7]. Therefore, clinical trials to examine the effect of MCT supplementation with a regular diet (but not with an energy-restricted diet) on energy expenditure were informative. Repeating a single meal (1000 kcal) containing MCTs (60% C8:0 and 30% C10:0, 400 kcal in each meal) on a regular diet for 5 days showed a two-fold postprandial energy expenditure than that containing LCTs, suggesting that under conditions favorable for fat synthesis (i.e., higher insulin concentrations), MCTs have a greater thermic effect of food than LCTs [48]. In addition, a greater thermic effect of food following a single meal (400–900 kcal) containing MCTs was also observed compared with a meal containing LCTs [49,50]. Interestingly, a respiratory chamber study showed that the intake of MCTs (15–30 g per day) as part of a habitual diet enhanced daily energy expenditure by about 500 KJ/day, increasing 24-h levels of urinary noradrenaline but not adrenaline and dopamine [51].

In the animal studies that examined the effects of C8:0 and C10:0, an increase in norepinephrine concentration was observed in mice fed an MCFA-enriched diet (MCFAs corresponding to 4% of total energy intake and 10% of total fat intake, a dose similar to that of clinical trials with frail older adults) compared with mice fed an LCFA-enriched diet, which leads to increased lipolysis in white and brown adipose tissues and thermogenesis in brown adipose tissue with reductions in body weight and fat mass [52,53].

These results strongly suggest that, compared to LCT supplementation, MCT supplementation taken with a meal increases energy expenditure during the postprandial period by activating the SNS.

4.2. Activation of SNS May Ameliorate Muscle Atrophy in Primary Sarcopenia

Increasing the PGC-1alpha protein in skeletal muscle is one promising strategy for ameliorating primary sarcopenia. Although exercise is recommended for this purpose, it is challenging for frail older adults. However, SNS activation can increase the PGC-1alpha protein without exercise [15].

As described in Section 1, a beta2-AR activation markedly increases the expression of isoforms of PGC-1 alpha mRNA in skeletal muscle via the alternative promoter [13,17,54]. SNS-induced isoforms of PGC-1alpha (i.e., PGC-1alpha-b and PGC-1alpha-c) mRNA are expressed in skeletal muscles, and they are functional [41,55,56]. Of 795 amino acids in the total canonical PGC-1alpha protein, only the N-terminal 16 amino acids differed from those in PGC-1alpha-b or PGC-1alpha-c [13]. Recently, to examine the effects of the PGC-1alpha derived from the alternative promotor on whole-body metabolism, mice disrupting the alternative exon 1 were produced [57]. As expected, these mice failed to upregulate muscle PGC-1alpha mRNAs in response to exercise and cold exposure, and eventually developed obesity and hyperinsulinemia. These data suggest that the PGC-1alpha-b and PGC-1alpha-c isoforms respond to external stimuli and are crucial for energy metabolism in the whole body.

In animal and human studies, chronic administrations of beta2-AR agonists, which are prescribed as bronchodilators for asthmatic patients, increased skeletal muscle mass and decreased fat mass [16,58]. Their mechanisms were elucidated as follows: activating the AKT/mTORC1 signaling pathway [59,60], attenuating the myostatin signaling pathway [61], and reducing calpain activity [60]. The muscle hypertrophy requires a stimulatory G-protein subunit [62] and beta-arrestin 1 [63]. However, in some studies, chronic treatment of a beta2-adrenergic receptor agonist (clenbuterol) resulted in decreases in mitochondrial contents and fatty acid oxidation [64,65], which were associated with a reduction of PGC-1alpha mRNA and protein [64]. Therefore, it is conceivable that beta2-AR agonists could increase muscle mass independent of the PGC-1alpha protein [16]. The acute effects of beta2-agonists may fundamentally differ from their chronic effects. The mechanisms by which the acute effects of a beta2-AR agonist are transient are unclear but may be partly due to a reduction in beta2-AR concentration in skeletal muscle by a long-time beta2-AR agonist administration (so-called "desensitization of beta2-AR signaling") [66].

4.3. Conclusions

MCT supplementation activates the SNS, which may prevent the decreased muscle mass and function observed in primary sarcopenia. A signal transduction pathway of beta2-AR to increase transcription of the PGC-1alpha isoform is shown in Figure 1. Note that the subcutaneous injection of clenbuterol in rats phosphorylates and activates AKT in skeletal muscles, by which muscle protein synthesis is promoted via PKA-independent signaling pathways [59]. Charlot et al. reported that, under normal conditions, 12-week-old male C57BL/6J mice fed a C8:0-rich diet (C8:0 corresponding to 18% of total energy intake and 80% of total fat intake) for 6 weeks presented higher spontaneous activity and endurance capacities, along with promoted mitochondrial biogenesis in skeletal muscle with increased PGC-1alpha mRNA and protein levels, relative to mice fed a chow (LCFAs-rich) diet [67]. However, SNS activity and body composition in each group of mice were not measured.

To prove the contributions of the SNS to the effect of MCTs on the prevention of sarcopenia, as well as animal studies examining the mechanisms using beta2-AR knockout mice, several clinical studies of MCT supplements targeted at older adults are needed to examine the chronic effects of MCTs on the association between postprandial SNS activity and muscle mass. If the results were positive, the tissues or soluble factors responsible for SNS activation might be sought. As suggested by cell culture experiments [40], MCFA might directly activate muscle fibers via beta2-adrenergic receptors. Recently, it was reported that the activation of the neurons in the ventromedial hypothalamic nucleus expressing steroidogenic factor-1 increases the sympathoadrenal activity and isoforms of PGC-1alpha in skeletal muscles via multiple downstream nodes in mice [68]. Therefore, MCT supplementation may activate the central nervous system to activate the SNS, increasing muscle mass and function. In addition, clinical trials to determine a suitable cessation interval for MCT supplementations might be necessary to avoid the possible desensitization of beta2-AR signaling due to the continuous activation of the SNS.

5. Acyl-Ghrelin

Although many fatty acids exist in nature, MCFAs (C6:0, C8:0, and C10:0) are specific; they are part of acyl-ghrelin (active ghrelin or simply ghrelin), an orexigenic hormone produced by the stomach [23]. Without MCFAs, no acyl-ghrelin is produced. Acyl-ghrelin stimulates GH release, and GH increases muscle mass (the MCTs/Ghrelin/GH hypothesis).

5.1. Clinical Trials to Estimate Muscle Mass in Response to MCTs, Acyl-Ghrelin, and GHSR Agonists

In clinical trials examining the effects of MCTs, acyl-ghrelin, and GHSR agonists on muscle mass, GHSR agonists consistently showed increased muscle mass [69–71]. However, clinical trials of MCTs were insufficient as the intervention periods were very short, and no negative control (i.e., LCTs) was provided.

Acyl-ghrelin (mostly C8:0-ghrelin) produced in the stomach enters the blood circulation or activates the gastric afferent vagal nerve and stimulates GH release in the pituitary [72]. GH can transmit its anabolic signal directly through its receptor or indirectly through insulin-like growth factor 1 (IGF-1) stimulation. Primary GH and IGF-1 receptor downstream signaling, which increases muscle mass, occurs via AKT [73] (Figure 1). To investigate this hypothesis, blood acyl-ghrelin concentrations after MCT supplementation were measured, and all studies showed increases in acyl-ghrelin concentration from its baseline value [74–77]. In one of these studies, body weight was significantly increased for a 2-week treatment: a single oral ingestion of 3 g of MCT (100% C8:0) with an EN formula (400 kcal in 400 mL) after breakfast in cachectic patients increased plasma acyl-ghrelin, and the 2-week administration of MCTs with the formula increased appetite score, body weight, and serum albumin and IGF-1 concentrations [74]. However, these increases after MCT supplementation were relative to their baseline values but not those in the LCTs-supplemented formula group. The increased body weight may be due to increased energy intake from the EN formula. The other three studies also did not make comparisons with an LCT group (isocaloric negative control) [75–77].

It is well known that acute acyl-ghrelin injection increases GH concentration, appetite, and food intake and that chronic acyl-ghrelin injection increases body weight [72]. However, few studies have shown increases in muscle mass, possibly due to the shorter intervention periods. Chronic effects of acyl-ghrelin over years are mimicked by GHSR agonists, which have been reported in three clinical trials [69–71], all of which showed increases in body weight and fat-free mass with no change in fat mass. Note that GH has a strong lipolytic effect on adipose tissues [78]; therefore, the GHSR agonist should have decreased fat mass. The reason for this discrepancy is unclear at present.

5.2. Animal Study to Support the MCTs/Ghrelin/GH Hypothesis

In an additional experiment by Zhang et al., to further examine the effects of acyl-ghrelin on LPS-induced cachexia (see Section 3), a specific inhibitor of GOAT (Go-CoA-Tat) was added to C8:0-rich EN [39]. The protective effects of C8:0-rich EN against cachexia were abolished mostly by the administration of Go-CoA-Tat, along with a decrease in serum acyl-ghrelin concentration. These results suggested that increased serum acyl-ghrelin causes the protective effects of dietary C8:0 against cachexia.

5.3. Conclusions

Applying the results from LPS-injected animals [39], "increased acyl-ghrelin concentration" appeared to be a cause of MCTs' effects in ameliorating primary sarcopenia. However, the effects of MCTs in animal models of primary sarcopenia, using inhibitors of GOAT or mice with knockout of GOAT, GHSR, or ghrelin are needed. Postprandial plasma acyl-ghrelin concentrations after MCT supplementation should be measured in frail older adults, who eventually develop increased muscle mass and function in 3-month intervention periods.

6. Nitrogen-Sparing Effect of Ketone Bodies

In a clinical trial of frail older adults (mean age 86.6 years) in a nursing home, LCT (6 g/day) supplementation with L-leucine (1.2 g/day), L-isoleucine (0.3 g/day), L-valine (0.3 g/day), and other essential amino acids and vitamins including a vitamin D (20 μg/day)-enriched supplement combined with a regular diet for 3 months did not significantly increase muscle mass and function relative to no supplementation [8]. The branched-chain amino acids (leucine, isoleucine, and valine) in circulation were preferentially used by muscle for energy production [79]. Leucine has been well known to play a pivotal role in the protein-sparing effects of amino acids [80]. Also, leucine located in the cytosol in muscle fiber indirectly activates the mTORC1, which is a protein kinase that phosphorylates substrates to potentiate anabolic processes and inhibits catabolic ones [81].

This result suggested that the participants in this nursing home might not be severely malnourished. Larger amounts of leucine and other essential amino acids or longer feeding periods (more than 3 months) may be required to obtain significant increases in muscle mass and function. In other words, even under this condition (i.e., non-significant effects of 1.2 g/day of leucine supplementation), 6 g MCTs/day of supplementation with a meal can effectively increase muscle mass and function. Because MCTs induce ketogenesis (Section 2), ketone bodies might be the molecules responsible for increases in muscle mass and function.

6.1. Clinical Trials Supporting the Nitrogen-Sparing Effect of MCT Relative to LCT Supplementation

Krotkiewski reported that in obese women (n = 22 in each group, overall mean age 43 years) receiving a very low-calorie diet (579 kcal/day), the MCT group (9.9 g MCTs/day) increased the rate of decline of body fat (MCT −5.8 kg vs. LCT −4.9 kg after 4 weeks, $p < 0.05$) but decreased the rate of decline of fat-free mass (MCT −2.7 kg vs. LCT −3.2 kg after 4 weeks, $p < 0.05$) relative to the LCT groups [43]. The author suggested that the increased plasma ketone body concentrations caused by dietary MCTs may provide additional energy to muscles and prevent the breakdown of muscle tissues caused by malnutrition (nitrogen-sparing effects of ketone bodies) [43]. In another study of clinically ill patients with sepsis and trauma receiving total parenteral nutrition (TPN), an improving effect of a structured MCT-containing emulsion (n = 9, mean age 52 years, 1.5 g MCTs/kg body weight/day) on the nitrogen balance was observed for at least 3 days of administration compared to an LCT emulsion (n = 11, mean age 59 years, 1.5 g LCTs/kg body weight/day). Thus, it appears that the nitrogen-sparing effect of ketone bodies was observed with long-term low-calorie diets or protein-energy malnutrition. The frail older adults who increased muscle mass and function after MCT supplementation might be experiencing some malnutrition; therefore, the nitrogen-sparing effect of ketone bodies might be in operation.

6.2. Can Increased Ketone Bodies after MCT Supplementation Promote Muscle Protein Synthesis?

Postprandial plasma ketone body concentrations after MCT supplementation with a meal have not been measured in clinical trials in frail older adults [1,8] but have been estimated by similar studies. For example, in healthy young volunteers, after oral intake of 10 g MCT (60% C8:0 and 40% C10:0) in a complete liquid diet mimicking a standard meal, high plasma levels of ketone bodies were produced over 2 h with a maximum plasma total ketone bodies concentration of 409 μM (150 μM at baseline) [82]. In another study, a supplementation with 10 g MCTs (60% C8:0 and 40% C10:0)/day at breakfast increased blood ketone body concentrations from 49–97 μM (baseline) to 455 μM (maximum) 30 min after MCT supplementation, followed by a return to baseline levels after 3 h [9]. Therefore, we estimated that 6 g MCT supplementation with a meal might result in mild ketosis (~0.5 mM) during a 3-h postprandial period.

The nitrogen-sparing effect of ketone bodies in skeletal muscle may depend on the acute (with or without a meal) and chronic (lean or obese) energy status in the whole body. Several clinical studies examined the acute effect of exogenous ketone bodies on muscle protein synthesis under different feeding conditions. Without a meal, which favors ketogenesis, more than 1 mM concentrations of ketone bodies seemed necessary to observe a meaningful nitrogen-sparing effect [83,84]. A 3-h infusion of BHB revealed nitrogen conservation in the whole body without a meal when total blood ketone body concentrations were 1.1–1.2 mM [83]. In another study, after an overnight fast, an 8-h infusion of BHB without a meal increased the incorporation of a tracer C^{13} leucine into skeletal muscles (protein synthesis) by 5–17% and decreased the appearance of the C^{13} in the expired CO_2 (leucine oxidation) by 18–41% compared to that during normal saline infusion, during which plasma BHB concentrations were 1–2.5 mM [84].

Very recently, the effects of acute ingestion of exogenous ketone monoester with and without dietary protein (10 g whey protein) co-ingestion on postprandial myofibrillar protein synthesis rates have been examined in healthy young males [85]. In a 1-h postprandial period, blood BHB concentrations reached 3.2–3.4 mM in the ketone-only and the ketone plus protein groups, whereas the concentration in the protein-only group was 0.6 mM. In all three groups, the postprandial rate of muscle protein synthesis was higher than that in basal conditions, with no difference between the groups. This result indicated that a large amount of plasma ketone bodies did not promote an increase in the rate of muscle protein synthesis either in the absence or presence of a protein-rich meal.

Taken together, 6 g MCTs/day supplementation with a regular meal, which might induce mild ketosis (~0.5 mM), may not exhibit the nitrogen-sparing effect of ketone bodies. However, the chronic effects of ketone bodies have yet to be examined. Repeated small increases in ketone bodies at each meal may contribute to the chronic increase in muscle mass in frail older adults.

6.3. Possible Molecular Mechanisms of the Protein-Sparing Effects of Ketone Bodies

The molecular mechanisms are speculated as follows. Under malnutrition, due to a shortage of ATP and increased adenosine monophosphate (AMP) in skeletal muscle, AMP-activated protein kinase (AMPK), the sensor of intracellular energy levels and glucose availability, is activated [86]. Then, AMPK reduces mTORC1 activity and inhibits protein synthesis [87]. Increases in ketone bodies could reverse these steps by providing ATP to skeletal muscles (Figure 1). However, it is worth noting that if a large amount of ketone bodies and fatty acids were given, AMPK in skeletal muscle may be activated due to a shortage of glucose.

Sirtuin 1 (SIRT1), an enzyme of class III deacetylase, may relate to ketone bodies' protein-sparing effects. In cells, it is well known that SIRT1 is activated by an increased nicotinamide adenine dinucleotide $(NAD)^+/NADH$ ratio, which plays an important role in cell survival and longevity under caloric restriction [88]. That may also apply to humans. When blood glucose declines due to a low-carbohydrate diet, the inhibited glycolysis increases the amount of NAD^+ and then activates SIRT1. In skeletal muscle, to utilize fatty acids instead of glucose as a fuel, SIRT1 deacetylates and activates the PGC-1alpha protein and promotes mitochondrial biogenesis [89]. SIRT1 also increases mitochondrial biogenesis by increasing the expression of mitochondrial transcription factor A (TFAM) in a PGC-1alpha-independent manner [90]. In addition, SIRT1 deacetylates and inactivates FOXO and inhibits muscle atrophy [91]. Therefore, SIRT1 activation might increase muscle mass and function [92].

Ketone bodies may activate SIRT1 in some tissues. The use of ^{31}P phosphorus magnetic resonance spectroscopy (MRS) in humans showed that a 10 g MCT (60% C8:0 and 40% C10:0) supplementation increased the amount of NAD^+ in the brain concomitant with an increase in plasma BHB concentration [82]. Increased NAD^+ may activate SIRT1. Cockayne syndrome is a progressive neurodegenerative disorder with mitochondrial dysfunction caused by mutations in genes encoding DNA repair proteins. In a mouse model of this condition, a high-fat diet increased blood and brain BHB concentrations, activated brain SIRT1, and rescued their phenotypes, including hearing loss but not muscle function [93]. In addition, in this cell model, BHB treatment activated SIRT1 and rescued its associated phenotypes [93]. Therefore, BHB can activate SIRT1 in the brain and change the phenotypes. However, whether BHB activates SIRT1 in skeletal muscle and whether it increases muscle mass and function are still being determined.

6.4. Conclusions

To prove the nitrogen-sparing effect of ketone bodies in frail older adults, the chronic effects of supplementation with a low dose of ketone monoester on muscle protein synthesis rate and muscle mass should be investigated in this population. Positive results indicate the need to clarify the molecular mechanisms of the nitrogen-sparing effects of ketone bodies.

7. Limitation

The mechanisms of MCTs' ameliorating effects on primary sarcopenia described here were mostly based on experiments using young animals and tissues. It is conceivable that young cells and older cells respond to MCTs differently. Therefore, comparative studies of MCTs targeted at young and older cells, tissues, animals, and humans are recommended.

8. Conclusions

Based on recent progress in understanding the etiology of primary sarcopenia and the discovery of SNS-sensitive isoforms of PGC-1alpha, three possible extracellular signals causing increases in muscle mass and function after MCT supplementation in frail older adults were discussed, as follows: (1) activation of the SNS, (2) an increased plasma acyl-ghrelin concentration, and (3) a nitrogen-sparing effect of ketone bodies. Each hypothesis is possible but will require further research that includes clinical trials. Studies using appropriate knockout mice and specific inhibitors will also be necessary to clarify the causality.

Funding: This research received no external funding.

Conflicts of Interest: The authors declare no conflict of interest.

Abbreviations

AMP: adenosine monophosphate; AMPK, AMP-activated protein kinase; AR, adrenergic receptor; BHB, beta-hydroxybutyrate; BMI, body mass index; CREB, cAMP-response element-binding protein; EN, enteral nutrition; FOXO, forkhead box protein O; GH, growth hormone; GHSR, growth hormone secretagogues receptor; GIP, gastric inhibitory peptide; GOAT, ghrelin O-acyltransferase; IGF, insulin-like growth factor; LCFAs, long-chain fatty acids; LCT, long-chain triglycerides; LPS, lipopolysaccharide; MCFAs, medium-chain fatty acids; MCT, medium-chain triglyceride; mTORC, mechanistic (mammalian) target of rapamycin complex; MuRF, muscle ring-finger protein; NAD, nicotinamide adenine dinucleotide; PGC, peroxisomal proliferator-activated receptor gamma coactivator; PKA, protein kinase A; SNS, sympathetic nervous system.

References

1. Abe, S.; Ezaki, O.; Suzuki, M. Medium-chain triglycerides (8:0 and 10:0) are promising nutrients for sarcopenia: A randomized controlled trial. *Am. J. Clin. Nutr.* **2019**, *110*, 652–665. [CrossRef] [PubMed]
2. Ezaki, O.; Abe, S. Medium-chain triglycerides (8:0 and 10:0) increase muscle mass and function in frail older adults: A combined data analysis of clinical trials. *Front. Nutr.* **2023**, *10*, 1284497. [CrossRef] [PubMed]
3. Bach, A.C.; Ingenbleek, Y.; Frey, A. The usefulness of dietary medium-chain triglycerides in body weight control: Fact or fancy? *J. Lipid Res.* **1996**, *37*, 708–726. [CrossRef]
4. Mumme, K.; Stonehouse, W. Effects of medium-chain triglycerides on weight loss and body composition: A meta-analysis of randomized controlled trials. *J. Acad. Nutr. Diet.* **2015**, *115*, 249–263. [CrossRef]
5. Bueno, N.B.; de Melo, I.V.; Florencio, T.T.; Sawaya, A.L. Dietary Medium-Chain Triacylglycerols versus Long-Chain Triacylglycerols for Body Composition in Adults: Systematic Review and Meta-analysis of Randomized Controlled Trials. *J. Am. Coll. Nutr.* **2015**, *34*, 175–183. [CrossRef]
6. Watanabe, S.; Tsujino, S. Applications of Medium-Chain Triglycerides in Foods. *Front. Nutr.* **2022**, *9*, 802805. [CrossRef]
7. Fernandez-Verdejo, R.; Mey, J.T.; Ravussin, E. Effects of ketone bodies on energy expenditure, substrate utilization, and energy intake in humans. *J. Lipid Res.* **2023**, *64*, 100442. [CrossRef]
8. Abe, S.; Ezaki, O.; Suzuki, M. Medium-Chain Triglycerides in Combination with Leucine and Vitamin D Increase Muscle Strength and Function in Frail Elderly Adults in a Randomized Controlled Trial. *J. Nutr.* **2016**, *146*, 1017–1026. [CrossRef] [PubMed]
9. Vandenberghe, C.; St-Pierre, V.; Fortier, M.; Castellano, C.A.; Cuenoud, B.; Cunnane, S.C. Medium Chain Triglycerides Modulate the Ketogenic Effect of a Metabolic Switch. *Front. Nutr.* **2020**, *7*, 3. [CrossRef] [PubMed]
10. Sandri, M.; Lin, J.; Handschin, C.; Yang, W.; Arany, Z.P.; Lecker, S.H.; Goldberg, A.L.; Spiegelman, B.M. PGC-1α protects skeletal muscle from atrophy by suppressing FoxO3 action and atrophy-specific gene transcription. *Proc. Natl. Acad. Sci. USA* **2006**, *103*, 16260–16265. [CrossRef]
11. Marzetti, E.; Calvani, R.; Coelho-Junior, H.J.; Landi, F.; Picca, A. Mitochondrial Quantity and Quality in Age-Related Sarcopenia. *Int. J. Mol. Sci.* **2024**, *25*, 2052. [CrossRef]
12. Miao, Y.; Xie, L.; Song, J.; Cai, X.; Yang, J.; Ma, X.; Chen, S.; Xie, P. Unraveling the causes of sarcopenia: Roles of neuromuscular junction impairment and mitochondrial dysfunction. *Physiol. Rep.* **2024**, *12*, e15917. [CrossRef]
13. Miura, S.; Kai, Y.; Kamei, Y.; Ezaki, O. Isoform-specific increases in murine skeletal muscle peroxisome proliferator-activated receptor-gamma coactivator-1α (PGC-1α) mRNA in response to beta2-adrenergic receptor activation and exercise. *Endocrinology* **2008**, *149*, 4527–4533. [CrossRef]
14. Yoshioka, T.; Inagaki, K.; Noguchi, T.; Sakai, M.; Ogawa, W.; Hosooka, T.; Iguchi, H.; Watanabe, E.; Matsuki, Y.; Hiramatsu, R.; et al. Identification and characterization of an alternative promoter of the human PGC-1α gene. *Biochem. Biophys. Res. Commun.* **2009**, *381*, 537–543. [CrossRef]
15. Chinsomboon, J.; Ruas, J.; Gupta, R.K.; Thom, R.; Shoag, J.; Rowe, G.C.; Sawada, N.; Raghuram, S.; Arany, Z. The transcriptional coactivator PGC-1α mediates exercise-induced angiogenesis in skeletal muscle. *Proc. Natl. Acad. Sci. USA* **2009**, *106*, 21401–21406. [CrossRef] [PubMed]
16. Lynch, G.S.; Ryall, J.G. Role of beta-adrenoceptor signaling in skeletal muscle: Implications for muscle wasting and disease. *Physiol. Rev.* **2008**, *88*, 729–767. [CrossRef]

17. Tadaishi, M.; Miura, S.; Kai, Y.; Kawasaki, E.; Koshinaka, K.; Kawanaka, K.; Nagata, J.; Oishi, Y.; Ezaki, O. Effect of exercise intensity and AICAR on isoform-specific expressions of murine skeletal muscle PGC-1α mRNA: A role of β_2-adrenergic receptor activation. *Am. J. Physiol. Endocrinol. Metab.* **2011**, *300*, E341–E349. [CrossRef]
18. Norrbom, J.; Sallstedt, E.K.; Fischer, H.; Sundberg, C.J.; Rundqvist, H.; Gustafsson, T. Alternative splice variant PGC-1α-b is strongly induced by exercise in human skeletal muscle. *Am. J. Physiol. Endocrinol. Metab.* **2011**, *301*, E1092–E1098. [CrossRef]
19. Azevedo Voltarelli, V.; Coronado, M.; Goncalves Fernandes, L.; Cruz Campos, J.; Jannig, P.R.; Batista Ferreira, J.C.; Fajardo, G.; Chakur Brum, P.; Bernstein, D. β_2-Adrenergic Signaling Modulates Mitochondrial Function and Morphology in Skeletal Muscle in Response to Aerobic Exercise. *Cells* **2021**, *10*, 146. [CrossRef] [PubMed]
20. Ryall, J.G.; Schertzer, J.D.; Lynch, G.S. Attenuation of age-related muscle wasting and weakness in rats after formoterol treatment: Therapeutic implications for sarcopenia. *J. Gerontol. A Biol. Sci. Med. Sci.* **2007**, *62*, 813–823. [CrossRef] [PubMed]
21. Liao, T.H.; Hamosh, P.; Hamosh, M. Fat digestion by lingual lipase: Mechanism of lipolysis in the stomach and upper small intestine. *Pediatr. Res.* **1984**, *18*, 402–409. [CrossRef]
22. Nishi, Y.; Hiejima, H.; Hosoda, H.; Kaiya, H.; Mori, K.; Fukue, Y.; Yanase, T.; Nawata, H.; Kangawa, K.; Kojima, M. Ingested medium-chain fatty acids are directly utilized for the acyl modification of ghrelin. *Endocrinology* **2005**, *146*, 2255–2264. [CrossRef]
23. Kojima, M.; Hosoda, H.; Date, Y.; Nakazato, M.; Matsuo, H.; Kangawa, K. Ghrelin is a growth-hormone-releasing acylated peptide from stomach. *Nature* **1999**, *402*, 656–660. [CrossRef] [PubMed]
24. Thomas, A.S.; Sassi, M.; Angelini, R.; Morgan, A.H.; Davies, J.S. Acylation, a Conductor of Ghrelin Function in Brain Health and Disease. *Front. Physiol.* **2022**, *13*, 831641. [CrossRef]
25. Hosoda, H.; Kojima, M.; Mizushima, T.; Shimizu, S.; Kangawa, K. Structural divergence of human ghrelin. Identification of multiple ghrelin-derived molecules produced by post-translational processing. *J. Biol. Chem.* **2003**, *278*, 64–70. [CrossRef]
26. Murata, Y.; Harada, N.; Yamane, S.; Iwasaki, K.; Ikeguchi, E.; Kanemaru, Y.; Harada, T.; Sankoda, A.; Shimazu-Kuwahara, S.; Joo, E.; et al. Medium-chain triglyceride diet stimulates less GIP secretion and suppresses body weight and fat mass gain compared with long-chain triglyceride diet. *Am. J. Physiol. Endocrinol. Metab.* **2019**, *317*, E53–E64. [CrossRef] [PubMed]
27. Murata, Y.; Harada, N.; Kishino, S.; Iwasaki, K.; Ikeguchi-Ogura, E.; Yamane, S.; Kato, T.; Kanemaru, Y.; Sankoda, A.; Hatoko, T.; et al. Medium-chain triglycerides inhibit long-chain triglyceride-induced GIP secretion through GPR120-dependent inhibition of CCK. *iScience* **2021**, *24*, 102963. [CrossRef]
28. Yamane, S.; Harada, N. Gastric inhibitory polypeptide/glucose-dependent insulinotropic polypeptide signaling in adipose tissue. *J. Diabetes Investig.* **2019**, *10*, 3–5. [CrossRef]
29. Papamandjaris, A.A.; MacDougall, D.E.; Jones, P.J. Medium chain fatty acid metabolism and energy expenditure: Obesity treatment implications. *Life Sci.* **1998**, *62*, 1203–1215. [CrossRef]
30. Binnert, C.; Pachiaudi, C.; Beylot, M.; Hans, D.; Vandermander, J.; Chantre, P.; Riou, J.P.; Laville, M. Influence of human obesity on the metabolic fate of dietary long- and medium-chain triacylglycerols. *Am. J. Clin. Nutr.* **1998**, *67*, 595–601. [CrossRef]
31. St-Pierre, V.; Vandenberghe, C.; Lowry, C.M.; Fortier, M.; Castellano, C.A.; Wagner, R.; Cunnane, S.C. Plasma Ketone and Medium Chain Fatty Acid Response in Humans Consuming Different Medium Chain Triglycerides During a Metabolic Study Day. *Front. Nutr.* **2019**, *6*, 46. [CrossRef] [PubMed]
32. Augustin, K.; Khabbush, A.; Williams, S.; Eaton, S.; Orford, M.; Cross, J.H.; Heales, S.J.R.; Walker, M.C.; Williams, R.S.B. Mechanisms of action for the medium-chain triglyceride ketogenic diet in neurological and metabolic disorders. *Lancet Neurol.* **2018**, *17*, 84–93. [CrossRef] [PubMed]
33. Shcherbakova, K.; Schwarz, A.; Apryatin, S.; Karpenko, M.; Trofimov, A. Supplementation of Regular Diet With Medium-Chain Triglycerides for Procognitive Effects: A Narrative Review. *Front. Nutr.* **2022**, *9*, 934497. [CrossRef] [PubMed]
34. Newman, J.C.; Verdin, E. Ketone bodies as signaling metabolites. *Trends Endocrinol. Metab.* **2014**, *25*, 42–52. [CrossRef] [PubMed]
35. Cotter, D.G.; Schugar, R.C.; Crawford, P.A. Ketone body metabolism and cardiovascular disease. *Am. J. Physiol. Heart Circ. Physiol.* **2013**, *304*, H1060–H1076. [CrossRef] [PubMed]
36. Sekine, S.; Terada, S.; Aoyama, T. Medium-chain triacylglycerol suppresses the decrease of plasma albumin level through the insulin-Akt-mTOR pathway in the livers of malnourished rats. *J. Nutr. Sci. Vitaminol.* **2013**, *59*, 123–128. [CrossRef] [PubMed]
37. Nishimura, S.; Inai, M.; Takagi, T.; Nonaka, Y.; Urashima, S.; Honda, K.; Aoyama, T.; Terada, S. Preventive Effects of the Dietary Intake of Medium-chain Triacylglycerols on Immobilization-induced Muscle Atrophy in Rats. *J. Oleo Sci.* **2017**, *66*, 917–924. [CrossRef] [PubMed]
38. Hirabayashi, T.; Tanaka, M.; Matsumoto, T.; Maeshige, N.; Kondo, H.; Fujino, H. Preventive effects of medium-chain triglycerides supplementation on the oxidative capacity in skeletal muscle under cachectic condition. *Biomed. Res.* **2020**, *41*, 179–186. [CrossRef]
39. Zhang, Z.; Li, X.; Cao, C. Octanoic acid-rich enteral nutrition attenuated hypercatabolism through the acylated ghrelin-POMC pathway in endotoxemic rats. *Nutrition* **2023**, *119*, 112329. [CrossRef]
40. Montgomery, M.K.; Osborne, B.; Brown, S.H.; Small, L.; Mitchell, T.W.; Cooney, G.J.; Turner, N. Contrasting metabolic effects of medium- versus long-chain fatty acids in skeletal muscle. *J. Lipid Res.* **2013**, *54*, 3322–3333. [CrossRef]
41. Tadaishi, M.; Miura, S.; Kai, Y.; Kano, Y.; Oishi, Y.; Ezaki, O. Skeletal muscle-specific expression of PGC-1α-b, an exercise-responsive isoform, increases exercise capacity and peak oxygen uptake. *PLoS ONE* **2011**, *6*, e28290. [CrossRef] [PubMed]
42. Miura, S.; Tomitsuka, E.; Kamei, Y.; Yamazaki, T.; Kai, Y.; Tamura, M.; Kita, K.; Nishino, I.; Ezaki, O. Overexpression of peroxisome proliferator-activated receptor gamma co-activator-1α leads to muscle atrophy with depletion of ATP. *Am. J. Pathol.* **2006**, *169*, 1129–1139. [CrossRef] [PubMed]

43. Krotkiewski, M. Value of VLCD supplementation with medium chain triglycerides. *Int. J. Obes. Relat. Metab. Disord.* **2001**, *25*, 1393–1400. [CrossRef] [PubMed]
44. Tsuji, H.; Kasai, M.; Takeuchi, H.; Nakamura, M.; Okazaki, M.; Kondo, K. Dietary medium-chain triacylglycerols suppress accumulation of body fat in a double-blind, controlled trial in healthy men and women. *J. Nutr.* **2001**, *131*, 2853–2859. [CrossRef] [PubMed]
45. St-Onge, M.P.; Jones, P.J. Greater rise in fat oxidation with medium-chain triglyceride consumption relative to long-chain triglyceride is associated with lower initial body weight and greater loss of subcutaneous adipose tissue. *Int. J. Obes. Relat. Metab. Disord.* **2003**, *27*, 1565–1571. [CrossRef] [PubMed]
46. St-Onge, M.P.; Ross, R.; Parsons, W.D.; Jones, P.J. Medium-chain triglycerides increase energy expenditure and decrease adiposity in overweight men. *Obes. Res.* **2003**, *11*, 395–402. [CrossRef]
47. St-Onge, M.P.; Bosarge, A. Weight-loss diet that includes consumption of medium-chain triacylglycerol oil leads to a greater rate of weight and fat mass loss than does olive oil. *Am. J. Clin. Nutr.* **2008**, *87*, 621–626. [CrossRef]
48. Hill, J.O.; Peters, J.C.; Yang, D.; Sharp, T.; Kaler, M.; Abumrad, N.N.; Greene, H.L. Thermogenesis in humans during overfeeding with medium-chain triglycerides. *Metabolism* **1989**, *38*, 641–648. [CrossRef] [PubMed]
49. Flatt, J.P.; Ravussin, E.; Acheson, K.J.; Jequier, E. Effects of dietary fat on postprandial substrate oxidation and on carbohydrate and fat balances. *J. Clin. Investig.* **1985**, *76*, 1019–1024. [CrossRef]
50. Seaton, T.B.; Welle, S.L.; Warenko, M.K.; Campbell, R.G. Thermic effect of medium-chain and long-chain triglycerides in man. *Am. J. Clin. Nutr.* **1986**, *44*, 630–634. [CrossRef]
51. Dulloo, A.G.; Fathi, M.; Mensi, N.; Girardier, L. Twenty-four-hour energy expenditure and urinary catecholamines of humans consuming low-to-moderate amounts of medium-chain triglycerides: A dose-response study in a human respiratory chamber. *Eur. J. Clin. Nutr.* **1996**, *50*, 152–158. [PubMed]
52. Liu, Y.H.; Zhang, Y.; Xu, Q.; Yu, X.M.; Zhang, X.S.; Wang, J.; Xue, C.; Yang, X.Y.; Zhang, R.X.; Xue, C.Y. Increased norepinephrine by medium-chain triglyceride attributable to lipolysis in white and brown adipose tissue of C57BL/6J mice. *Biosci. Biotechnol. Biochem.* **2012**, *76*, 1213–1218. [CrossRef] [PubMed]
53. Zhang, Y.; Xu, Q.; Liu, Y.H.; Zhang, X.S.; Wang, J.; Yu, X.M.; Zhang, R.X.; Xue, C.; Yang, X.Y.; Xue, C.Y. Medium-Chain Triglyceride Activated Brown Adipose Tissue and Induced Reduction of Fat Mass in C57BL/6J Mice Fed High-fat Diet. *Biomed. Environ. Sci.* **2015**, *28*, 97–104. [PubMed]
54. Wen, X.; Wu, J.; Chang, J.S.; Zhang, P.; Wang, J.; Zhang, Y.; Gettys, T.W.; Zhang, Y. Effect of exercise intensity on isoform-specific expressions of NT-PGC-1 α mRNA in mouse skeletal muscle. *BioMed Res. Int.* **2014**, *2014*, 402175. [CrossRef] [PubMed]
55. Hatazawa, Y.; Tadaishi, M.; Nagaike, Y.; Morita, A.; Ogawa, Y.; Ezaki, O.; Takai-Igarashi, T.; Kitaura, Y.; Shimomura, Y.; Kamei, Y.; et al. PGC-1α-mediated branched-chain amino acid metabolism in the skeletal muscle. *PLoS ONE* **2014**, *9*, e91006. [CrossRef] [PubMed]
56. Hatazawa, Y.; Senoo, N.; Tadaishi, M.; Ogawa, Y.; Ezaki, O.; Kamei, Y.; Miura, S. Metabolomic Analysis of the Skeletal Muscle of Mice Overexpressing PGC-1α. *PLoS ONE* **2015**, *10*, e0129084. [CrossRef] [PubMed]
57. Nomura, K.; Kinoshita, S.; Mizusaki, N.; Senga, Y.; Sasaki, T.; Kitamura, T.; Sakaue, H.; Emi, A.; Hosooka, T.; Matsuo, M.; et al. Adaptive gene expression of alternative splicing variants of PGC-1α regulates whole-body energy metabolism. *Mol. Metab.* **2024**, *86*, 101968. [CrossRef]
58. Hostrup, M.; Jacobson, G.A.; Jessen, S.; Lemminger, A.K. Anabolic and lipolytic actions of β_2 -agonists in humans and antidoping challenges. *Drug Test. Anal.* **2020**, *12*, 597–609. [CrossRef] [PubMed]
59. Kline, W.O.; Panaro, F.J.; Yang, H.; Bodine, S.C. Rapamycin inhibits the growth and muscle-sparing effects of clenbuterol. *J. Appl. Physiol.* **2007**, *102*, 740–747. [CrossRef]
60. Koopman, R.; Gehrig, S.M.; Leger, B.; Trieu, J.; Walrand, S.; Murphy, K.T.; Lynch, G.S. Cellular mechanisms underlying temporal changes in skeletal muscle protein synthesis and breakdown during chronic β-adrenoceptor stimulation in mice. *J. Physiol.* **2010**, *588*, 4811–4823. [CrossRef]
61. Pearen, M.A.; Ryall, J.G.; Lynch, G.S.; Muscat, G.E. Expression profiling of skeletal muscle following acute and chronic β2-adrenergic stimulation: Implications for hypertrophy, metabolism and circadian rhythm. *BMC Genom.* **2009**, *10*, 448. [CrossRef] [PubMed]
62. Hagg, A.; Colgan, T.D.; Thomson, R.E.; Qian, H.; Lynch, G.S.; Gregorevic, P. Using AAV vectors expressing the β2-adrenoceptor or associated Galpha proteins to modulate skeletal muscle mass and muscle fibre size. *Sci. Rep.* **2016**, *6*, 23042. [CrossRef] [PubMed]
63. Kim, J.; Grotegut, C.A.; Wisler, J.W.; Li, T.; Mao, L.; Chen, M.; Chen, W.; Rosenberg, P.B.; Rockman, H.A.; Lefkowitz, R.J. β-arrestin 1 regulates β2-adrenergic receptor-mediated skeletal muscle hypertrophy and contractility. *Skelet. Muscle* **2018**, *8*, 39. [CrossRef] [PubMed]
64. Hoshino, D.; Yoshida, Y.; Holloway, G.P.; Lally, J.; Hatta, H.; Bonen, A. Clenbuterol, a β2-adrenergic agonist, reciprocally alters PGC-1 α and RIP140 and reduces fatty acid and pyruvate oxidation in rat skeletal muscle. *Am. J. Physiol. Regul. Integr. Comp. Physiol.* **2012**, *302*, R373–R384. [CrossRef] [PubMed]
65. Meister, J.; Bone, D.B.J.; Knudsen, J.R.; Barella, L.F.; Velenosi, T.J.; Akhmedov, D.; Lee, R.J.; Cohen, A.H.; Gavrilova, O.; Cui, Y.; et al. Clenbuterol exerts antidiabetic activity through metabolic reprogramming of skeletal muscle cells. *Nat. Commun.* **2022**, *13*, 22. [CrossRef] [PubMed]

66. Torgan, C.E.; Etgen, G.J., Jr.; Brozinick, J.T., Jr.; Wilcox, R.E.; Ivy, J.L. Interaction of aerobic exercise training and clenbuterol: Effects on insulin-resistant muscle. *J. Appl. Physiol.* **1993**, *75*, 1471–1476. [CrossRef] [PubMed]
67. Charlot, A.; Morel, L.; Bringolf, A.; Georg, I.; Charles, A.L.; Goupilleau, F.; Geny, B.; Zoll, J. Octanoic Acid-Enrichment Diet Improves Endurance Capacity and Reprograms Mitochondrial Biogenesis in Skeletal Muscle of Mice. *Nutrients* **2022**, *14*, 2721. [CrossRef] [PubMed]
68. Yoshida, T.; Fujitani, M.; Farmer, S.; Harada, A.; Shi, Z.; Lee, J.J.; Tinajero, A.; Singha, A.K.; Fujikawa, T. VMHdm/c(SF-1) neuronal circuits regulate skeletal muscle PGC1-α via the sympathoadrenal drive. *Mol. Metab.* **2023**, *77*, 101792. [CrossRef] [PubMed]
69. Nass, R.; Pezzoli, S.S.; Oliveri, M.C.; Patrie, J.T.; Harrell, F.E., Jr.; Clasey, J.L.; Heymsfield, S.B.; Bach, M.A.; Vance, M.L.; Thorner, M.O. Effects of an oral ghrelin mimetic on body composition and clinical outcomes in healthy older adults: A randomized trial. *Ann. Intern. Med.* **2008**, *149*, 601–611. [CrossRef]
70. Katakami, N.; Uchino, J.; Yokoyama, T.; Naito, T.; Kondo, M.; Yamada, K.; Kitajima, H.; Yoshimori, K.; Sato, K.; Saito, H.; et al. Anamorelin (ONO-7643) for the treatment of patients with non-small cell lung cancer and cachexia: Results from a randomized, double-blind, placebo-controlled, multicenter study of Japanese patients (ONO-7643-04). *Cancer* **2018**, *124*, 606–616. [CrossRef]
71. White, H.K.; Petrie, C.D.; Landschulz, W.; MacLean, D.; Taylor, A.; Lyles, K.; Wei, J.Y.; Hoffman, A.R.; Salvatori, R.; Ettinger, M.P.; et al. Effects of an oral growth hormone secretagogue in older adults. *J. Clin. Endocrinol. Metab.* **2009**, *94*, 1198–1206. [CrossRef] [PubMed]
72. Yanagi, S.; Sato, T.; Kangawa, K.; Nakazato, M. The Homeostatic Force of Ghrelin. *Cell Metab.* **2018**, *27*, 786–804. [CrossRef] [PubMed]
73. Yoshida, T.; Delafontaine, P. Mechanisms of IGF-1-Mediated Regulation of Skeletal Muscle Hypertrophy and Atrophy. *Cells* **2020**, *9*, 1970. [CrossRef] [PubMed]
74. Ashitani, J.; Matsumoto, N.; Nakazato, M. Effect of octanoic acid-rich formula on plasma ghrelin levels in cachectic patients with chronic respiratory disease. *Nutr. J.* **2009**, *8*, 25. [CrossRef] [PubMed]
75. Kawai, K.; Nakashima, M.; Kojima, M.; Yamashita, S.; Takakura, S.; Shimizu, M.; Kubo, C.; Sudo, N. Ghrelin activation and neuropeptide Y elevation in response to medium chain triglyceride administration in anorexia nervosa patients. *Clin. Nutr. ESPEN* **2017**, *17*, 100–104. [CrossRef] [PubMed]
76. Yoshimura, Y.; Shimazu, S.; Shiraishi, A.; Nagano, F.; Tominaga, S.; Hamada, T.; Kudo, M.; Yamasaki, Y.; Noda, S.; Bise, T. Ghrelin activation by ingestion of medium-chain triglycerides in healthy adults: A pilot trial. *J. Aging Res. Clin. Pract.* **2018**, *7*, 42–46. [CrossRef]
77. Miki, K.; Kitada, S.; Miki, M.; Hui, S.P.; Shrestha, R.; Yoshimura, K.; Tsujino, K.; Kagawa, H.; Oshitani, Y.; Kida, H.; et al. A phase II, open-label clinical trial on the combination therapy with medium-chain triglycerides and ghrelin in patients with chronic obstructive pulmonary disease. *J. Physiol. Sci.* **2019**, *69*, 969–979. [CrossRef]
78. Kopchick, J.J.; Berryman, D.E.; Puri, V.; Lee, K.Y.; Jorgensen, J.O.L. The effects of growth hormone on adipose tissue: Old observations, new mechanisms. *Nat. Rev. Endocrinol.* **2020**, *16*, 135–146. [CrossRef]
79. Odessey, R.; Khairallah, E.A.; Goldberg, A.L. Origin and possible significance of alanine production by skeletal muscle. *J. Biol. Chem.* **1974**, *249*, 7623–7629. [CrossRef]
80. Buse, M.G.; Reid, S.S. Leucine. A possible regulator of protein turnover in muscle. *J. Clin. Investig.* **1975**, *56*, 1250–1261. [CrossRef]
81. Wolfson, R.L.; Sabatini, D.M. The Dawn of the Age of Amino Acid Sensors for the mTORC1 Pathway. *Cell Metab.* **2017**, *26*, 301–309. [CrossRef]
82. Xin, L.; Ipek, O.; Beaumont, M.; Shevlyakova, M.; Christinat, N.; Masoodi, M.; Greenberg, N.; Gruetter, R.; Cuenoud, B. Nutritional Ketosis Increases NAD^+/NADH Ratio in Healthy Human Brain: An in Vivo Study by (31)P-MRS. *Front. Nutr.* **2018**, *5*, 62. [CrossRef]
83. Sherwin, R.S.; Hendler, R.G.; Felig, P. Effect of ketone infusions on amino acid and nitrogen metabolism in man. *J. Clin. Investig.* **1975**, *55*, 1382–1390. [CrossRef]
84. Nair, K.S.; Welle, S.L.; Halliday, D.; Campbell, R.G. Effect of β-hydroxybutyrate on whole-body leucine kinetics and fractional mixed skeletal muscle protein synthesis in humans. *J. Clin. Investig.* **1988**, *82*, 198–205. [CrossRef]
85. Hannaian, S.J.; Lov, J.; Hawley, S.E.; Dargegen, M.; Malenda, D.; Gritsas, A.; Gouspillou, G.; Morais, J.A.; Churchward-Venne, T.A. Acute ingestion of a ketone monoester, whey protein, or their co-ingestion in the overnight postabsorptive state elicit a similar stimulation of myofibrillar protein synthesis rates in young males: A double-blind randomized trial. *Am. J. Clin. Nutr.* **2024**, *119*, 716–729. [CrossRef]
86. Lin, S.C.; Hardie, D.G. AMPK: Sensing Glucose as well as Cellular Energy Status. *Cell Metab.* **2018**, *27*, 299–313. [CrossRef]
87. Panwar, V.; Singh, A.; Bhatt, M.; Tonk, R.K.; Azizov, S.; Raza, A.S.; Sengupta, S.; Kumar, D.; Garg, M. Multifaceted role of mTOR (mammalian target of rapamycin) signaling pathway in human health and disease. *Signal Transduct. Target. Ther.* **2023**, *8*, 375. [CrossRef]
88. Cohen, H.Y.; Miller, C.; Bitterman, K.J.; Wall, N.R.; Hekking, B.; Kessler, B.; Howitz, K.T.; Gorospe, M.; de Cabo, R.; Sinclair, D.A. Calorie restriction promotes mammalian cell survival by inducing the SIRT1 deacetylase. *Science* **2004**, *305*, 390–392. [CrossRef]
89. Gerhart-Hines, Z.; Rodgers, J.T.; Bare, O.; Lerin, C.; Kim, S.H.; Mostoslavsky, R.; Alt, F.W.; Wu, Z.; Puigserver, P. Metabolic control of muscle mitochondrial function and fatty acid oxidation through SIRT1/PGC-1α. *EMBO J.* **2007**, *26*, 1913–1923. [CrossRef]

90. Gomes, A.P.; Price, N.L.; Ling, A.J.; Moslehi, J.J.; Montgomery, M.K.; Rajman, L.; White, J.P.; Teodoro, J.S.; Wrann, C.D.; Hubbard, B.P.; et al. Declining NAD^+ induces a pseudohypoxic state disrupting nuclear-mitochondrial communication during aging. *Cell* **2013**, *155*, 1624–1638. [CrossRef]
91. Lee, D.; Goldberg, A.L. SIRT1 protein, by blocking the activities of transcription factors FoxO1 and FoxO3, inhibits muscle atrophy and promotes muscle growth. *J. Biol. Chem.* **2013**, *288*, 30515–30526. [CrossRef]
92. Yang, L.; Liu, D.; Jiang, S.; Li, H.; Chen, L.; Wu, Y.; Essien, A.E.; Opoku, M.; Naranmandakh, S.; Liu, S.; et al. SIRT1 signaling pathways in sarcopenia: Novel mechanisms and potential therapeutic targets. *Biomed. Pharmacother.* **2024**, *177*, 116917. [CrossRef]
93. Scheibye-Knudsen, M.; Mitchell, S.J.; Fang, E.F.; Iyama, T.; Ward, T.; Wang, J.; Dunn, C.A.; Singh, N.; Veith, S.; Hasan-Olive, M.M.; et al. A high-fat diet and NAD^+ activate Sirt1 to rescue premature aging in cockayne syndrome. *Cell Metab.* **2014**, *20*, 840–855. [CrossRef] [PubMed]

nutrients

MDPI

Systematic Review

The Role of Nutrition in the Treatment of Sarcopenia in Old Patients: From Restoration of Mitochondrial Activity to Improvement of Muscle Performance, a Systematic Review

Camille Cochet [1], Giulia Belloni [2], Ilaria Buondonno [3], Francesco Chiara [4] and Patrizia D'Amelio [1,3,*]

1 Service of Geriatric Medicine and Geriatric Rehabilitation, Department of Internal Medicine, University of Lausanne Hospital Centre (CHUV), 1011 Lausanne, Switzerland; camilleanais.cochet@gmail.com
2 Department of Epidemiology and Health Systems, Center for Primary Care and Public Health (Unisanté), University of Lausanne, 1011 Lausanne, Switzerland
3 Laboratory of Osteobiology and Aging Diseases, Department of Medical Sciences, University of Turin, Via Santena 5 bis, 10124 Turin, Italy
4 Clinical Biochemistry Laboratory, City of Health and Science University Hospital, 10043 Turin, Italy
* Correspondence: patrizia.damelio@chuv.ch

Abstract: Sarcopenia is an age-related disease characterized by loss of muscle strength, mass and performance. Malnutrition contributes to sarcopenia pathogenesis. The aim of this systematic review is to analyze existing evidence on the efficacy of nutritional supplementation on muscle and mitochondrial health among sarcopenic or malnourished older adults. We included randomized controlled trials (RCTs) assessing the effect of branched-chain amino acid (BCAA), vitamin D and/or omega-3 polyunsaturated fatty acid (PUFA) on muscle mass, strength and performance and/or on mitochondrial activity and redox state in older sarcopenic and/or malnourished adults. The literature search was on MEDLINE, Embase and Cochrane Central, restricted to articles published in the last 10 years (2012–2022). Twelve RCTs with a total of 1337 subjects were included. BCAA with vitamin D significantly ameliorates appendicular muscle mass (4 RCTs), hand grip strength (4 RCTs), gait speed (3 RCTs), short physical performance battery (3 RCTs) or chair stand test (3 RCTs) among six out of nine RCTs. BCAA alone (2 RCTs) or PUFA (1 RCT) were not effective in improving muscle health. Mitochondrial function was significantly improved by the administration of BCAA alone (1 RCT) or in association with vitamin D (1 RCT). In conclusion, BCAA in association with vitamin D may be useful in the treatment of sarcopenia and boost mitochondrial bioenergetic and redox activity. PROSPERO CRD42022332288.

Keywords: sarcopenia; malnutrition; mitochondrial bioenergetic; redox activity; older adults; BCAA; vitamin D; omega-3 PUFA; geriatrics

1. Introduction

Sarcopenia is a condition characterized by loss of muscle strength, mass and performance. Sarcopenia reduces independency, increases risk of falls and hospital admissions, reduces mobility and increases mortality. This disease is frequently associated with aging and plays a major role in the development of frailty syndrome [1]. The worldwide prevalence of sarcopenia in people older than 60 years is estimated between 10% and 27%, the use of different diagnostic criteria [2] contributes to the under-diagnosis and under treatment of this disease. The ancient criteria essentially considered muscle mass reduction, whereas the more recent criteria take into account mainly the reduction of muscle strength and performance; different diagnostic criteria are summarized in Table 1.

Table 1. Different criteria for the diagnosis of sarcopenia.

Criteria	Muscle Mass	Muscle Strength	Muscle Performance	Summary Definition
European Working Group on Sarcopenia in Older People (EWGSOP1, 2010) [3]	2 SD < mean reference value	Grip Strength: <30 kg	Gait Speed < 0.8 m/s.	Sarcopenia: Low muscle mass + low muscle strength OR low performance. Severe sarcopenia: All 3 criteria.
European Working Group on Sarcopenia in Older People (EWGSOP2, 2019) [4]	ASM M < 20 kg; F < 15 kg ASM/height^2 M < 7.0 kg/m^2; F < 5.5 kg/m^2	Grip Strength: M < 27 kg, F < 16 kg Chair stand: >15 s for five rises	Gait Speed < 0.8 m/s SPPB < 8 TUG < 20 s 400 m walk test > 6 min or non-completion	Assess sarcopenia with low muscle strength confirmed sarcopenia with low muscle mass. Class Sarcopenia's severity with performance. Severe sarcopenia: All 3 criteria.
Foundation for the National Institutes of Health (FNIH, 2014) [5]	ALM/BMI M < 0.789; F < 0.512	Grip Strength: M < 26 kg; F < 16 kg	Gait speed < 0.8 ms	Sarcopenia: Low muscle mass and low muscle strength. Class Sarcopenia's severity with performance severe sarcopenia: All 3 criteria.
International Working Group (2011) [6]	ALM//height^2 M < 7.23 kg/m^2 F < 5.67 kg/m^2	NA	Gait Speed < 1.0 m/s	Sarcopenia: Low muscle mass and low muscle performance.
Asian Working Group for Sarcopenia (ASIA, 2019) [7]	ASM//height^2 DXA: M < 7.0 kg/m^2; F < 5.4 kg/m^2 or BIA: M < 7.0 kg/m; F < 5.7 kg/m^2	Grip Strength: M < 28 kg; F < 18 kg	Gait Speed < 1 m/s. 6-metre walk < 1.0 m/s 5-time CST $\geq$ 12 s SPPB $\leq$ 9	Sarcopenia: Low muscle mass + low muscle strength OR Low physical performance. Severe sarcopenia: All 3 criteria.

ASM: Appendicular Skeletal Muscle Mass; ALM: Appendicular Lean Mass; BIA: Bioelectrical impedance analysis; CST: chair stand test; F: female; M: male; SD: Standard Deviation; SMM: Skeletal Muscle Mass; SPPB: short physical performance battery; TUG: Timed up and go.

The aging of the population will lead to a significant increase in the prevalence of sarcopenia, increasing the associated socio-economic burden [4]. Therefore, an early diagnosis, an adapted and individualized treatment are essential to counteract the spreading of the disease.

The physiopathology of sarcopenia is complex; several lifestyle factors contribute to its development, including malnutrition and physical inactivity [8–10] that are frequent amongst older persons [2]. The muscle being a high-energy-demanding tissue, mitochondrial dysfunction [11,12] associated with aging [13], malnutrition [14] and physical inactivity [14] has been proposed as an important contributor to the pathogenesis of the disease [15]. Aging is associated with a decrease in mitochondrial biogenesis and performance; the impairment of mitochondrial function has been proposed as one of the causes of unhealthy aging [13].

Even though healthy mitochondria reduce oxidative stress levels, their respiratory chain generates reactive oxygen species (ROS) within the organelle; the consequent increase in oxidative stress is counteracted by antioxidant enzymes. Although the production of ROS is physiological at a cellular level, if the balance between ROS and the antioxidant system exceeds a certain threshold, oxidative stress increases and mitochondrial DNA is damaged; this damage induces a mitochondrial protein degradation and impairs mitochondrial function [13].

The impairment of mitochondrial function contributes to the increase in oxidative stress and to the development of immune senescence and immune system dysfunction [16]. The immune system dysfunction associated with unhealthy aging causes an increase in chronic, sterile inflammation, named "inflammaging". This pro-inflammatory status favors muscle catabolism [17,18] and, hence, sarcopenia. The increase in inflammation and the decrease in mitochondrial function at cellular and molecular levels are associated with unhealthy aging at a clinical level and play a paramount role in the development of sarcopenia [15].

Amongst clinical factors associated with sarcopenia, malnutrition plays an important role. Malnutrition is a complex syndrome with multifactorial causes due to both a quantitative and qualitative reduction of dietary intake and an impairment of gut absorption, and it is frequent amongst older adults. This condition has been defined as an involuntary weight loss of 5% in less than 6 months or 10% in more than 6 months, or a BMI of less than 22, in patients over 70 years old, associated with reduced energy intake or acute injury or chronic-related disease [19]. Several screening tools for the diagnosis of malnutrition have been proposed; in older persons, the more frequently used screening tools are the Mini Nutritional Assessment (MNA) or its short form (MNA-SF) [20,21] and the Nutritional Risk Screening Score (NRS) [22,23]. These tools are user-friendly questionnaires. The three questionnaires focus on BMI, reduced food intake, weight loss and the presence of acute and severe illness. Within the MNA and MNA-SF, the presence of cognitive impairment or depression is considered as well as the patient's mobility. The full form of the MNA also takes into account dietary habits, self-perceived health, medication, living environment, skin wounds, and calf and arm circumference. The decreased intake/absorption of nutrients essential for muscle health, such as protein, amino acids, vitamins and minerals in malnourished subjects contributes to the development of sarcopenia [15,24].

The negative impact of reduced dietary intake of essential amino acid on muscle synthesis [25] in older age is enhanced by the reduced muscular protein synthesis during aging [14,26], this phenomenon has been named "anabolic resistance" of aging. The impairment in mRNA translation and in the activation of the mammalian target of rapamycin (mTOR) pathway, are amongst the underlying causes of this anabolic resistance. Higher protein intake can counteract the anabolic resistance [27]. Indeed, branched-chain amino acids (BCAAs), particularly leucine, isoleucine and valine, have the capacity to activate the mTOR pathway [28], promoting mitochondrial biogenesis, and boosting the PGC 1 alpha pathway which increases oxidative resistance [14,15,29]. Omega-3 PUFA also contributes to the activation of the mTOR pathway, leading to increased protein mitochondrial synthesis in muscle cells [30]. Furthermore, omega-3 PUFA have an anti-inflammatory effect [31] and can decrease oxidative stress [32] thus counteracting the catabolic effect of inflammaging.

Vitamin D plays a role in maintaining muscle health and, in particular, it influences muscle cells differentiation, maturation and growth [33]. Aging is associated with reduced expression of muscle vitamin D receptor and lower vitamin D level, and these factors contribute to the reduction of muscle mass [34]. Furthermore, at the cellular level, some evidence shows that the active form of vitamin D, 1–25 dihydroxyvitamin D3 or calcitriol, increases mitochondrial biogenesis and function [35], and regulates the expression of genes involved in this process in muscles [36].

Given the role of malnutrition in the pathogenesis of sarcopenia, nutritional interventions targeting mitochondrial health and oxidative stress imbalance are emerging as possible strategies to treat this disease.

The aim of this systematic review is to analyze existing evidence on the efficacy of nutritional supplementation on muscle and mitochondrial health among sarcopenic and/or malnourished older adults. This review is divided into two parts: a clinical section focused on the role of nutritional supplementation on muscle strength, mass and performance and a biological section focused on the role of nutritional supplementation on mitochondrial health.

2. Methods

2.1. Study Design

We conducted a systematic review according to the Preferred Reporting Items for Systematic Reviews and Meta-Analyses (PRISMA) statement [37], and Testing Methodological Guidance on the Conduct of Narrative Synthesis in Systematic Reviews [38], as well as Synthesis without meta-analysis reporting guidelines [39]. The review protocol is published on the PROSPERO database (registration number CRD42022332288).

2.2. Inclusion and Exclusion Criteria

We included randomized controlled trials (RCTs) and meta-analyses which assessed the effect of nutritional supplementation the clinical parameters for the diagnosis of sarcopenia or the mitochondrial activity in older adults (65 years and older) affected by sarcopenia and/or malnutrition. The nutritional supplements analyzed were BCAA and/or vitamin D and/or omega-3 PUFA.

The research question was formulated by the following PICOS (participants, intervention, comparison, outcomes, study) strategy.

- Participants: Adults aged 65 years and older affected by sarcopenia and/or malnutrition clinically diagnosed. Sarcopenia was defined using recognized diagnostic criteria, such as the European Working Group on Sarcopenia in Older People (EWGSOP-1 [3] and EWGSOP-2 [4]), FNIH [5], International Working Group [6] or Asian working group on sarcopenia (ASIA) [7,40]. The diagnosis of malnutrition or the presence of a malnutrition risk was defined according to a validated tool such as MNA, MNA-SF or NRS score [23].
- Intervention: BCAA and/or omega-3 PUFA and/or vitamin D supplementation, associated or not to physical exercise.
- Comparison: Standard clinical treatment, placebo, or physical exercise without nutritional supplementation.
- Outcomes: Measurement of the parameters of muscle mass and/or muscle strength and/or muscle performance or the mitochondrial activity and/or the oxidative stress.
- Studies: RCT and meta-analysis.

We excluded trials whose participants were affected by sarcopenia due to the following diseases: kidney or hepatic failure, cancer, neuromuscular disease, AIDS, COVID-19, recent surgery or transplants, or severe neurological disorders. We also excluded clinical trials assessing the effect of any kind of protein, (soy, casein, whey proteins or unknown) alone without BCAA, vitamin D or omega-3 PUFA.

2.3. Search Strategy

MEDLINE, Embase, Cochrane Central Register of Controlled Trials (CENTRAL) were used for the literature search; the final search was run on 30 July 2022. A re-run was carried out on 9 August 2023 and retrieved two additional papers, that are discussed in the discussion section. The search was restricted to the last ten years of publications (2012–2022), to studies performed in humans and written in English.

The search string for the clinical part was as follows: ((("Sarcopenia"[MeSH Terms] OR "Malnutrition"[MeSH Terms]) AND "amino acids, branched chain"[MeSH Terms]) OR "fatty acids, omega 3"[MeSH Terms] OR "Vitamin D"[MeSH Terms]) AND ((y_10[Filter]) AND (meta-analysis[Filter] OR randomizedcontrolledtrial[Filter]) AND (humans[Filter]) AND (english[Filter]) AND (aged[Filter] OR 80 andover[Filter])). The search strategy is publicly available at: (((("Sarcopenia"[Mesh]) OR "Malnutrition"[Mesh]) AND "Amino Acids, Branched-Chain"[Mesh]) OR "Fatty Acids, Omega-3"[Mesh]) OR "Vitamin D"[Mesh]—Search Results—PubMed (nih.gov), and https://www.cochranelibrary.com/advanced-search/search-manager?search=6982594 (accessed on 9 August 2023).

The search string for the biological part was as follows: ((("amino acids, branched chain"[MeSH Terms] OR "fatty acids, omega 3"[MeSH Terms] OR "Vitamin D"[MeSH Terms]) AND "Mitochondria"[MeSH Terms]) OR "Oxidative Stress"[MeSH Terms]) AND ((y_10[Filter]) AND (meta-analysis[Filter] OR randomizedcontrolledtrial[Filter]) AND (humans[Filter]) AND (english[Filter]) AND (80 andover[Filter] OR aged[Filter])). The search strategy is publicly available at: (((("Amino Acids, Branched-Chain"[Mesh]) OR "Fatty Acids, Omega-3"[Mesh]) OR "Vitamin D"[Mesh]) AND "Mitochondria"[Mesh]) OR "Oxidative Stress"[Mesh]—Search Results—PubMed (nih.gov) and https://www.cochranelibr

ary.com/web/cochrane/advanced-search/search-manager?search=6982595 (accessed on 9 August 2023).

A second, less restrictive, search strategy for the biological part was as follows: (('mitochondria/exp OR 'mitochondrion' OR 'oxidative stress'/exp OR 'oxidative stress') AND ('vitamin d'/exp OR 'vitamin d') OR 'omega 3 fatty acid'/exp OR 'omega 3 fatty acid' OR 'branched chain amino acid'/exp OR 'branched chain amino acid') AND [randomized controlled trial]/lim AND([aged]/lim OR [very elderly]/lim) AND [humans]/lim AND [2012–2022]/py. The search strategy is publicly available at: https://pubmed.ncbi.nlm.nih.gov/?term=%28%28%28%28%22Amino+Acids%2C+Branched-Chain%22%5BMesh%5D%29+OR+%22Fatty+Acids%2C+Omega-3%22%5BMesh%5D%29+OR+%22Vitamin+D%22%5BMesh%5D%29+AND+%22Mitochondria%22%5BMesh%5D%29+OR+%22Oxidative+Stress%22%5BMesh%5D&filter=pubt.clinicaltrial&filter=pubt.randomizedcontrolledtrial&filter=datesearch.y_10&filter=hum_ani.humans&filter=lang.english&filter=age.aged&filter=age.80andover&show_snippets=off&sort=date (accessed on 9 August 2023).

2.4. Study Selection

Two teams of reviewers worked, according to their expertise, on the clinical part (CC and GB) or on the biological part (FC and IB). The reviewers of each team worked independently on the studies retrieved by the search to evaluate the inclusion of each study; discrepancies between the two reviewers were solved through discussion and if necessary, by the third reviewer (PDA for both parts of the review). The Rayyan® tool was used to detect duplicates and abstracts published before 2012 having escaped our filter, and to a systemized article selection process between the authors (available at Rayyan—Intelligent Systematic Review—Rayyan). The bibliography of included articles was screened to detect other eligible RCTs and subjected to the same process as the RCT detected by the search-string strategy.

2.5. Data Extraction

From each study, we extracted the following data: authors, year of publication, location, number and characteristic of participants (mean age and gender), diagnosis at baseline (sarcopenia vs. malnutrition), type and duration of intervention, type of control group used, type of physical exercise, inclusion and exclusion criteria, and sample size and measured outcomes. If applicable, information on falls, length of hospital stay and mortality was also recorded.

2.6. Quality Assessment

The risk of bias was assessed by the RoB 2 tool according to Cochrane (https://methods.cochrane.org/risk-bias-2 (accessed on 14 July 2022)). As for the study selection, two reviewers independently evaluated the bias of each study; discrepancies between the reviewers were solved by discussion and consensus or by the third reviewer (PDA, for both parts of the review).

2.7. Data Synthesis and Analysis

The aim of our work was to write a systematic narrative review, without meta-analysis. To compare the different studies, we grouped them according to the type of intervention. Therefore, three subgroups were identified: 1. Vitamin D + BCAA or Whey protein; 2. BCAA alone; 3. Omega-3 PUFA.

3. Results

3.1. Study Selection

For the clinical part, we retrieved 4348 abstracts after searching the following databases: Medline (n = 1761), Cochrane Central (n = 1744) and Embase (n = 843). We removed 1585 duplicates and 382 papers were marked as ineligible by the Rayyan tool (abstracts dating before 2012). We identified 17 supplementary records through the bibliographies of the included articles [41–57]. Hence, forty-six full-text articles were reviewed as previously described. After reading the full text article, we excluded twenty-two articles for violation of inclusion criteria. Twelve articles were excluded due to the following reasons: wrong study design (*n* = 3), wrong outcome (*n* = 4), wrong study population (*n* = 1), wrong intervention (*n* = 2), full paper not-available (*n* = 2). Twelve articles were included in the clinical part of the review [45,47,50,57–65].

For the biological part, with the first search strip, we retrieved 1542 abstracts, after searching the following databases: Medline (*n* = 776), Cochrane Central (*n* = 193) and Embase (*n* = 573). We removed 196 duplicates. A less restrictive search strategy was re-run and retrieved 652 more abstracts. We excluded 16 articles for violation of inclusion criteria and four articles were excluded due to unavailability of the full paper. Hence, two full-text articles were reviewed as previously described. Two articles were included in the biological part of this review and were also eligible for the clinical part [49,57,58,65]. All the included articles were RCTs; no meta-analyses were included.

Figure 1 shows the study's flowcharts for both the clinical and biological part.

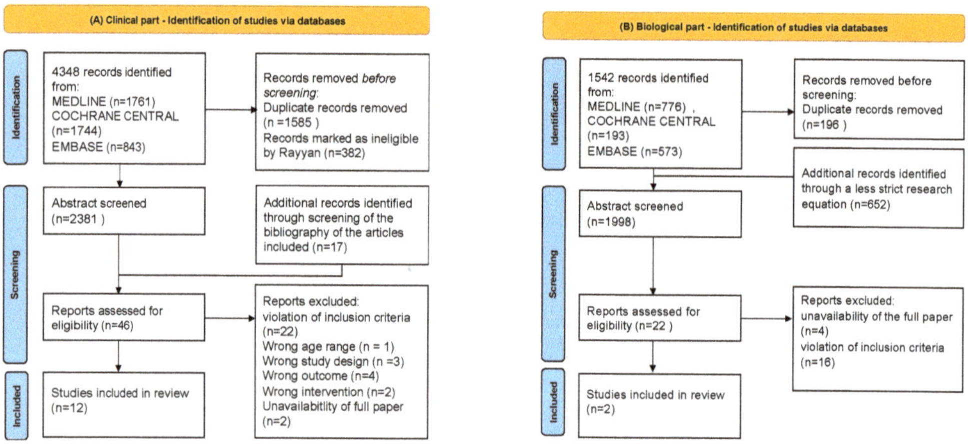

Figure 1. Study's flow charts. Panel (**A**) shows the flow chart of the clinical part and panel (**B**) shows the flow chart of the biological part.

3.2. Study Characteristics

Amongst the 12 studies included in the clinical part of the review, seven evaluated nutritional supplementations with BCAA mixture in association with vitamin D [45,50,57,60–62,64], one BCAA mixture alone [65], one leucine alone [58], two evaluated whey protein or proteins in association with vitamin D [59,63] and one omega-3 PUFA [47]. All the characteristics of nutritional supplementation are specified in Table 1. Nine studies had a similar timing of intervention between 8 and 12 weeks; one was run for 30 days [64] and two considered a longer intervention period: six months [59] and one year [58]. Amongst the included studies, four included a physical training program; in two [60,61], physical exercise intervention including resistance training with gait and balance training was proposed both in the intervention and the control group. Two studies [45,63] were a four-arm trials comparing nutrition alone, exercise alone, nutrition plus exercise, versus no intervention [63] or health education [45]; exercise consists in resistance training [63] or in resistance and aerobic training [45]. All characteristics of the studies are summarized in Table 2.

Table 2. Characteristics of the RCT included within the clinical and biological part.

Authors Year Location	Number of Partici-pants	Mean Age (Years)	Gender (Women %)	Diagnosis (Used Criteria)	Duration of the Intervention	Type of Intervention	Control	Physical Exercice	Outcomes Measured
BCAA or Whey Protein with Vitamin D									
Rondanelli 2016 [60] Italy	130	80.4	59	sarcopenia (EWGSOP1)	12 weeks	Once daily EAA: Leucine 4 g, Isoleucine 1 g, Valine 1 g, L-Lysine, 1.5 g L-Threonine 1.1 g L-Tryptophane 0.3 g, L-Valine 1.0 g NEAA: DL-Met 0.6 g, L-Cys 0.4 g, L-Phe 0.5 g, L-Tyr 0.5 g, Asp 1.8 g, Ser 0.8 g, Glu 5.2 g, Pro 1.0 g, Gly 0.3 g, Ala 0.8 g, Arg 0.8 g Vitamin D3 312 IU Whey protein 68.9 g	Placebo	Yes (RT, gait and balance training)	MM: Relative skeletal muscle mass; MS: Hand grip
Rondanelli 2020 [61] Italy	140	81	63	sarcopenia (EWGSOP1)	8 weeks	Twice daily Leucine 2.8 g Vitamin D 800 IU Whey proteins 20 g	Placebo	Yes (RT, gait and balance training)	MM: Skeletal muscle index, appendicular muscle mass; MS: Hand grip; MP: Gait speed, chair stand test, TUG, SPPB
Lin 2020 [50] Taiwan	56	73.1	28.6	sarcopenia (ASIA 2019)	12-weeks	Once daily Leucine 1.2 g Vitamin D 120 IU Whey protein 8.5 g +Diet advice	Diet advice; instructed to consume 1.5 g pro-tein/kg/BW/day	No	MM: Appendicular muscle mass index; MS: Hand grip; MP: Gait speed
Bauer 2015 [62] Germany	380	77.7	65.5	sarcopenia (EWGSOP1)	13 weeks	Twice daily Leucine 3 g Vitamin D3 800 IU Whey protein 20 g	Placebo	No	MM: Skeletal muscle index; MS: Hand grip; MP: SPPB
Kim 2016 [45] Japan	139	81.1	NI	sarcopenic obesity	3 months	Once daily EAA: Leucine 1.20 g, lysine HCL 0.50 g, valine 0.33 g, isoleucine 0.32 g, threonine 0.28 g, phenylalanine 0.20 g and other 0.17 g Vitamin D 800 IU Tea catechin 540 mg	Exercise Exercise + Nutrition Health education	Yes (RT and AT)	MM: Skeletal muscle mass index; MS: Hand grip, knee extension strength; MP: Gait speed
Grootswagers 2021 [57] The Netherlands	9 (malnour-ish)	74.1	20.5	malnutrition (MNA-sf)	12 weeks	Twice daily free BCAA 7 g Vitamin D3 432 IU Urosalic acid 206 mg Whey protein 11 g Casein protein 11 g	Twice daily Oral standard nutritional sup-plementation (with Vit D3 172 IU)	No	MM: Appendicular muscle mass index; MS: Hand grip, knee extension and flexion; MP: SPPB mRNA expression of mitochondrial activity and biogenesis (PGC1-alpha, AMPK, TFAM, redox activity)

Table 2. *Cont.*

Authors Year Location	Number of Participants	Mean Age (Years)	Gender (Women %)	Diagnosis (Used Criteria)	Duration of the Intervention	Type of Intervention	Control	Physical Exercice	Outcomes Measured
Ekinci 2016 [64] Turkey	62	82.6	100	malnutrition (NRS)	30 days	Twice daily Vitamin D 1000 IU CaHMB 3 g Protein 36 g (unknown origin)	Standard nutrition	No	MM: Calf and arm circumference, triceps skinfold thickness; MS: Hand grip
Bo 2019 [59] China	60	74	55	sarcopenia (ASIA 2014)	6 months	Twice daily Vitamin D 702 UI Whey proteins 22 g	Placebo	No	MM: Relative muscle mass index; MS: Hand grip; MP: Gait speed, TUG, chair stand test
Yamada 2019 [63] Japan	34 (sarcopenic)	No data for sarcopenic patient	No data for sarcopenic patient	sarcopenia and dynapenia (ASIA 2014)	12 weeks	Once daily Vitamin D 800 IU Whey protein 10 g	Exercise Exercise + Nutrition Nothing	Yes (RT)	MM: Appendicular muscle mass, MS: Hand grip, knee extension; MP: Gait speed, chair stand test, one-leg stand test
						BCAA alone			
Buondonno 2020 [65] Italy	155	83	72.5	malnutrition (MNA)	2 months	Twice daily EAA:Leucine 1.25 g, Lysine 0.65 g, Isoleucine 0.625 g, Valine 0.625 g, Threonine 0.35 g, Histidine 0.15 g, Phenylalanine 0.01 g, Methionine 0.05 g, Tryptophan 0.02 g, NEAA: Cystine 0.15 g, Tyrosine 0.03 g Vitamin B 6 0.1 mg, Vitamin B1 0.15 mg	Nutritional conseilling	No	MM: Calf and arm circumference; MS: Hand grip; MP: Gait speed, TUG, Tinetti, chair stand test Mitochondrial activity (ATP, electron flux); Mitochondiral biogenesis (COX-1, COX-4, TFAM, NRF-1, MFN-1, MFN-2); redoc activity (TBARs)
Achison 2022 [58] UK	145	78.4	54	sarcopenia (EWGSOP1)	12 months	3 times daily Leucine 2.5 g	Placebo	No	MM: Appendicular muscle mass index; MS: Hand grip, quadriceps strentgh; MP: Gait speed, SPPB, Chair stand test, 6 min walk
						Omega-3			
Krzymińska-Siemaszko 2015 [47] Poland	27	75.8	59	sarcopenia (EWGSOP1)	12 weeks	Once daily n-3 PUFA 1.3 g with vitamine E	Once daily vitamin E 11 mg	No	MM: Appendicular muscle mass index; MS: Hand grip; MP: Gait speed, TUG

Essential amino acid (EAA): histidine, isoleucine, leucine, lysine, methionine, phenylalanine, threonine, tryptophane, and valine. Non-essential amino acid (NEAA): alanine, arginine, asparagine, aspartic acid, cysteine, glutamic acid, glutamine, glycine, proline, serine, and tyrosine. MM: muscle mass; MS: muscle strength; MP: muscle performance. Resistance training (RT), aerobic training (AT), short physical performance battery (SPPB) which consists of (1) a 4-m usual gait speed walking test, (2) a repeated chair rise test and (3) a balance test. TUG: timed up and go test. COX-1 and COX-4: Cytochrome C oxidase-1 and 4; TFAM: Mitochondrial Transcription Factor A; MFN-: Mitofusin-1; MFN-2: Mitofusin-2; TBARs: Thiobarbituric Acid-Reactive Substances. MNA-sf: Mini Nutritional Assessment Scale Short Form; NRS: nutritional risk score; CaHMB: Calcium β-Hydroxy-β-Methylbutyrate; EWGSOP-1: European Working Group on Sarcopenia in Older People; ASIA: Asian Working Group for Sarcopenia.

3.3. Quality Assessment

Amongst the included studies, 16.6% raised some concerns or were evaluated at high risk of biases in the randomization process, 8.3% were at high risk from intended interventions, 16.6% were at high risk for missing outcome data, and 25% raised some concerns or were at high risk for reported result. Overall, 25% of the studies have high risk of bias and 8.3% raised some concerns (Figure 2).

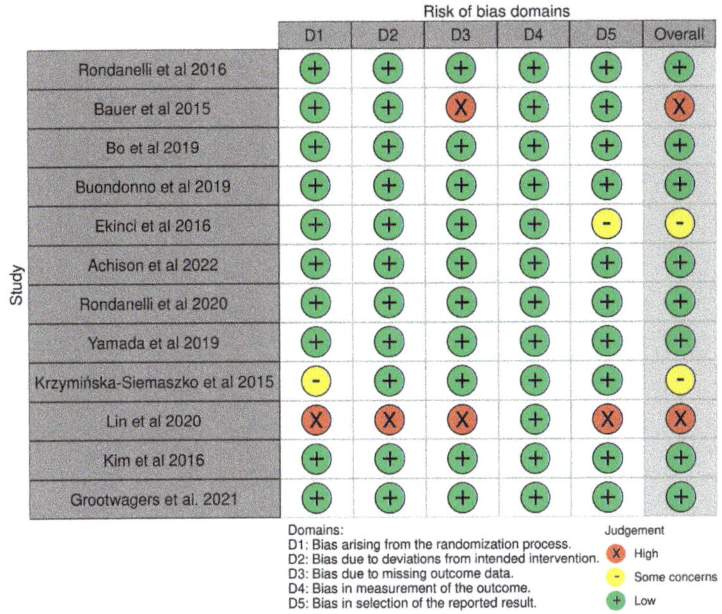

Figure 2. Details of the evaluation of bias risk for each study [45,47,50,57–65].

3.4. Outcomes of the Included RCT

Five clinical studies out of nine testing vitamin D and whey protein or BCAA supplementation [59–62,64] found a significant improvement of muscle mass and/or on muscle strength in supplemented patients. Moreover, three studies [57,61,62] out of seven showed a significant effect of this intervention in the amelioration of physical performance; the studies that did not reach statistical significance showed a positive trend in the intervention group. Studies focusing on nutritional supplementation with vitamin D with whey proteins or BCAA administered with physical exercise [45,63] demonstrated a significant effect of the combined intervention on muscle mass, strength or performance. Regarding the biological findings, one study [65] assessing nutritional supplementation with a BCAA mixture showed a significant improvement of mitochondrial bioenergetics and redox activity. The second study showed a positive trend on mitochondrial activity with nutritional supplementation including a vitamin D, whey protein and BCAA mixture. Detailed results of intervention's studies are summarized in Table 3 for the clinical outcomes and Table 4 for the biological outcomes.

Table 3. Clinical Outcomes.

Authors Year Location	Overall Risk of Bias	Mucsle Mass	Muscle Strength	Muscle Performance
		BCAA or Whey Protein with Vitamin D		
Rondanelli 2016 [60] Italy	Low	RSMM (kg/m^2) Treatment effect (mean difference): 0.27, 95% CI (0.07–0.47), p 0.009	Handgrip strength (kg) Treatment effect (mean difference): 3.68, 95% CI (2.55–4.81), $p < 0.001$	NA
Rondanelli 2020 [61] Italy	Low	SMMI (kg/m^2/month) Crude between-group difference: 0.40, 95% CI (0.06 to 0.73), p 0.023	Handgrip strength (kg/month) Crude between-group difference: 5.45, 95% CI (4.51 to 6.38), $p < 0.001$	4 m gait speed (m/s/month) Crude between-group difference: 0.062, 95% CI (0.043 to 0.082), $p < 0.001$ Chair stand test (s/month) Crude between-group difference: 12.64, 95% CI (10.84 to 14.44), $p < 0.001$ Timed up and go (s/month) Crude between-group difference: 3.71, 95% CI (3.09 to 4.33), $p < 0.001$ SPPB (score/month) crude Between-group difference: 2.27, 95% CI (1.88 to 2.68), $p < 0.001$
Lin 2020 [50] Taiwan	High	AMMI (kg/m^2) Within group change in the intervention group from baseline (6.1 $\pm$ 0.64) to 12 weeks (6.56 $\pm$ 0.95), $p < 0.001$ Within group change in the control group from baseline (6.27 $\pm$ 0.68) to 12 weeks (6.61 $\pm$ 0.76), $p < 0.001$	Handgrip strength (kg) Within group change in the intervention group from baseline (25.3 $\pm$ 10.1) to 12 weeks (26.1 $\pm$ 7.76), p 0.19 Within group change in the control group from baseline (26.3 $\pm$ 6.95) to 12 weeks (27.6 $\pm$ 7.0), p 0.03	Gait speed (m/s) Within group change in the intervention group from baseline (0.98 $\pm$ 0.14) to 12 weeks (0.97 $\pm$ 0.13) p 0.016 Within group change in the control group from baseline (0.98 $\pm$ 0.14) to 12 weeks (0.97 $\pm$ 0.13), p 0.58
Bauer 2015 [62] Germany	High	SMI (kg) Treatment effect (mean difference): 0.17, 95% CI (0.004–0.338), p 0.045	Handgrip strength (kg) Treatment effect (mean difference): 0.30, 95% CI (−0.46–1.05), p 0.44	Gait speed (m/s) Treatment effect (mean difference): 0.01, 95% CI (−0.02–0.04), p 0.46 SPPB (score) Treatment effect (mean difference): 0.11, 95% CI (−0.21–0.42), p 0.51 Chair stand test (s) Treatment effect (mean difference): −1.01, 95% CI (−1.77–0.19), p 0.018

Table 3. *Cont.*

Authors Year Location	Overall Risk of Bias	Mucsle Mass	Muscle Strength	Muscle Performance
Kim 2016 [45] Japan	Low	SMI (kg/m^2) Odds Ratio for Changes compared to HE group: NU: 0.78 (0.25–2.21) EX: 0.83 (0.29–2.41) EXNU: 0.67 (0.23–1.93)	Handgrip and knee extension strength (kg) Odds Ratio for changes compared to HE group: NU: 2.71 (0.96–7.64) EX: 3.72 (1.24–11.17) EXNU: 3.69 (1.28–10.71)	Gait speed (m/s) Odds Ratio for Changes compared to HE group: NU: 1.53 (0.52–4.55) EX: 2.06 (0.67–6.29) EXNU: 3.05 (1.01–9.19)
Grootswagers 2021 [57] The Netherlands	Low	ALMI (kg/m^2) Treatment*time interaction $p > 0.05$	Non-dominant knee extension (Newton) Mean change in the intervention group: 8 ± 12, p 1.000 Mean change in the control group: 38 ± 10, p 0.003 Treatment*time interaction p 0.058 Dominant knee extension (Newton) Mean change in the intervention group: -2 ± 14, p 1.000 Mean change in the control group: 26 ± 21, p 1.000 Treatment*time interaction p 0.145 Dominant knee flexion (Newton) Mean change in the intervention group: 12 ± 9, p 1.000 Mean change in the control group: 23 ± 8, p 0.036 Treatment*time interaction p 0.351 Handgrip strength (dominant hand, kg) Mean change in the intervention group: 0 ± 1, p 1.000 Mean change in the control group: 0 ± 1, p 1.000 Treatment*time interaction p 0.948	400 m walk test (s) Mean change in the intervention group: -7.4 ± 8.7, p 1.000 Mean change in the control group: 17.6 ± 7.8, p 0.172 Treatment*time interaction p 0.038 4 m walk test (s) Mean change in the intervention group: -0.4 ± 0.1, p 0.047 Mean change in the control group: 0.0 ± 0.1, p 1.000 Treatment*time interaction p 0.048 Chair rise test (s) Mean change in the intervention group: 0.0 ± 0.5, p 1.000 Mean change in the control group: -0.3 ± 0.4, p 1.000 Treatment*time interaction p 0.634 SPPB (score) Mean change in the intervention group: 0.1 ± 0.2, p 1.000 Mean change in the control group: 0.3 ± 0.2, p 0.523 Treatment*time interaction p 0.355

Table 3. *Cont.*

Authors Year Location	Overall Risk of Bias	Mucsle Mass	Muscle Strength	Muscle Performance
Ekinci 2016 [64] Turkey	Some concerns	Arm circumference (cm) Change in the intervention group from baseline (24.00 ± 2.57) to 30 days (24.84 ± 2.00), p 0.001 Change in the control group from baseline (25.30 ± 2.53) to 30 days (24.87 ± 2.62), p 0.320 Difference between group p 0.969 Calf circumference (cm) Change in the intervention group from baseline (41.13 ± 4.19) to 30 days (40.56 ± 3.78), p 0.672 Change in the control group from baseline (41.40 ± 3.04) to 30 days (41.77 ± 3.15), p 0.986 Difference between group p 0.180 Triceps skinfold thickness (TST) (mm) Change in the intervention group from baseline (12.47 ± 3.89) to 30 days (13.94 ± 3.81), $p < 0.001$ Change in the control group from baseline (13.53 ± 3.01) to 30 days (13.13 ± 3.66), p 0.999 Difference between group p 0.400	Handgrip strength (kg) Change in the intervention group from baseline (7.13 ± 4.01) to 30 days (8.63 ± 3.83), p 0.015 Change in the control group from baseline (5.53 ± 3.42) to 30 days (6.40 ± 3.86), p 0.157 Difference between group p 0.026	NA
Bo 2019 [59] China	Low	RSMi (kg/m^2) Treatment effect (mean difference): 0.18, 95% CI (0.01–0.35), p 0.040	Handgrip strength (kg) Treatment effect (mean difference): 2.68, 95% CI (0.71–4.65), p 0.009	6 m gait speed (walking at usual pace) (m/s) Mean change: Intervention group: 0.14 ± 0.15, $p < 0.001$ Control group: 0.08 ± 0.24, p 0.074 Treatment effect (mean difference): 0.05, 95% CI (−0.06 to 0.15), p 0.402 Timed up and go (s) Mean change: Intervention group: -1.36 ± 2.43, $p < 0.001$ Control group: −0.68 ± 3.29, p 0.267 Treatment effect (mean difference): −0.67, 95% CI (−2.20 to 0.86), p 0.383 Chair stand test (s) Mean change: Intervention group: −2.79 ± 3.73, p 0.005 Control group: −1.21 ± 6.28, p 0.507 Treatment effect (mean difference): −1.84, 95% CI (−4.53 to 0.85), p 0.176

Table 3. *Cont.*

Authors Year Location	Overall Risk of Bias	Mucsle Mass	Muscle Strength	Muscle Performance
Yamada 2019 [63] Japan	Low	Appendicular muscle mass (kg) Median change from baseline (IQR range): EXNU: 0.51 (0.04–1.24) EX: −0.30 (−0.64 to 1.07) NU: 0.11 (−0.29 to 0.40) Control group: −0.72 (−1.88 to 0.01) Differences between EXNU and control group p 0.02	Maximal isometric knee extension strength (Newton) Median change from baseline (IQR range): EXNU: 39.20 (31.36–45.20) EX: 1.84 (−6.13 to 10.05) NU: −4.41 (−10.17 to 10.41) Control group: 3.92 (−7.60 to 9.43) Differences between EXNU group and control group p 0.46 Handgrip strength (kg) Median change from baseline (IQR range): EXNU: 1.70 (−0.20 to 2.70) EX: −0.05 (−2.45 to 1.08) NU: −0.40 (−1.90 to 0.95) Control group: −0.67 (−3.18 to 0.73) Differences between EXNU group and control group p 0.07	5 m maximum walking time (s) Median change from baseline (IQR range): EXNU: −0.82 (−1.28 to −0.52) EX: −0.57 (−2.05–0.47) NU: −0.04 (−0.88 to 0.41) Control group: 0.44 (−0.26 to 1.11) Differences between EXNU group and control group p 0.01 Five-repetition chair stand test (s) Median change from baseline (IQR range): EXNU: −1.15 (−2.48 to −0.26) EX: −0.27 (−1.92 to 0.00) NU: −0.41 (−1.46 to 0.06) Control group: −0.37 (−1.68 to 0.77) Differences between EXNU group and control group p 0.47
			BCAA alone	
Buondonno 2020 [65] Italy	Low	Calf circumference (cm) Mean in intervention group at baseline 30.4 ± 0.35, 95% CI (−1.14 to −0.04) and 2 months 31.3 ± 0.39, 95% CI (−1.45 to −0.36) Mean in control group at baseline 30.7 ± 0.43, 95% CI (−1.08 to 0.02) and 2 months 31.19 ± 0.39, 95% CI (−1.07 to 0.03) Time difference p 0.0004, treatment difference p 0.8560, interaction difference p 0.4521 Arm circumference (cm) Mean in intervention group at baseline 22.7 ± 0.36, 95% CI (−0.75 to 0.23) and 2 months 23.3 ± 0.38, 95% CI (−1.03 to −0.04) Mean in control group at baseline 23.0 ± 0.40, 95% CI (−0.83 to 0.15) and 2 months 23.4 ± 0.44, 95% CI (−0.88 to −0.10) Time difference p 0.0045, treatment difference p 0.6754, interaction difference p 0.7351	Handgrip strength (kg) Mean in intervention group at baseline 17.9 ± 1.0, 95% CI (−2.01 to 0.47) and 2 months 18.3 ± 1.0, 95% CI (−2.4 to 0.10) Mean in control group at baseline 17.9 ± 1.0, 95% CI (−1.84 to 0.65) and 2 months 19.1 ± 1.0, 95% CI (−1.59 to 0.90) Time mean difference p 0.0474, treatment mean difference p 0.7796, Interaction mean difference p 0.5231	4 m gait speed (s) Mean in intervention group at baseline 8.2 ± 0.6, 95% CI (−0.3 to 1.7) and 2 months 7.2 ± 0.6, 95% CI (0.04 to 2.0) Mean in control group at baseline 9.8 ± 0.7, 95% CI (0.4 to 2.3) and 2 months 8.0 ± 0.7, 95% CI (0.8 to 2.8) Time mean difference $p < 0.001$, treatment mean difference: p 0.1685, Interaction mean difference p 0.3955 Timed up and go (s) Mean in intervention group at baseline 19.8 ± 2.14, 95% CI (1.5 to 7.6) and 2 months 15.1 ± 1.1, 95% CI (1.6 to 7.8) Mean in control group at baseline 20.5 ± 1.5, 95% CI (−1.2 to 4.9) and 2 months 17.7 ± 1.7, 95% CI (−0.3 to 5.9) Time mean difference p 0.0001, treatment mean difference) p 0.2780, interaction mean difference p 0.3215 30-s Chair to stand test (s) Mean in intervention group at baseline 6.8 ± 0.5, 95% CI (−2.6 to −0.7) and 2 months 8.5 ± 0.7, 95% CI (−2.7 to −0.7) Mean in control group at baseline 6.0 ± 0.5, 95% CI (−2.4 to −0.5) and 2 months 8.1 ± 0.6, 95% CI (−3.0 to −1.1) Time mean difference $p < 0.0001$, treatment mean difference p 0.3328, Interaction mean difference p 0.5810

Table 3. *Cont.*

Authors Year Location	Overall Risk of Bias	Mucsle Mass	Muscle Strength	Muscle Performance
				Balance test (Tinetti) Mean in intervention group at baseline 20.4 ± 0.8, 95% CI (−2.1 to −0.1) and 2 months 22.2 ± 0.7, 95% CI (−2.8—0.8) Mean in control group at baseline 18.3± 0.8, 95% CI (−3.2 to −1.1) and 2 months 20.7 ± 0.9, 95% CI (−3.4 to −1.4) Time mean difference < 0.0001, treatment mean difference: *p* 0.1503, interaction mean difference *p* 0.2076
Achison 2022 [58] UK	Low	RSMI (kg/m^2) Between-group difference over 12 months follow up: −0.3, 95% CI (−1.0, 0.4), *p* 0.47	Handgrip Strength (kg) Between-group difference over 12 months follow up: −0.3, 95% CI (−1.2, 0.7), *p* 0.55 Quadriceps strength (kg) Between-group difference over 12 months follow up: −1.0, 95% CI (−4.4, 2.4), *p* 0.55	4-m gait speed (m/s): Between-group difference over 12-month follow-up: 0.01, 95% CI (−0.18, 0.19), *p* 0.96 Six min walk (m): Between-group difference over 12-month follow-up: 17, 95% CI (−25, 59), *p* 0.43 SPPB (score) Between-group difference over 12-month follow-up: 0.1, 95% CI (−1.0, 1.1), *p* 0.90 Chair stand test (s) Between-group difference over 12-month follow-up: −3.1, 95% CI (−9.5, 3.3), *p* 0.34
			Omega-3	
Krzymińska-Siemaszko 2015 [47] Poland	High	ALMI (kg/m^2) Change in the intervention group from the baseline: 0.00 ± 0.30 Change in the control group from the baseline: 0.03 ± 0.36 Between-group difference *p* 0.53	Handgrip strength (kg) Change in the intervention group from the baseline: 0.68 ± 1.43 Change in the control group from the baseline: 0.54 ± 2.77 Between-group difference *p* 0.12	4-m walking test (s) Change in the intervention group from the baseline: 0.11 ± 0.26 Change in the control group from the baseline: 0.09 ± 0.13 Between-group difference *p* 0.06 Timed up and go (s) Change in the intervention group from the baseline: 0.05 ± 1.50 Change in the control group from the baseline: 0.42 ± 1.18 Between-group difference *p* 0.11

Results showing an efficacy of the nutritional intervention as compared to placebo are highlighted in shadow. Efficacity of treatment are reported for all the studies RSMM: Relative Skeletal Muscle Mass; SMI: skeletal muscle index; AMS: Appendicular Muscle Mass; AMMI; Appendicular Muscle Mass Index; ALMI: Appendicular Lean Muscle Index; RSMI: relative muscle mass index; SPPB: Short Performance Physical Battery; EXNU: exercise in addition to nutrition; EX: exercise alone; NU: nutrition alone; HE: health education, CI: confidence interval.

Table 4. Biological outcomes. Results showing an efficacy of the nutritional intervention as compared to placebo are highlighted in shadow. Efficacity of the treatment is reported for all the studies.

Authors Year Location	Overall Risk of Bias	Mitochondrial Bioenergetics	Mitochondrial Dynamics	Redox Activity
			BCAA and Whey Protein with Vitamin D	
Grootswagers 2021 [57] The Netherlands	Low	NA	mRNA PGC-1 alpha (Peroxisome proliferator-activated receptor-γ coactivator-1α) Fold change expression: Intervention group: 4.7 +/− 1.8; Control group: 2.2 +/− 0.6 Between treatment difference p 0.685 AMPK (5′adenosine monophosphate-activated protein kinase) Within-treatment difference baseline vs. 12 weeks in intervention group p 0.031 Within-treatment difference baseline vs. 12 weeks in control group p 0.125 TFAM (Mitochondrial Transcription Factor A) Between treatment difference p 0.603	NA
			BCAA alone	
Buondonno 2020 [65] Italy	Low	ATP Mean in intervention group at baseline:1.0 ± 0.0, 95% CI (−0.45 to −0.15) and 2 months:1.43 ± 0.10, 95% CI (−0.58 to −0.28) Mean in control group at baseline 1.0 ± 0.0, 95% CI (−0.13 to 0.17) and 2 months 0.99 ± 0.02, 95% CI (−0.14 to 0.16) Time difference p 0.0001, Treatment difference p 0.0005, Interaction difference p 0.0001 Electron flux Mean in intervention group at baseline 1.0 ± 0.0, 95% CI (−0.38 to −0.13) and 2 months 1.50 ± 0.09, 95% CI (−0.62 to −0.38) Mean in control group at baseline 1.0 ± 0.0, 95% CI (−0.13 to 0.13) and 2 months 1.01 ± 0.04, 95% CI (−0.14 to 0.12) Time difference $p < 0.0001$, Treatment difference $p < 0.0001$, Interaction difference $p < 0.0001$	COX-1 (Cytochrome C oxidase -1) Mean in intervention group at baseline 1.0 ± 0.0, 95% CI (−23.8 to −1.9) and 2 months 7.3 ± 3.6, 95% CI (−17.3 to 4.7); Mean in control group at baseline 1.0 ± 0.0, 95% CI (−11.4 to 10.6) and 2 months 3.7 ± 1.2, 95% CI (−13.6 to 8.3) Time difference p 0.1155, Treatment difference p 0.1967, Interaction difference p 0.1409 COX-4 (Cytochrome C oxidase-4) Mean in intervention group at baseline 1.0 ± 0.0, 95% CI (−2.3 to −0.10) and 2 months: 1.8 ± 0.5, 95% CI (−1.9 to 0.3); Mean in control group at baseline 1.0 ± 0.0, 95% CI (−1.39 to 0.76) and 2 months 1.3 ± 0.19, 95% CI (−1.4 to 0.73) Time difference p 0.0459, Treatment difference p 0.2373, Interaction difference p 0.3786	Thiobarbituric Acid Reactiv Substances (TBARs) mcg/M Mean in intervention group at baseline 2.3 ± 0.4, 95% CI (−2.8 to 1.2) and 2 months 3.2 ± 0.70, 95% CI (−3.1 to 0.85) Mean in control group at baseline 4.1 ± 0.7, 95% CI (−3.02 to 0.97) and 2 months 6.7 ± 1.3, 95% CI (−5.64 to −1.64) Time difference p 0.0007, Treatment difference p 0.0289, Interaction difference p 0.0332

Table 4. *Cont.*

Authors Year Location	Overall Risk of Bias	Mitochondrial Bioenergetics	Mitochondrial Dynamics	Redox Activity
			TFAM (Mitochondrial Transcription Factor A) Mean in intervention group at baseline 1.0 ± 0.0, 95% CI (−6.9 to −0.6) and 2 months 4.2 ± 1.15, 95% CI (−6.2 to 0.1); Mean in control group at baseline 1.0 ± 0.0, 95% CI (−3.8 to 2.5) and 2 months: 3.0 ± 1.5, 95% CI (−5.1 to 1.2) Time difference p 0.0178, Treatment difference p 0.0932, Interaction difference p 0.2235 NRF-1 (Nuclear Respiratory Factor-1) Mean in intervention group at baseline 1.0 ± 0.0, 95% CI (−20.2 to 3.5) and 2 months 11.6 ± 9.5, 95% CI (−22.4 to 1.3); Mean in control group at baseline 1.0 ± 0.0, 95% CI (−13.5 to 10.2) and 2 months:3.6 ± 1.2, 95% CI (−14.4 to 9.3) Time difference 0.6599, Treatment difference p 0.3507, Interaction difference p 0.2055 MFN-1 (Mitofusin-1) Mean in intervention group at baseline 1.0 ± 0.0, 95% CI (−22.4 to −2.1) and 2 months 10.1 ± 6.1, 95% CI (−19.3 to 1.0); Mean in control group at baseline 1.0 ± 0.0, 95% CI (−10.8 to 9.4) and 2 months 1.8 ± 0.3, 95% CI (−11.0 to 9.3) Time difference 0.0746, Treatment difference p 0.1648, Interaction difference p 0.1320 MFN-2 (Mitofusin-2) Mean in intervention group at baseline 1.0 ± 0.0 95% CI (−11.6 to −1.1) and 2 months 3.9 ± 1.6, 95% CI (−8.2 to 2.3); Mean in control group at baseline 1.0 ± 0.0, 95% CI (−6.0 to 4.5) and 2 months 2.4 ± 0.5, 95% CI (−6.6 to 3.9) Time difference p 0.0772, Treatment difference p 0.2046, Interaction difference p 0.1810	

Results showing an efficacy of the nutritional intervention as compared to placebo are highlighted in shadow. Efficacity of treatment are reported for all the studies.

4. Discussion

Sarcopenia is a frequent condition amongst older adults and malnutrition plays a causal role on its pathogenesis as well as mitochondrial dysfunction. The use of nutritional supplementation can be effective in improving mitochondrial function and in improving muscle health, thus treating sarcopenia; however, there is no clear recommendation on the type of nutritional supplements useful for treating this disease and ameliorating mitochondrial function.

Here, we suggest that nutritional supplementation with vitamin D combined with whey proteins or BCAA mixture can be effective in increasing muscle mass, muscle strength or performance, whereas supplementation with BCAA or omega-3 PUFA alone is not. All the studies on vitamin D used cholecalciferol in doses ranging between 100 and 1000 IU per day. Four out of nine studies on this combination demonstrated a significant gain in muscle mass [59–62] compared to placebo; however, no gain in muscle mass was observed if the control groups received nutritional counseling or a standard nutritional intervention. This result suggests that standard clinical treatment is sufficient to induce an improvement in muscle mass.

Nutritional supplementation with vitamin D combined with whey protein or BCAA may be effective in improving muscle strength [59–61,64], regardless of the addition of physical exercise. These findings are particularly relevant in geriatrics as, frequently, older patients are not willing to or are unable to follow an adequate physical activity program [66]; thus, we suggest to treat these patients with the combination of vitamin D and whey protein or BCAA, even without adding physical activity.

Three out of seven studies [57,61,62] showed a significant improvement of the muscular performance thanks to the nutritional intervention with vitamin D and whey protein or BCAA. The four studies that did not reach statistically significant results showed a trend towards improvement of muscle performance [45,50,59,63]. The included studies differed in the quantity of BCAA and whey proteins administered; the studies showing a significant improvement in muscle performance evaluated doses of total leucine higher than 3 g/day combined with 20 to 40 g of whey proteins/day, administered during the two main meals. Although there is no unique recommendation for the amount of BCAA required to trigger muscular protein synthesis in older patients [67], the literature suggests a dose-dependent correlation between BCAA supplements and muscular protein synthesis induction [68]. It is noteworthy that the only study that did not find significant effects of BCAA on muscle mass, strength or performance [58] administrates leucin without other EAA or protein supplementation, suggesting that, although leucin is necessary to boost muscle health, it is not sufficient and must be administered with other EAA or with whey proteins in order to achieve a clinical benefit. According to Wolfe [69] and Wilkison [70], leucine alone might contribute to an enhanced anabolic response and therefore increase muscle mass, However, all the EAA are needed to induce muscular proteins synthesis [71]. Hence, if only leucine is supplemented, the other needed EAA will be obtained from muscle protein breakdown, blunting the efficacy of leucine.

Although nutritional supplementation can boost muscular health even without physical exercise, adding aerobic or resistance training will significantly improve muscle mass, strength and performance [45,63]. Physical inactivity increases anabolic resistance [14] and hence contributes to muscle waste and sarcopenia; moreover, physical exercise enhances the effect of BCAA and protein supplementation as muscle protein breakdown during exercise releases EAA that boost protein synthesis [69]. A meta-analysis [72] published after the end of our research, including three RCTs, also included in our review [60–62], concludes that supplementation with whey protein, leucine and vitamin D, without physical exercise, increases muscle mass in sarcopenic patients; if nutritional supplementation is combined with physical activity, it also increases muscle strength and performances. Differently from Chang and Choo [72], we included in our systematic review not only supplementation with protein, leucine, and vitamin D in sarcopenic patients but also BCAA and omega-3 PUFA and widen the target population by also including malnourished patients. Despite these differences, our work supports their findings and contributes to generalization to a wider population.

Thus, we suggest adding appropriate and adapted training to nutritional supplementation whenever possible.

Only two studies evaluated the effect of nutritional supplements on mitochondrial activity: one study focused on BCAA alone [65], the other studied the effect of BCAA plus vitamin D [57]. BCAA were proven to be effective in ameliorating mitochondrial bioenergetics and function even without the addition of vitamin D by Buondonno et al. [65]. These authors showed an increase in the ability of mitochondria to produce ATP and an increased ability to reduce oxidative stress over two months. The authors also evaluated clinical outcomes on muscle mass, strength and performance, showing a positive trend on muscle performance; thus, they hypothesize that the improvement of mitochondrial activity via the mTOR signaling pathway and reduction of ROS may explain the improvement of muscle performance.

Grootswagers et al. [57] evaluated the effect of nutritional supplementation with BCAA and vitamin D on mitochondrial activity showing that this treatment is effective in improving mitochondrial bioenergetics and function, and that this improvement might explain the amelioration of muscular performance; the authors demonstrated that patients with improved mitochondrial biogenesis had a significant improvement in the gait speed, whereas the others did not.

The results from these two studies suggest that nutritional supplementation may play a role in improving age-associated mitochondrial dysfunction, paving the way for further studies on other conditions associated with unhealthy aging, such as cognitive impairment. In this regard, Buondonno et al. [65] showed an improvement not only in muscular, but also in cognitive performance in old malnourished patients treated with BCAA, this improvement correlated with the increase in ATP production by mitochondria.

Even if some papers suggest that omega-3 PUFA may be beneficial in sarcopenic patients [30], we cannot draw any conclusions on their efficacy as only one study was eligible for our review [47]. For instance, the results of this study were not significant; these authors tested omega-3 PUFA in addition to vitamin E that was also administered to the control group. After the conclusion of our search, an RCT tested the impact of supplementation with omega-3 PUFA (500 mg), leucine (2.5 g) and Lactobacillus paracasei PS23 in a population of sarcopenic patients [73]. This study demonstrated a significant improvement in muscle mass, strength and performance in patients supplemented, compared to the placebo group. Although this study shows positive results, the researchers used a mixture of supplements; thus, it is not possible to conclude in favor of the supplementation with the sole omega-3 PUFA.

A major limitation for this review is the small sample size of the studies included and the presence of several biases in the study designs that are often not double blind and placebo controlled. Moreover, in this review, we could not evaluate the effect of vitamin D alone. Even though several authors showed that active vitamin D supplementation (calcitriol) significantly improves mitochondrial function, biogenesis and decreases oxidative stress [35,36,74], and, clinically, hypovitaminosis D induces muscle atrophy, clinical trials on vitamin D supplementation were not effective in proving its effectiveness on muscle mass, strength and function [75].

One of the strengths of our review is to focus on BCAA rather than on whey proteins; indeed, even if whey proteins contain BCAA (approximately 11%) [14], their quality, quantity and bioavailability might not be sufficient in order to activate muscle anabolism.

Another significant strength of our review is to include only studies on patients suffering from sarcopenia or malnutrition. This restriction allows us to reduce the heterogeneity between the studies as concerns the targeted population.

To the best of our knowledge, this review is the first to evaluate the impact of targeted nutritional supplementation on both biological and clinical outcomes and to shed light on their relationship.

5. Conclusions

Although there is still room for well-designed studies on the optimal treatment strategy for older malnourished patients, the results of this systematic review suggest that the treatment for these patients is vitamin D supplementation associated with either BCAA or whey proteins together with appropriate aerobic or anaerobic training.

All the studies included in this review used cholecalciferol in doses ranging between 100 and 1000 IU per day, and even at the lower doses [60], the supplementation of vitamin D coupled with BCAA and whey proteins is effective in ameliorating muscle health; hence, we recommend to give cholecalciferol 1000 IU daily in patients at increased risk of vitamin D deficiency according to the recent guidelines from the Society of Clinical and Economical Aspects of Osteoporosis, Osteoarthritis and Musculoskeletal Diseases (ESCEO) working group [76].

As regards the BCAA to be used, a mixture of amino acids including leucin, valine and isoleucine with a 2:1:1 ratio [65,77] or 4:1:1 ratio [45,60] works better than the sole leucine that might be used at doses equal to or higher than 3 g/day [69]. If whey proteins are used, the intake needed to overcome anabolic resistance in older sarcopenic patients is at least 1.2 g/kg/day [78]. These doses are above the internationally recommended dietary allowance of 0.8 g/kg/day [14,15,78] and can cause gastrointestinal problems or be contraindicated in case of kidney failure. These limitations in the use of whey proteins must be considered in the clinical prescription of nutritional supplementation in older patients.

Regarding the timing of protein intake, we suggest distributing proteins and or BCAA supplements over the three main meals to boost protein synthesis after each meal [14].

We also suggest using BCAA with or without vitamin D to boost mitochondrial activity and potentially ameliorate other chronic conditions associated with aging, such as cognitive impairment.

As concerns omega-3 PUFA, we cannot, based on the results of this review, recommend their use in sarcopenic older adults.

Author Contributions: Conceptualization, C.C. and P.D.; methodology, C.C. and P.D.; validation, P.D.; investigation, C.C., G.B., F.C. and I.B.; data curation, C.C. and G.B.; writing—original draft, C.C.; review, G.B. and P.D.; supervision, P.D. All authors have read and agreed to the published version of the manuscript.

Funding: This research received no external funding.

Institutional Review Board Statement: Not applicable.

Informed Consent Statement: Not applicable.

Data Availability Statement: All the data are available to the public and presented in the manuscript.

Conflicts of Interest: The authors declare no conflict of interest.

References

1. Cattaneo, F.; Buondonno, I.; Cravero, D.; Sassi, F.; D'Amelio, P. Musculoskeletal Diseases Role in the Frailty Syndrome: A Case–Control Study. *Int. J. Environ. Res. Public Health* **2022**, *19*, 11897. [CrossRef] [PubMed]
2. Petermann-Rocha, F.; Balntzi, V.; Gray, S.R.; Lara, J.; Ho, F.K.; Pell, J.P.; Celis-Morales, C. Global Prevalence of Sarcopenia and Severe Sarcopenia: A Systematic Review and Meta-Analysis. *J. Cachexia. Sarcopenia Muscle* **2022**, *13*, 86–99. [CrossRef] [PubMed]
3. Cruz-Jentoft, A.J.; Baeyens, J.P.; Bauer, J.M.; Boirie, Y.; Cederholm, T.; Landi, F.; Martin, F.C.; Michel, J.P.; Rolland, Y.; Schneider, S.M.; et al. Sarcopenia: European Consensus on Definition and Diagnosis. *Age Ageing* **2010**, *39*, 412–423. [CrossRef] [PubMed]
4. Cruz-Jentoft, A.J.; Bahat, G.; Bauer, J.; Boirie, Y.; Bruyère, O.; Cederholm, T.; Cooper, C.; Landi, F.; Rolland, Y.; Sayer, A.A.; et al. Sarcopenia: Revised European Consensus on Definition and Diagnosis. *Age Ageing* **2019**, *48*, 16–31. [CrossRef] [PubMed]
5. Studenski, S.A.; Peters, K.W.; Alley, D.E.; Cawthon, P.M.; McLean, R.R.; Harris, T.B.; Ferrucci, L.; Guralnik, J.M.; Fragala, M.S.; Kenny, A.M.; et al. The FNIH Sarcopenia Project: Rationale, Study Description, Conference Recommendations, and Final Estimates. *J. Gerontol. Ser. A Biol. Sci. Med. Sci.* **2014**, *69*, 547–558. [CrossRef] [PubMed]
6. Fielding, R.A.; Vellas, B.; Evans, W.J.; Bhasin, S.; Morley, J.E.; Newman, A.B.; Abellan van Kan, G.; Andrieu, S.; Bauer, J.; Breuille, D.; et al. Sarcopenia: An Undiagnosed Condition in Older Adults. Current Consensus Definition: Prevalence, Etiology, and Consequences. International Working Group on Sarcopenia. *J. Am. Med. Dir. Assoc.* **2011**, *12*, 249–256. [CrossRef]

7. Chen, L.K.; Woo, J.; Assantachai, P.; Auyeung, T.W.; Chou, M.Y.; Iijima, K.; Jang, H.C.; Kang, L.; Kim, M.; Kim, S.; et al. Asian Working Group for Sarcopenia: 2019 Consensus Update on Sarcopenia Diagnosis and Treatment. *J. Am. Med. Dir. Assoc.* **2020**, *21*, 300–307.e2. [CrossRef]
8. Rom, O.; Kaisari, S.; Aizenbud, D.; Reznick, A.Z. Lifestyle and Sarcopenia—Etiology, Prevention and Treatment. *Rambam Maimonides Med. J.* **2012**, *3*, e0024. [CrossRef]
9. Marzetti, E.; Calvani, R.; Tosato, M.; Cesari, M.; Di Bari, M.; Cherubini, A.; Broccatelli, M.; Savera, G.; D'Elia, M.; Pahor, M.; et al. Physical Activity and Exercise as Countermeasures to Physical Frailty and Sarcopenia. *Aging Clin. Exp. Res.* **2017**, *29*, 35–42. [CrossRef]
10. Calvani, R.; Picca, A.; Coelho-Júnior, H.J.; Tosato, M.; Marzetti, E.; Landi, F. Diet for the Prevention and Management of Sarcopenia. *Metabolism* **2023**, *146*, 155637. [CrossRef]
11. Alizadeh Pahlavani, H.; Laher, I.; Knechtle, B.; Zouhal, H. Exercise and Mitochondrial Mechanisms in Patients with Sarcopenia. *Front. Physiol.* **2022**, *13*, 1040381. [CrossRef]
12. Hood, D.A.; Memme, J.M.; Oliveira, A.N.; Triolo, M. Maintenance of Skeletal Muscle Mitochondria in Health, Exercise, and Aging. *Annu. Rev. Physiol.* **2019**, *81*, 19–41. [CrossRef] [PubMed]
13. Cedikova, M.; Pitule, P.; Kripnerova, M.; Markova, M.; Kuncová, J. Multiple Roles of Mitochondria in Aging Processes. *Physiol. Res.* **2016**, *65*, S519–S531. [CrossRef] [PubMed]
14. Moore, D.R. Keeping Older Muscle "Young" through Dietary Protein and Physical Activity. *Adv. Nutr.* **2014**, *5*, 599S–607S. [CrossRef]
15. Romani, M.; Berger, M.M.; D'amelio, P. From the Bench to the Bedside: Branched Amino Acid and Micronutrient Strategies to Improve Mitochondrial Dysfunction Leading to Sarcopenia. *Nutrients* **2022**, *14*, 483. [CrossRef]
16. Buondonno, I.; Sassi, F.; Cattaneo, F.; D'Amelio, P. Association between Immunosenescence, Mitochondrial Dysfunction and Frailty Syndrome in Older Adults. *Cells* **2023**, *12*, 44. [CrossRef] [PubMed]
17. Prado, C.M.; Landi, F.; Chew, S.T.H.; Atherton, P.J.; Molinger, J.; Ruck, T.; Gonzalez, M.C. Advances in Muscle Health and Nutrition: A Toolkit for Healthcare Professionals. *Clin. Nutr.* **2022**, *41*, 2244–2263. [CrossRef]
18. Nelke, C.; Dziewas, R.; Minnerup, J.; Meuth, S.G.; Ruck, T. Skeletal Muscle as Potential Central Link between Sarcopenia and Immune Senescence. *EBioMedicine* **2019**, *49*, 381–388. [CrossRef] [PubMed]
19. Cederholm, T.; Jensen, G.L.; Correia, M.I.T.D.; Gonzalez, M.C.; Fukushima, R.; Higashiguchi, T.; Baptista, G.; Barazzoni, R.; Blaauw, R.; Coats, A.; et al. GLIM Criteria for the Diagnosis of Malnutrition—A Consensus Report from the Global Clinical Nutrition Community. *Clin. Nutr.* **2019**, *38*, 1–9. [CrossRef]
20. Garry, P.J.; Vellas, B.J. Practical and Validated Use of the Mini Nutritional Assessment in Geriatric Evaluation. *Nutr. Clin. Care* **1999**, *2*, 146–154. [CrossRef]
21. Rubenstein, L.Z.; Harker, J.O.; Salvà, A.; Guigoz, Y.; Vellas, B. Screening for Undernutrition in Geriatric Practice: Developing the Short-Form Mini-Nutritional Assessment (MNA-SF). *J. Gerontol. Ser. A Biol. Sci. Med. Sci.* **2001**, *56*, 366–372. [CrossRef]
22. Kondrup, J.; Ramussen, H.H.; Hamberg, O.; Stanga, Z.; Camilo, M.; Richardson, R.; Elia, M.; Allison, S.; Meier, R.; Plauth, M. Nutritional Risk Screening (NRS 2002): A New Method Based on an Analysis of Controlled Clinical Trials. *Clin. Nutr.* **2003**, *22*, 321–336. [CrossRef] [PubMed]
23. Cederholm, T.; Bosaeus, I.; Barazzoni, R.; Bauer, J.; Van Gossum, A.; Klek, S.; Muscaritoli, M.; Nyulasi, I.; Ockenga, J.; Schneider, S.M.; et al. Diagnostic Criteria for Malnutrition—An ESPEN Consensus Statement. *Clin. Nutr.* **2015**, *34*, 335–340. [CrossRef]
24. Norman, K.; Haß, U.; Pirlich, M. Malnutrition in Older Adults—Recent Advances and Remaining Challenges. *Nutrients* **2021**, *13*, 2764. [CrossRef] [PubMed]
25. Atherton, P.J.; Smith, K.; Etheridge, T.; Rankin, D.; Rennie, M.J. Distinct Anabolic Signalling Responses to Amino Acids in C2C12 Skeletal Muscle Cells. *Amino Acids* **2010**, *38*, 1533–1539. [CrossRef]
26. Rolland, Y.; Dupuy, C.; Abellan van Kan, G.; Gillette, S.; Vellas, B. Treatment Strategies for Sarcopenia and Frailty. *Med. Clin. N. Am.* **2011**, *95*, 427–438. [CrossRef] [PubMed]
27. Moore, D.R.; Churchward-Venne, T.A.; Witard, O.; Breen, L.; Burd, N.A.; Tipton, K.D.; Phillips, S.M. Protein Ingestion to Stimulate Myofibrillar Protein Synthesis Requires Greater Relative Protein Intakes in Healthy Older versus Younger Men. *J. Gerontol. Ser. A Biol. Sci. Med. Sci.* **2015**, *70*, 57–62. [CrossRef]
28. Le Couteur, D.G.; Solon-Biet, S.M.; Cogger, V.C.; Ribeiro, R.; de Cabo, R.; Raubenheimer, D.; Cooney, G.J.; Simpson, S.J. Branched Chain Amino Acids, Aging and Age-Related Health. *Ageing Res. Rev.* **2020**, *64*, 101198. [CrossRef]
29. D'Antona, G.; Ragni, M.; Cardile, A.; Tedesco, L.; Dossena, M.; Bruttini, F.; Caliaro, F.; Corsetti, G.; Bottinelli, R.; Carruba, M.O.; et al. Branched-Chain Amino Acid Supplementation Promotes Survival and Supports Cardiac and Skeletal Muscle Mitochondrial Biogenesis in Middle-Aged Mice. *Cell Metab.* **2010**, *12*, 362–372. [CrossRef] [PubMed]
30. Dupont, J.; Dedeyne, L.; Dalle, S.; Koppo, K.; Gielen, E. The Role of Omega-3 in the Prevention and Treatment of Sarcopenia. *Aging Clin. Exp. Res.* **2019**, *31*, 825–836. [CrossRef]
31. Domingues-Faria, C.; Vasson, M.P.; Goncalves-Mendes, N.; Boirie, Y.; Walrand, S. Skeletal Muscle Regeneration and Impact of Aging and Nutrition. *Ageing Res. Rev.* **2016**, *26*, 22–36. [CrossRef]
32. Lalia, A.Z.; Dasari, S.; Robinson, M.M.; Abid, H.; Morse, D.M.; Klaus, K.A.; Lanza, I.R. Influence of Omega-3 Fatty Acids on Skeletal Muscle Protein Metabolism and Mitochondrial Bioenergetics in Older Adults. *Aging* **2017**, *9*, 1096. [CrossRef] [PubMed]

33. D'amelio, P.; Quacquarelli, L. Hypovitaminosis d and Aging: Is There a Role in Muscle and Brain Health? *Nutrients* **2020**, *12*, 628. [CrossRef]
34. Bischoff-Ferrari, H.A.; Borchers, M.; Gudat, F.; Dürmüller, U.; Stähelin, H.B.; Dick, W. Vitamin D Receptor Expression in Human Muscle Tissue Decreases with Age. *J. Bone Miner. Res.* **2004**, *19*, 265–269. [CrossRef] [PubMed]
35. Ryan, Z.C.; Craig, T.A.; Folmes, C.D.; Wang, X.; Lanza, I.R.; Schaible, N.S.; Salisbury, J.L.; Nair, K.S.; Terzic, A.; Sieck, G.C.; et al. 1α,25-Dihydroxyvitamin D3 Regulates Mitochondrial Oxygen Consumption and Dynamics in Human Skeletal Muscle Cells. *J. Biol. Chem.* **2016**, *291*, 1514–1528. [CrossRef] [PubMed]
36. Salles, J.; Chanet, A.; Guillet, C.; Vaes, A.M.; Brouwer-Brolsma, E.M.; Rocher, C.; Giraudet, C.; Patrac, V.; Meugnier, E.; Montaurier, C.; et al. Vitamin D Status Modulates Mitochondrial Oxidative Capacities in Skeletal Muscle: Role in Sarcopenia. *Commun. Biol.* **2022**, *5*, 1288. [CrossRef] [PubMed]
37. Page, M.J.; McKenzie, J.E.; Bossuyt, P.M.; Boutron, I.; Hoffmann, T.C.; Mulrow, C.D.; Shamseer, L.; Tetzlaff, J.M.; Akl, E.A.; Brennan, S.E.; et al. The PRISMA 2020 Statement: An Updated Guideline for Reporting Systematic Reviews. *BMJ* **2021**, *372*, n71. [CrossRef]
38. Rodgers, M.; Sowden, A.; Petticrew, M.; Arai, L.; Roberts, H.; Britten, N.; Popay, J. Testing Methodological Guidance on the Conduct of Narrative Synthesis in Systematic Reviews: Effectiveness of Interventions to Promote Smoke Alarm Ownership and Function. *Evaluation* **2009**, *15*, 49–73. [CrossRef]
39. Campbell, M.; McKenzie, J.E.; Sowden, A.; Katikireddi, S.V.; Brennan, S.E.; Ellis, S.; Hartmann-Boyce, J.; Ryan, R.; Shepperd, S.; Thomas, J.; et al. Synthesis without Meta-Analysis (SWiM) in Systematic Reviews: Reporting Guideline. *BMJ* **2020**, *368*, l6890. [CrossRef]
40. Chen, L.K.; Liu, L.K.; Woo, J.; Assantachai, P.; Auyeung, T.W.; Bahyah, K.S.; Chou, M.Y.; Chen, L.Y.; Hsu, P.S.; Krairit, O.; et al. Sarcopenia in Asia: Consensus Report of the Asian Working Group for Sarcopenia. *J. Am. Med. Dir. Assoc.* **2014**, *15*, 95–101. [CrossRef]
41. Amasene, M.; Cadenas-Sanchez, C.; Echeverria, I.; Sanz, B.; Alonso, C.; Tobalina, I.; Irazusta, J.; Labayen, I.; Besga, A. Effects of Resistance Training Intervention along with Leucine-Enriched Whey Protein Supplementation on Sarcopenia and Frailty in Post-Hospitalized Older Adults: Preliminary Findings of a Randomized Controlled Trial. *J. Clin. Med.* **2022**, *11*, 97. [CrossRef]
42. Cheng, H.; Kong, J.; Underwood, C.; Petocz, P.; Hirani, V.; Dawson, B.; O'Leary, F. Systematic Review and Meta-Analysis of the Effect of Protein and Amino Acid Supplements in Older Adults with Acute or Chronic Conditions. *Br. J. Nutr.* **2018**, *119*, 527–542. [CrossRef]
43. Guo, Y.; Fu, X.; Hu, Q.; Chen, L.; Zuo, H. The Effect of Leucine Supplementation on Sarcopenia-Related Measures in Older Adults: A Systematic Review and Meta-Analysis of 17 Randomized Controlled Trials. *Front. Nutr.* **2022**, *9*, 929891. [CrossRef]
44. Ispoglou, T.; White, H.; Preston, T.; McElhone, S.; McKenna, J.; Hind, K. Double-Blind, Placebo-Controlled Pilot Trial of L-Leucine-Enriched Amino-Acid Mixtures on Body Composition and Physical Performance in Men and Women Aged 65–75 Years. *Eur. J. Clin. Nutr.* **2016**, *70*, 182–188. [CrossRef]
45. Kim, H.; Kim, M.; Kojima, N.; Fujino, K.; Hosoi, E.; Kobayashi, H.; Somekawa, S.; Niki, Y.; Yamashiro, Y.; Yoshida, H. Exercise and Nutritional Supplementation on Community-Dwelling Elderly Japanese Women with Sarcopenic Obesity: A Randomized Controlled Trial. *J. Am. Med. Dir. Assoc.* **2016**, *17*, 1011–1019. [CrossRef] [PubMed]
46. Kim, H.K.; Suzuki, T.; Saito, K.; Yoshida, H.; Kobayashi, H.; Kato, H.; Katayama, M. Effects of Exercise and Amino Acid Supplementation on Body Composition and Physical Function in Community-Dwelling Elderly Japanese Sarcopenic Women: A Randomized Controlled Trial. *J. Am. Geriatr. Soc.* **2012**, *60*, 16–23. [CrossRef]
47. Krzymińska-Siemaszko, R.; Czepulis, N.; Lewandowicz, M.; Zasadzka, E.; Suwalska, A.; Witowski, J.; Wieczorowska-Tobis, K. The Effect of a 12-Week Omega-3 Supplementation on Body Composition, Muscle Strength and Physical Performance in Elderly Individuals with Decreased Muscle Mass. *Int. J. Environ. Res. Public Health* **2015**, *12*, 10558–10574. [CrossRef]
48. Kemmler, W.; Grimm, A.; Bebenek, M.; Kohl, M.; von Stengel, S. Effects of Combined Whole-Body Electromyostimulation and Protein Supplementation on Local and Overall Muscle/Fat Distribution in Older Men with Sarcopenic Obesity: The Randomized Controlled Franconia Sarcopenic Obesity (FranSO) Study. *Calcif. Tissue Int.* **2018**, *103*, 266–277. [CrossRef] [PubMed]
49. Martínez-Arnau, F.M.; Fonfría-Vivas, R.; Buigues, C.; Castillo, Y.; Molina, P.; Hoogland, A.J.; van Doesburg, F.; Pruimboom, L.; Fernández-Garrido, J.; Cauli, O. Effects of Leucine Administration in Sarcopenia: A Randomized and Placebo-Controlled Clinical Trial. *Nutrients* **2020**, *12*, 932. [CrossRef] [PubMed]
50. Lin, C.C.; Shih, M.H.; Chen, C.D.; Yeh, S.L. Effects of Adequate Dietary Protein with Whey Protein, Leucine, and Vitamin D Supplementation on Sarcopenia in Older Adults: An Open-Label, Parallel-Group Study. *Clin. Nutr.* **2021**, *40*, 1323–1329. [CrossRef]
51. Mathieu, L.; Maltais, J.P.; Ladouceur, A.I.J.D. Effect of Specific Resistance Training. *J. Strength Cond. Res.* **2016**, *27*, 229–235.
52. Murphy, C.H.; Flanagan, E.M.; De Vito, G.; Susta, D.; Mitchelson, K.A.J.; De Marco Castro, E.; Senden, J.M.G.; Goessens, J.P.B.; Mikłosz, A.; Chabowski, A.; et al. Does Supplementation with Leucine-Enriched Protein Alone and in Combination with Fish-Oil-Derived n-3 PUFA Affect Muscle Mass, Strength, Physical Performance, and Muscle Protein Synthesis in Well-Nourished Older Adults? A Randomized, Double-Blind, Placebo. *Am. J. Clin. Nutr.* **2021**, *113*, 1411–1427. [CrossRef] [PubMed]
53. Stow, R.; Ives, N.; Smith, C.; Rick, C.; Rushton, A. A Cluster Randomised Feasibility Trial Evaluating Nutritional Interventions in the Treatment of Malnutrition in Care Home Adult Residents. *Trials* **2015**, *16*, 433. [CrossRef] [PubMed]

54. Takeuchi, I.; Yoshimura, Y.; Shimazu, S.; Jeong, S.; Yamaga, M.; Koga, H. Effects of Branched-Chain Amino Acids and Vitamin D Supplementation on Physical Function, Muscle Mass and Strength, and Nutritional Status in Sarcopenic Older Adults Undergoing Hospital-Based Rehabilitation: A Multicenter Randomized Controlled Trial. *Geriatr. Gerontol. Int.* **2019**, *19*, 12–17. [CrossRef]
55. Dimori, S.; Leoni, G.; Fior, L.; Gasparotto, F. Clinical Nutrition and Physical Rehabilitation in a Long-Term Care Setting: Preliminary Observations in Sarcopenic Older Patients. *Aging Clin. Exp. Res.* **2018**, *30*, 951–958. [CrossRef]
56. Englund, D.A.; Kirn, D.R.; Koochek, A.; Zhu, H.; Travison, T.G.; Reid, K.F.; Von Berens, Å.; Melin, M.; Cederholm, T.; Gustafsson, T.; et al. Nutritional Supplementation with Physical Activity Improves Muscle Composition in Mobility-Limited Older Adults, the VIVE2 Study: A Randomized, Double-Blind, Placebo-Controlled Trial. *J. Gerontol. Ser. A Biol. Sci. Med. Sci.* **2018**, *73*, 95–101. [CrossRef]
57. Grootswagers, P.; Smeets, E.; Oteng, A.B.; de Groot, L. A Novel Oral Nutritional Supplement Improves Gait Speed and Mitochondrial Functioning Compared to Standard Care in Older Adults with (or at Risk of) Undernutrition: Results from a Randomized Controlled Trial. *Aging* **2021**, *13*, 9398–9418. [CrossRef]
58. Achison, M.; Adamson, S.; Akpan, A.; Aspray, T.; Avenell, A.; Band, M.M.; Bashir, T.; Burton, L.A.; Cvoro, V.; Donnan, P.T.; et al. Effect of Perindopril or Leucine on Physical Performance in Older People with Sarcopenia: The LACE Randomized Controlled Trial. *J. Cachexia. Sarcopenia Muscle* **2022**, *13*, 858–871. [CrossRef]
59. Bo, Y.; Liu, C.; Ji, Z.; Yang, R.; An, Q.; Zhang, X.; You, J.; Duan, D.; Sun, Y.; Zhu, Y.; et al. A High Whey Protein, Vitamin D and E Supplement Preserves Muscle Mass, Strength, and Quality of Life in Sarcopenic Older Adults: A Double-Blind Randomized Controlled Trial. *Clin. Nutr.* **2019**, *38*, 159–164. [CrossRef]
60. Rondanelli, M.; Klersy, C.; Terracol, G.; Talluri, J.; Maugeri, R.; Guido, D.; Faliva, M.A.; Solerte, B.S.; Fioravanti, M.; Lukaski, H.; et al. Whey Protein, Amino Acids, and Vitamin D Supplementation with Physical Activity Increases Fat-Free Mass and Strength, Functionality, and Quality of Life and Decreases Inflammation in Sarcopenic Elderly. *Am. J. Clin. Nutr.* **2016**, *103*, 830–840. [CrossRef]
61. Rondanelli, M.; Cereda, E.; Klersy, C.; Faliva, M.A.; Peroni, G.; Nichetti, M.; Gasparri, C.; Iannello, G.; Spadaccini, D.; Infantino, V.; et al. Improving Rehabilitation in Sarcopenia: A Randomized-Controlled Trial Utilizing a Muscle-Targeted Food for Special Medical Purposes. *J. Cachexia. Sarcopenia Muscle* **2020**, *11*, 1535–1547. [CrossRef] [PubMed]
62. Bauer, J.M.; Verlaan, S.; Bautmans, I.; Brandt, K.; Donini, L.M.; Maggio, M.; McMurdo, M.E.T.; Mets, T.; Seal, C.; Wijers, S.L.; et al. Effects of a Vitamin D and Leucine-Enriched Whey Protein Nutritional Supplement on Measures of Sarcopenia in Older Adults, the PROVIDE Study: A Randomized, Double-Blind, Placebo-Controlled Trial. *J. Am. Med. Dir. Assoc.* **2015**, *16*, 740–747. [CrossRef] [PubMed]
63. Yamada, M.; Kimura, Y.; Ishiyama, D.; Nishio, N.; Otobe, Y.; Tanaka, T.; Ohji, S.; Koyama, S.; Sato, A.; Suzuki, M.; et al. Synergistic Effect of Bodyweight Resistance Exercise and Protein Supplementation on Skeletal Muscle in Sarcopenic or Dynapenic Older Adults. *Geriatr. Gerontol. Int.* **2019**, *19*, 429–437. [CrossRef] [PubMed]
64. Ekinci, O.; Yanlk, S.; Terzioğlu Bebitoğlu, B.; Yllmaz Akyüz, E.; Dokuyucu, A.; Erdem, Ş. Effect of Calcium β-Hydroxy-β-Methylbutyrate (CaHMB), Vitamin D, and Protein Supplementation on Postoperative Immobilization in Malnourished Older Adult Patients with Hip Fracture. *Nutr. Clin. Pract.* **2016**, *31*, 829–835. [CrossRef]
65. Buondonno, I.; Sassi, F.; Carignano, G.; Dutto, F.; Ferreri, C.; Pili, F.G.; Massaia, M.; Nisoli, E.; Ruocco, C.; Porrino, P.; et al. From Mitochondria to Healthy Aging: The Role of Branched-Chain Amino Acids Treatment: MATeR a Randomized Study. *Clin. Nutr.* **2020**, *39*, 2080–2091. [CrossRef]
66. VanDerVeer, S.; Markert, R.; Bickford, B.; Yuhas, J.; Pikman, P.; Wall, T.; Burtson, K. Increasing Exercise Adherence among Elderly Patients with Chronic Disease in Primary Care: A Prospective Cohort Study. *BMC Geriatr.* **2021**, *21*, 616. [CrossRef]
67. Bauer, J.; Biolo, G.; Cederholm, T.; Cesari, M.; Cruz-Jentoft, A.J.; Morley, J.E.; Phillips, S.; Sieber, C.; Stehle, P.; Teta, D.; et al. Evidence-Based Recommendations for Optimal Dietary Protein Intake in Older People: A Position Paper from the Prot-Age Study Group. *J. Am. Med. Dir. Assoc.* **2013**, *14*, 542–559. [CrossRef]
68. Pennings, B.; Groen, B.; de Lange, A.; Gijsen, A.P.; Zorenc, A.H.; Senden, J.M.G.; van Loon, L.J.C. Amino Acid Absorption and Subsequent Muscle Protein Accretion Following Graded Intakes of Whey Protein in Elderly Men. *Am. J. Physiol. Endocrinol. Metab.* **2012**, *302*, 992–999. [CrossRef]
69. Wolfe, R.R. Branched-Chain Amino Acids and Muscle Protein Synthesis in Humans: Myth or Reality? *J. Int. Soc. Sports Nutr.* **2017**, *14*, 30. [CrossRef]
70. Wilkinson, D.J.; Hossain, T.; Hill, D.S.; Phillips, B.E.; Crossland, H.; Williams, J.; Loughna, P.; Churchward-Venne, T.A.; Breen, L.; Phillips, S.M.; et al. Effects of Leucine and Its Metabolite β-Hydroxy-β-Methylbutyrate on Human Skeletal Muscle Protein Metabolism. *J. Physiol.* **2013**, *591*, 2911–2923. [CrossRef]
71. Gielen, E.; Beckwée, D.; Delaere, A.; De Breucker, S.; Vandewoude, M.; Bautmans, I.; Bautmans, I.; Beaudart, C.; Beckwée, D.; Beyer, I.; et al. Nutritional Interventions to Improve Muscle Mass, Muscle Strength, and Physical Performance in Older People: An Umbrella Review of Systematic Reviews and Meta-Analyses. *Nutr. Rev.* **2021**, *79*, 121–147. [CrossRef] [PubMed]
72. Chang, M.C.; Choo, Y.J. Effects of Whey Protein, Leucine, and Vitamin D Supplementation in Patients with Sarcopenia: A Systematic Review and Meta-Analysis. *Nutrients* **2023**, *15*, 521. [CrossRef] [PubMed]

73. Rondanelli, M.; Gasparri, C.; Barrile, G.C.; Battaglia, S.; Cavioni, A.; Giusti, R.; Mansueto, F.; Moroni, A.; Nannipieri, F.; Patelli, Z.; et al. Effectiveness of a Novel Food Composed of Leucine, Omega-3 Fatty Acids and Probiotic Lactobacillus Paracasei PS23 for the Treatment of Sarcopenia in Elderly Subjects: A 2-Month Randomized Double-Blind Placebo-Controlled Trial. *Nutrients* **2022**, *14*, 4566. [CrossRef] [PubMed]
74. Dzik, K.P.; Kaczor, J.J. Mechanisms of Vitamin D on Skeletal Muscle Function: Oxidative Stress, Energy Metabolism and Anabolic State. *Eur. J. Appl. Physiol.* **2019**, *119*, 825–839. [CrossRef]
75. Beaudart, C.; Dawson, A.; Shaw, S.C.; Harvey, N.C.; Kanis, J.A.; Binkley, N.; Reginster, J.Y.; Chapurlat, R.; Chan, D.C.; Bruyère, O.; et al. Nutrition and Physical Activity in the Prevention and Treatment of Sarcopenia: Systematic Review. *Osteoporos. Int.* **2017**, *28*, 1817–1833. [CrossRef]
76. Chevalley, T.; Brandi, M.L.; Cashman, K.D.; Cavalier, E.; Harvey, N.C.; Maggi, S.; Cooper, C.; Al-Daghri, N.; Bock, O.; Bruyère, O.; et al. Role of Vitamin D Supplementation in the Management of Musculoskeletal Diseases: Update from an European Society of Clinical and Economical Aspects of Osteoporosis, Osteoarthritis and Musculoskeletal Diseases (ESCEO) Working Group. *Aging Clin. Exp. Res.* **2022**, *34*, 2603–2623. [CrossRef]
77. Mantuano, P.; Boccanegra, B.; Bianchini, G.; Conte, E.; De Bellis, M.; Sanarica, F.; Camerino, G.M.; Pierno, S.; Cappellari, O.; Allegretti, M.; et al. BCAAs and Di-Alanine Supplementation in the Prevention of Skeletal Muscle Atrophy: Preclinical Evaluation in a Murine Model of Hind Limb Unloading. *Pharmacol. Res.* **2021**, *171*, 105798. [CrossRef]
78. Baum, J.I.; Kim, I.Y.; Wolfe, R.R. Protein Consumption and the Elderly: What Is the Optimal Level of Intake? *Nutrients* **2016**, *8*, 359. [CrossRef]

nutrients

MDPI

Brief Report

Associations between Vitality/Nutrition and the Other Domains of Intrinsic Capacity Based on Data from the INSPIRE ICOPE-Care Program

Luc Gaussens [1], Emmanuel González-Bautista [1], Marc Bonnefoy [2], Marguerite Briand [1], Neda Tavassoli [1], Philipe De Souto Barreto [1,3], Yves Rolland [1,3,*] and on behalf of the GEGN Group [†]

[1] Gérontopôle de Toulouse, Institut du Vieillissement, Centre Hospitalo-Universitaire de Toulouse, 31300 Toulouse, France
[2] Service de Médecine Gériatrique, CHU Lyon, Groupement Hospitalier Sud, 69495 Pierre-Bénite, France
[3] CERPOP UMR 1295, University of Toulouse III, Inserm, UPS, 31062 Toulouse, France
* Correspondence: rolland.y@chu-toulouse.fr
† And the French Groupe Expert Nutrition en Gériatrie (GEGN) sous L'égide de la Société Française de Gériatrie et Gérontologie (SFGG): Berrut G (Nantes, France), Bonnefoy M (Lyon, France), Bruyère O (Liège, Belgique), Guerin O (Nice, France), Haine M (Villefranche-sur-Saône, France), Jeandel C (Montpellier, France), Lahaye C (Clermont-Ferrand, France), Paillaud P (Paris, France), Raynaud A (Paris, France), Sanchez M (Paris, France).

Abstract: Background: The vitality domain of intrinsic capacity (IC) represents the synthesis of biological interactions and metabolism. As part of the Integrated Care for Older People (ICOPE) program developed by the World Health Organization (WHO), vitality focuses on the nutritional status of older adults. The objective of this work was to describe the vitality domain of IC in community-dwelling older people and to examine the associations of the vitality components (appetite loss and weight loss) with the other IC domains assessed within the framework of ICOPE. Methods: Cross-sectional data were obtained between January 2020 and February 2022 through the INSPIRE-ICOPE-Care program, a real-life ICOPE implementation initiative developed in the Occitania region of France. Participants were men and women aged 60 and older, looking for primary care services within the French healthcare system. Results: Appetite loss was reported by 14.0% (2013) of the participants, and weight loss by 12.4% (1788). A total of 863 participants (6.01%) declaring weight loss also suffered from appetite loss. In total, 2910 participants (20.27%) screened positive for the domain of vitality. Appetite loss was significantly associated with positive screenings for the domains of cognition (OR = 2.14 [1.84;2.48]), vision (OR = 1.51 [1.28;1.79]), hearing (OR = 1.18 [1.01;1.37]), psychology (OR = 3.95 [3.46;4.52]), and locomotion 'OR = 2.19 [1.91;2.51]). We found significant associations of weight loss with the IC domains of cognition (OR = 1.65 [1.42;1.93]), psychology (OR = 1.80 [1.56;2.07]), locomotion (OR = 1.64 [1.41;1.91]), vision (OR = 1.24 [1.04;1.47]), and hearing (OR = 1.32 [1.12;1.55]). People reporting simultaneous appetite and weight loss showed higher odds of screening positive for psychological (OR = 5.33 [4.53;6.27]) and locomotion impairments (OR = 3.38 [2.88;3.98]). Conclusions: Appetite and weight loss are common among older people and are related to other potential IC impairments, especially psychological and locomotion. Further studies are needed to explore the longitudinal associations of vitality with the incidence of clinically meaningful declines in the other IC domains.

Keywords: ICOPE; vitality; appetite; weight loss; prevention; intrinsic capacity; older people

1. Background

The World Health Organization (WHO) has launched the Integrated Care for Older People (ICOPE) program to prevent disability in older adults [1]. ICOPE uses intrinsic capacity (IC)—i.e., the aggregate of all mental and physical capacities that people draw upon as they age—to monitor early functional decline [2–6]. IC's domains are vitality, cognition, locomotion, sensory (vision and hearing), and psychology.

"Vitality" has been recently defined as a physiological state (due to normal or accelerated biological aging processes) resulting from the interaction between multiple physiological systems, reflected in (the level of) energy and metabolism, neuromuscular function, and immune and stress response functions of the body [7].

However, there is no consensus on how to operationalize vitality in clinical settings [7–10]. The WHO, in its description of the ICOPE program [11], suggests vitality be screened through appetite and weight loss, which are malnutrition hallmarks. Malnutrition is a major determinant of frailty and adverse health events [1,2,6,7,12–15], yet its link with the other IC domains in a real-life population of users of the healthcare system is unclear. Although several studies investigating the associations of weight loss and cognition [16,17], mobility [13,18], and mood [19,20] have already been shown previously, the associations between appetite loss in older people and those functions are not well-known. Furthermore, such associations have never been investigated in a real-life population of users of the healthcare system using the ICOPE screening tool, which is a more practical instrument (few items, feasible to use in clinical practice, high specificity) than full-length scales [21,22]. Moreover, studies examining the associations of appetite and weight loss with the sensory domain in older adults are scarce.

The objective of this study was to describe the vitality domain of IC in a real-life population of older adult users of primary care services and to examine the associations of the vitality screening components of appetite loss and weight loss with the other IC domains assessed within the framework of ICOPE.

2. Materials and Methods

2.1. INSPIRE ICOPE-CARE Cohort

This study uses the ICOPE MONITOR database containing the data of the INSPIRE ICOPE-CARE cohort (implementation of ICOPE in a real-life population of users of primary care services in the Occitania region, Southwest, France) [23]. Data have been collected between 1 January 2020 and 18 November 2021. The first step (Step 1) of the ICOPE program is to screen for potential impairments in IC using the ICOPE screening tool. If an older person screens positive for an IC impairment, the system will send an alert to the surveillance team of nurses for further follow-up [24].

The French Ethical Committee residing in Rennes (CPP Ouest V) validated the study protocol in October 2019. The protocol is registered on the site http://clinicaltrials.gov (accessed on 26 February 2023) (ID NCT04224038). All data are stored in the Gerontopole ICOPE MONITOR database and received the authorization of the French "National Commission for Data Protection" on 13 April 2017 (Ref. Nb. MMS/OSS/NDT171027, authorization request Nb. 19141154). All patients gave either their oral consent or consented via the ICOPE MONITOR app.

2.2. Participants

Participants are community-dwelling men and women aged 60 and over who sought care within the French healthcare system. Health professionals (mainly community nurses, but also primary care physicians, pharmacists, and physical therapists) assessed IC using the ICOPE MONITOR mobile app. In this cross-sectional study, we are using the first Step 1 screening for each participant.

Exclusion criteria were: people younger than 60 years old; institutionalized subjects.

2.3. Vitality

Vitality was screened using two self-reported questions related to appetite and weight loss: "Have you experienced appetite loss?"; "Have you unintentionally lost more than 3 kg over the last three months?". Responding "Yes" to any of these questions determined a positive screening for vitality [11]. We further classified participants into four mutually exclusive groups: only appetite loss, only weight loss, both, and none.

2.4. Other IC Domains

A positive screening in the other IC domains was defined according to the ICOPE Step 1 as follows:

Cognition: failure to recall three words or an error in the current date (year, month, date of the month, and day of the week) or report memory or orientation problems (such as not knowing where one is or what day it is).

Locomotion: inability to complete five chair rises without using arms in less than 14 s.

Vision: self-reported difficulty in seeing far, reading, eye diseases, or currently under medical treatment (e.g., diabetes, high blood pressure).

Hearing: failing the whisper test (when assessed by a health professional) or self-reported hearing loss in the past four months.

Psychology: answering yes to "feeling down, depressed or hopeless?" or "little interest or pleasure in doing things?".

2.5. Covariates

Age was estimated from the date of birth and was used as a categorical variable (per decade). Sex and weight were self-reported.

2.6. Statistical Analysis

Means, standard deviations, and percentages were used to describe the study population. We used logistic regression models to estimate the cross-sectional associations of vitality (and its screening components appetite and weight loss, alone or combined) with positive screening in other IC domains. The models were adjusted for age, sex, and weight. A p-value of ≤ 0.05 defined statistical significance. All analyses were carried out using Stata 16.

3. Results

Table 1 displays the characteristics of the 14,572 participants. Their mean age was 76.7 (SD = 8.8), 62% were female, and the mean weight was 70 kg (SD = 15.5).

Table 1. Description of the population aged 60 and over with professional assessment. (n, % unless otherwise noted).

Demographics			
Age (total, mean, SD)	14,572	76.7	8.8
Weight (total, mean, SD)	14,027	70.0	15.5
Female (total, number, percentage)	14,572	9039	62.0
Self-reported appetite and weight loss	total	n	%
Appetite loss	14,406	2013	14.0
Weight loss	14,403	1788	12.4
Only appetite loss	14,358	1132	7.9
Only weight loss	14,358	915	6.4
Both	14,358	863	6.0
None	14,358	11,448	79.7
Domain alerts from ICOPE Care			
Cognition	14,338	8708	60.7
Nutrition	14,386	2938	20.4
Vision	13,316	9981	75.0
Hearing	12,209	7721	63.2
Psychological	14,516	5750	39.6
Locomotion	13,967	5169	37.0

Appetite loss was reported by 14.0% (2013) of the participants, and weight loss by 12.4% (1788). A total of 863 participants (6.01%) declaring weight loss also suffered from appetite loss. In total, 2910 participants (20.27%) screened positive for vitality impairment.

The results of the associations of vitality and its components with the other IC domains are presented in Table 2. Both appetite and weight loss were significantly associated with positive screenings of the other five IC domains (i.e., cognition, psychology, locomotion, vision, and hearing) when adjusting for covariates. When examining the associations of appetite loss without weight loss, appetite was associated with all IC domains. Weight loss without appetite loss was associated with all IC domains. The strongest association of a positive screening for vitality impairment was found for a positive screening in the locomotion (OR 3.38, 95%CI 2.88; 3.98) and psychological domains (OR 5.33, 95%CI 4.53; 6.27) among participants who reported simultaneous appetite and weight loss.

Table 2. Odds ratio for the association between appetite loss and weight loss with positive screening for IC impairments.

		n	aOR *	95%CI		*p*
		Appetite Loss				
Cognition		13,822	2.16	1.92	2.42	<0.001
Vision		12,796	1.39	1.23	1.58	<0.001
Hearing		11,859	1.22	1.09	1.38	0.001
Psychological		13,852	4.24	3.81	4.72	<0.001
Locomotion		13,356	2.52	2.27	2.81	<0.001
		Weight Loss				
Cognition		13,818	1.83	1.63	2.06	<0.001
Vision		12,785	1.23	1.08	1.39	0.002
Hearing		11,848	1.31	1.16	1.49	<0.001
Psychological		13,847	2.54	2.29	2.83	<0.001
Locomotion		13,347	2.15	1.92	2.41	<0.001
		Mutually Exclusive Groups				
Cognition		13,782				
	Only appetite loss		2.14	1.84	2.48	<0.001
	Only weight loss		1.65	1.42	1.93	<0.001
	Both		2.36	1.98	2.81	<0.001
Vision		12,762				
	Only appetite loss		1.51	1.28	1.79	<0.001
	Only weight loss		1.24	1.04	1.47	0.014
	Both		1.31	1.09	1.58	0.004
Hearing		11,831				
	Only appetite loss		1.18	1.01	1.37	0.035
	Only weight loss		1.32	1.12	1.55	0.001
	Both		1.35	1.13	1.61	0.001
Psychological		13,812				
	Only appetite loss		3.95	3.46	4.52	<0.001
	Only weight loss		1.80	1.56	2.07	<0.001
	Both		5.33	4.53	6.27	<0.001
Locomotion		13,321				
	Only appetite loss		2.19	1.91	2.51	<0.001
	Only weight loss		1.64	1.41	1.91	<0.001
	Both		3.38	2.88	3.98	<0.001

* Adjusted for age (per decade), sex, and weight.

4. Discussion

One-fifth of the INSPIRE ICOPE-CARE cohort screened positive for vitality impairment (20.4%). Almost one person in seven declared appetite loss (14.0%) and almost one in eight reported weight loss (12.4%). Both appetite and weight loss were associated with an increased likelihood of having positive screenings in all the other IC domains (i.e., cognition, psychology, locomotion, vision, and hearing). People with both appetite and weight loss have a higher risk of screening positive with the locomotion and psychology domains than people with only one of the two vitality items.

The prevalence of vitality impairment in our study is quite similar to the prevalence found in other cross-sectional studies among Mexicans (12,459 community-dwelling older people, mean age of 71.2 years, prevalence of vitality impairment of 27.5%) [6] or Chinese (304 participants living in their own homes, mean age of 78 years, prevalence of vitality impairment of 18.1%) [21]. The higher vitality alert in the Mexican study could be explained by their broader definition (using a two-year recall instead of three months for appetite and weight loss), whereas the Chinese study used the same items for vitality as our study.

In a meta-analysis [25], the prevalence of weight loss among older people ranged between 3.1% (in community-dwelling) to 29.4% (in rehabilitation/sub-acute care). The prevalence of weight loss in studies of community-dwelling older people is critically dependent on age, autonomy, or definition of weight loss [26–30]. The prevalence of weight loss in our study may be higher than in other studies because our detection threshold is lower than in most studies (a weight loss of 3 kg represents approximately 4% of an average weight of 70 kg whereas many studies have a threshold of 5% weight loss). However, any comparison with other studies should take into account that our population is more prone to having some health issues than the average community-dwelling people because our participants are already seeking care.

Our results on the prevalence of appetite loss are consistent with some studies [31,32]. Our low value compared to other studies could be explained by our population which was slightly younger, for example, the average age of Landi et al.'s study population [33] was 80.4 years for more than 25% appetite loss. Our definition of appetite loss could also explain some differences; for example, Vázquez–Valdez's study (with a prevalence of appetite loss of 30.1%) included any appetite loss in the last 12 months [34].

Although appetite loss is a major cause of weight loss [35], only one in two people suffer from both appetite and weight loss, as has already been suggested in previous studies [14,28,33,36].

Weight loss was associated with positive screenings in all the other IC domains. Weight loss is known to impact locomotor function through sarcopenia [37–39]. Appetite loss could also influence locomotor function, independently of weight loss, through changes in diet. Indeed, older people with appetite loss tend to reduce their protein and micronutrient intake, with a potential increase in carbohydrate intake, which can lead to impaired muscle function without weight loss [40–42].

We found a strong cross-sectional association between having a positive screening for the vitality and the psychological domain. It is important to highlight that loss of weight or appetite make part of the diagnostic criteria for depression according to the Diagnostic and Statistical Manual of Mental Disorders (5th ed.; DSM–5) [43]. Mood disorders are one of the main causes of nutritional disorders [44,45] and treatments for depression can also affect both weight and appetite [46,47]. There is also evidence suggesting nutritional deficiencies may promote mood disorders [20,48,49]. Some proinflammatory cytokines and hormones (such as ghrelin, leptin, orexin, or insulin), involved in appetite and weight regulation, are cited among the neurobiological substrates of depression [50–52]. Vitality seems to be an interesting domain to better understand causes and consequences of mood disorders.

Malnutrition is a critical factor in the complications of cognitive disorders [17,53,54] and appetite loss can be an early sign of neurodegenerative pathologies [55]. However, some studies suggest energy metabolism and micronutrient deficiencies could be precursors in the genesis of cognitive disorders [17,41,55,56]. Further investigations are needed to understand the influence of vitality on the development and aging of neurological tissues.

Weight could be a specific factor in hearing loss. Indeed, an increase in hearing loss has been observed in the Korean population in cases of underweight and severe obesity [57]. Certain micronutrient deficiencies (such as vitamins A, B, C, D, and E or iron and magnesium) and a poor-quality diet (low protein intake or high cholesterol intake) could affect hearing function [58]. Micronutrients are already known to be an important factor in vision [41] but there may be more complex molecular interactions between energetic metabolism and the retina [59]. Nutritionally dependent biological processes may affect the sensory domain and require further investigation.

Our results showed five-fold higher odds of psychological and three-fold higher odds of locomotion-positive screening among participants with both weight and appetite loss simultaneously, strongly suggesting that the overlapping of these conditions should be identified as a "red flag" to identify older adults with critical low IC levels.

The biological processes regulating appetite might influence other age-related mechanisms as seen in the GDF-15 studies [60]. It is not yet known to what extent appetite loss is a cause or a consequence of other medical conditions, but it could represent a target for early treatment.

Our study has strengths, for example, our data were collected in a large real-life population seeking care, and our study is among the first to provide an in-depth approach to the screening for vitality impairment and its relationship with other intrinsic capacity domains among the ICOPE CARE cohort. On the other hand, there are limitations, for example, we could not estimate a causal effect of vitality over the other IC domains given that the IC domain impairment was not confirmed (not all positive screenings imply a true impairment). Furthermore, given that objectively measured body composition was not available in the database, we used self-reported weight. We could not use Body Mass Index (BMI) without a measure of height. Yet, studies have found that self-reported weight in older adults is quite close to the actual weight (to within 2 kg), although overestimation of weight increases with age, cognitive impairments, poor health, and underweight [61,62].

Further studies with verified IC domain impairments and preferably longitudinal follow-up are needed to better understand the specific biological processes linking the domains. Population monitoring and further assessment of impairment through the INSPIRE ICOPE-CARE program will provide more data.

5. Conclusions

Appetite and weight loss are common among older people and are related to other potential IC impairments, especially psychological and locomotion. Further studies are needed to explore the longitudinal associations of vitality with the incidence of clinically meaningful declines in the other IC domains.

Author Contributions: Formal analysis, E.G.-B.; Writing—original draft, L.G.; Writing—review & editing, M.B. (Marc Bonnefoy), M.B. (Marguerite Briand), N.T. and GEGN Group; Supervision, P.D.S.B. and Y.R. All authors have read and agreed to the published version of the manuscript.

Funding: The Inspire Program is supported by grants from the Region Occitanie/Pyrénées-Méditerranée (Reference number: 1901175), the European Regional Development Fund (ERDF) (Project number: MP0022856), MSD Avenir, and the Inspire Chairs of Excellence funded by: Alzheimer Prevention in Occitania and Catalonia (APOC), EDENIS, KORIAN, Pfizer, Pierre-Fabre.

Institutional Review Board Statement: The French Ethical Committee residing in Rennes (CPP Ouest V) validated the study protocol in October 2019. The protocol is registered on the site http://clinicaltrials.gov (accessed on 26 February 2023) (ID NCT04224038). All data are stored in the Gerontopole ICOPE MONITOR database and received the authorization of the French "National Commission for Data Protection" on 13 April 2017 (Ref. Nb. MMS/OSS/NDT171027, authorization request Nb. 19141154.

Informed Consent Statement: All patients gave either their oral consent or consented via the ICOPE MONITOR app.

Data Availability Statement: Data may be made available under reasonable request to the author at barreto.p@chu-toulouse.fr.

Conflicts of Interest: The authors declare no conflict of interest.

References

1. WHO. *Integrated Care for Older People: Guidelines on Community-Level Interventions to Manage Declines in Intrinsic Capacity*; World Health Organization: Geneva, Switzerland, 2017. (WHO Guidelines Approved by the Guidelines Review Committee). Available online: http://www.ncbi.nlm.nih.gov/books/NBK488250/ (accessed on 9 January 2023).
2. Sanchez-Rodriguez, D.; Piccard, S.; Dardenne, N.; Giet, D.; Annweiler, C.; Gillain, S. Implementation of the Integrated Care of Older People (ICOPE) App and ICOPE Monitor in Primary Care: A study protocol. *J. Frailty Aging* **2021**, *10*, 290–296. [CrossRef] [PubMed]
3. González-Bautista, E.; de Souto Barreto, P.; Andrieu, S.; Rolland, Y.; Vellas, B.; MAPT/DSA Group. Screening for intrinsic capacity impairments as markers of increased risk of frailty and disability in the context of integrated care for older people: Secondary analysis of MAPT. *Maturitas* **2021**, *150*, 1–6. [CrossRef] [PubMed]
4. Beard, J.R.; Jotheeswaran, A.T.; Cesari, M.; Araujo de Carvalho, I. The structure and predictive value of intrinsic capacity in a longitudinal study of ageing. *BMJ Open* **2019**, *9*, e026119. [CrossRef] [PubMed]
5. Stolz, E.; Mayerl, H.; Freidl, W.; Roller-Wirnsberger, R.; Gill, T.M. Intrinsic Capacity Predicts Negative Health Outcomes in Older Adults. *J. Gerontol. Ser. A Biol. Sci. Med. Sci.* **2022**, *77*, 101–105. [CrossRef]
6. Gutiérrez-Robledo, L.M.; García-Chanes, R.E.; Pérez-Zepeda, M.U. Screening intrinsic capacity and its epidemiological characterization: A secondary analysis of the Mexican Health and Aging Study. *Rev. Panam. Salud Pública* **2021**, *45*, e121. [CrossRef] [PubMed]
7. Bautmans, I.; Knoop, V.; Thiyagarajan, J.A.; Maier, A.B.; Beard, J.R.; Freiberger, E.; Belsky, D.; Aubertin-Leheudre, M.; Mikton, C.; Cesari, M.; et al. WHO working definition of vitality capacity for healthy longevity monitoring. *Lancet Healthy Longev.* **2022**, *3*, e789–e796. [CrossRef]
8. Cesari, M.; De Carvalho, I.A.; Thiyagarajan, J.A.; Cooper, C.; Martin, F.C.; Reginster, J.-Y.; Vellas, B.; Beard, J.R. Evidence for the Domains Supporting the Construct of Intrinsic Capacity. *J. Gerontol. Ser. A Biol. Sci. Med. Sci.* **2018**, *73*, 1653–1660. [CrossRef]
9. George, P.P.; Lun, P.; Ong, S.P.; Lim, W.S. A Rapid Review of the Measurement of Intrinsic Capacity in Older Adults. *J. Nutr. Health Aging* **2021**, *25*, 774–782. [CrossRef]
10. Gonzalez-Bautista, E.; Andrieu, S.; Gutiérrez-Robledo, L.M.; García-Chanes, R.E.; de Souto Barreto, P. In the quest of a Standard Index of Intrinsic Capacity. A Critical Literature Review. *J. Nutr. Health Aging* **2020**, *24*, 959–965. [CrossRef]
11. WHO. Guidance on Person-Centred Assessment and Pathways in Primary Care [Internet]. February 2022. Available online: https://www.who.int/publications/i/item/WHO-FWC-ALC-19.1 (accessed on 26 February 2023).
12. Stuck, A.E.; Walthert, J.M.; Nikolaus, T.; Büla, C.J.; Hohmann, C.; Beck, J.C. Risk factors for functional status decline in community-living elderly people: A systematic literature review. *Soc. Sci. Med.* **1999**, *48*, 445–469. [CrossRef]
13. Ni Lochlainn, M.; Cox, N.J.; Wilson, T.; Hayhoe, R.P.G.; Ramsay, S.E.; Granic, A.; Isanejad, M.; Roberts, H.C.; Wilson, D.; Welch, C.; et al. Nutrition and Frailty: Opportunities for Prevention and Treatment. *Nutrients* **2021**, *13*, 2349. [CrossRef]
14. Landi, F.; Russo, A.; Liperoti, R.; Tosato, M.; Barillaro, C.; Pahor, M.; Bernabei, R.; Onder, G. Anorexia, Physical Function, and Incident Disability Among the Frail Elderly Population: Results From the ilSIRENTE Study. *J. Am. Med. Dir. Assoc.* **2010**, *11*, 268–274. [CrossRef] [PubMed]
15. Sanford, A.M. Anorexia of aging and its role for frailty. *Curr. Opin. Clin. Nutr. Metab. Care* **2017**, *20*, 54–60. [CrossRef] [PubMed]
16. Giudici, K.V.; Guyonnet, S.; Morley, J.E.; Nguyen, A.D.; Aggarwal, G.; Parini, A.; Li, Y.; Bateman, R.J.; Vellas, B.; Barreto, P.D.S.; et al. Interactions Between Weight Loss and Plasma Neurodegenerative Markers for Determining Cognitive Decline Among Community-Dwelling Older Adults. *J. Gerontol. Ser. A Biol. Sci. Med. Sci.* **2022**, *77*, 1159–1168. [CrossRef] [PubMed]
17. Minaglia, C.; Giannotti, C.; Boccardi, V.; Mecocci, P.; Serafini, G.; Odetti, P.; Monacelli, F. Cachexia and advanced dementia. *J. Cachexia Sarcopenia Muscle* **2019**, *10*, 263–277. [CrossRef] [PubMed]
18. Ramsey, K.A.; Meskers, C.G.M.; Trappenburg, M.C.; Verlaan, S.; Reijnierse, E.M.; Whittaker, A.; Maier, A.B. Malnutrition is associated with dynamic physical performance. *Aging Clin. Exp. Res.* **2020**, *32*, 1085–1092. [CrossRef]
19. Hajek, A.; Bock, J.O.; König, H.H. Psychosocial correlates of unintentional weight loss in the second half of life in the German general population. *PLoS ONE* **2017**, *12*, e0185749. [CrossRef]
20. Koster, A.; van Gool, C.H.; Kempen, G.I.; Penninx, B.W.; Lee, J.S.; Rubin, S.M.; Tylavsky, F.A.; Yaffe, K.; Newman, A.B.; Harris, T.B.; et al. Late-life depressed mood and weight change contribute to the risk of each other. *Am. J. Geriatr. Psychiatry Off. J. Am. Assoc. Geriatr. Psychiatry* **2010**, *18*, 236–244. [CrossRef]
21. Leung, A.Y.M.; Su, J.J.; Lee, E.S.H.; Fung, J.T.S.; Molassiotis, A. Intrinsic capacity of older people in the community using WHO Integrated Care for Older People (ICOPE) framework: A cross-sectional study. *BMC Geriatr.* **2022**, *22*, 304. [CrossRef]

22. Rojano i Luque, X.; Blancafort-Alias, S.; Prat Casanovas, S.; Forné, S.; Martín Vergara, N.; Fabregat Povill, P.; Royo, M.V.; Serrano, R.; Sanchez-Rodriguez, D.; Saldaña, M.V.; et al. Identification of decreased intrinsic capacity: Performance of diagnostic measures of the ICOPE Screening tool in community dwelling older people in the VIMCI study. *BMC Geriatr.* **2023**, *23*, 106. [CrossRef]
23. Tavassoli, N.; Piau, A.; Berbon, C.; De Kerimel, J.; Lafont, C.; Barreto, P.D.S.; Guyonnet, S.; Takeda, C.; Carrie, I.; Angioni, D.; et al. Framework Implementation of the INSPIRE ICOPE-CARE Program in Collaboration with the World Health Organization (WHO) in the Occitania Region. *J. Frailty Aging* **2021**, *10*, 103–109. [CrossRef]
24. Tavassoli, N.; Barreto, P.D.S.; Berbon, C.; Mathieu, C.; de Kerimel, J.; Lafont, C.; Takeda, C.; Carrie, I.; Piau, A.; Jouffrey, T.; et al. Implementation of the WHO integrated care for older people (ICOPE) programme in clinical practice: A prospective study. *Lancet Healthy Longev.* **2022**, *3*, e394–e404. [CrossRef] [PubMed]
25. Cereda, E.; Pedrolli, C.; Klersy, C.; Bonardi, C.; Quarleri, L.; Cappello, S.; Turri, A.; Rondanelli, M.; Caccialanza, R. Nutritional status in older persons according to healthcare setting: A systematic review and meta-analysis of prevalence data using MNA®. *Clin. Nutr.* **2016**, *35*, 1282–1290. [CrossRef] [PubMed]
26. Granic, A.; Mendonça, N.; Hill, T.R.; Jagger, C.; Stevenson, E.J.; Mathers, J.C.; Sayer, A.A. Nutrition in the Very Old. *Nutrients* **2018**, *10*, 269. [CrossRef] [PubMed]
27. Norman, K.; Haß, U.; Pirlich, M. Malnutrition in Older Adults—Recent Advances and Remaining Challenges. *Nutrients* **2021**, *13*, 2764. [CrossRef]
28. Takehara, S.; Hirani, V.; Wright, F.A.C.; Naganathan, V.; Blyth, F.M.; Le Couteur, D.G.; Waite, L.M.; Seibel, M.J.; Handelsman, D.J.; Cumming, R.G. Appetite, oral health and weight loss in community-dwelling older men: An observational study from the Concord Health and Ageing in Men Project (CHAMP). *BMC Geriatr.* **2021**, *21*, 255. [CrossRef]
29. Newman, A.B.; Yanez, D.; Harris, T.; Duxbury, A.; Enright, P.L.; Fried, L.P.; Cardiovascular Study Research Group. Weight change in old age and its association with mortality. *J. Am. Geriatr. Soc.* **2001**, *49*, 1309–1318. [CrossRef]
30. Alibhai, S.M.H.; Greenwood, C.; Payette, H. An approach to the management of unintentional weight loss in elderly people. *Can. Med. Assoc. J.* **2005**, *172*, 773–780. [CrossRef]
31. Tsutsumimoto, K.; Doi, T.; Makizako, H.; Hotta, R.; Nakakubo, S.; Makino, K.; Suzuki, T.; Shimada, H. Aging-related anorexia and its association with disability and frailty. *J. Cachexia Sarcopenia Muscle* **2018**, *9*, 834–843. [CrossRef]
32. Donini, L.M.; Poggiogalle, E.; Piredda, M.; Pinto, A.; Barbagallo, M.; Cucinotta, D.; Sergi, G. Anorexia and Eating Patterns in the Elderly. *PLoS ONE* **2013**, *8*, e63539. [CrossRef]
33. Landi, F.; Liperoti, R.; Lattanzio, F.; Russo, A.; Tosato, M.; Barillaro, C.; Bernabei, R.; Onder, G. Effects of anorexia on mortality among older adults receiving home care: An observational study. *J. Nutr. Health Aging* **2012**, *16*, 79–83. [CrossRef] [PubMed]
34. Vázquez-Valdez, O.E.; Aguilar-Navarro, S.; Avila-Funes, J.A. Association between anorexia of aging and disability in older community-dwelling Mexicans. *J. Am. Geriatr. Soc.* **2010**, *58*, 2044–2046. [CrossRef]
35. Rolland, Y.; Perrin, A.; Gardette, V.; Filhol, N.; Vellas, B. Screening older people at risk of malnutrition or malnourished using the Simplified Nutritional Appetite Questionnaire (SNAQ): A comparison with the Mini-Nutritional Assessment (MNA) tool. *J. Am. Med. Dir. Assoc.* **2012**, *13*, 31–34. [CrossRef]
36. van der Meij, B.S.; Wijnhoven, H.A.; Lee, J.S.; Houston, D.K.; Hue, T.; Harris, T.B.; Kritchevsky, S.B.; Newman, A.B.; Visser, M. Poor appetite and dietary intake in community-dwelling older adults. *J. Am. Geriatr. Soc.* **2017**, *65*, 2190–2197. [CrossRef] [PubMed]
37. Hai, S.; Gao, Q.; Gwee, X.; Chua, D.Q.L.; Yap, K.B.; Ng, T.P. Malnutrition Risk, Physical Function Decline and Disability in Middle-Aged and Older Adults Followed Up in the Singapore Longitudinal Ageing Study. *Clin. Interv. Aging* **2021**, *16*, 1527–1539. [CrossRef]
38. Murphy, R.A.; Patel, K.V.; Kritchevsky, S.B.; Houston, D.K.; Newman, A.B.; Koster, A.; Simonsick, E.M.; Tylvasky, F.A.; Cawthon, P.M.; Harris, T.B. Weight change, body composition, and risk of mobility disability and mortality in older adults: A population-based cohort study. *J. Am. Geriatr. Soc.* **2014**, *62*, 1476–1483. [CrossRef] [PubMed]
39. Cruz-Jentoft, A.J.; Bahat, G.; Bauer, J.; Boirie, Y.; Bruyère, O.; Cederholm, T.; Cooper, C.; Landi, F.; Rolland, Y.; Sayer, A.A.; et al. Sarcopenia: Revised European consensus on definition and diagnosis. *Age Ageing* **2019**, *48*, 16–31. [CrossRef]
40. de Souto Barreto, P.; Cesari, M.; Morley, J.E.; Roberts, S.; Landi, F.; Cederholm, T.; Rolland, Y.; Vellas, B.; Fielding, R. Appetite Loss and Anorexia of Aging in Clinical Care: An ICFSR Task Force Report. *J. Frailty Aging* **2022**, *11*, 129–134. [CrossRef] [PubMed]
41. Bruins, M.J.; Van Dael, P.; Eggersdorfer, M. The Role of Nutrients in Reducing the Risk for Noncommunicable Diseases during Aging. *Nutrients* **2019**, *11*, 85. [CrossRef]
42. Landi, F.; Liperoti, R.; Russo, A.; Giovannini, S.; Tosato, M.; Barillaro, C.; Capoluongo, E.D.; Bernabei, R.; Onder, G. Association of anorexia with sarcopenia in a community-dwelling elderly population: Results from the ilSIRENTE study. *Eur. J. Nutr.* **2013**, *52*, 1261–1268. [CrossRef]
43. American Psychiatric Association Publishing; American Psychiatric Association. *Diagnostic and Statistical Manual of Mental Disorders*, 5th ed.; American Psychiatric Association: Washington, DC, USA, 2022. [CrossRef]
44. Landi, F.; Calvani, R.; Tosato, M.; Martone, A.M.; Ortolani, E.; Savera, G.; Sisto, A.; Marzetti, E. Anorexia of Aging: Risk Factors, Consequences, and Potential Treatments. *Nutrients* **2016**, *8*, 69. [CrossRef] [PubMed]
45. McMinn, J.; Steel, C.; Bowman, A. Investigation and management of unintentional weight loss in older adults. *BMJ* **2011**, *342*, d1732. [CrossRef] [PubMed]

46. Oliva, V.; Lippi, M.; Paci, R.; Del Fabro, L.; Delvecchio, G.; Brambilla, P.; De Ronchi, D.; Fanelli, G.; Serretti, A. Gastrointestinal side effects associated with antidepressant treatments in patients with major depressive disorder: A systematic review and meta-analysis. *Prog. Neuropsychopharmacol. Biol. Psychiatry* **2021**, *109*, 110266. [CrossRef] [PubMed]
47. Gill, H.; Gill, B.; El-Halabi, S.; Chen-Li, D.; Lipsitz, O.; Rosenblat, J.D.; Van Rheenen, T.E.; Rodrigues, N.B.; Mansur, R.B.; Majeed, A.; et al. Antidepressant Medications and Weight Change: A Narrative Review. *Obes. Silver Spring Md.* **2020**, *28*, 2064–2072. [CrossRef] [PubMed]
48. Jin, Y.; Cho, J.; Lee, I.; Park, S.; Kim, D.; Kong, J.; Kang, H. Involuntary Weight Loss and Late-Life Depression in Korean Older Adults. *Iran. J. Public Health* **2020**, *49*, 637–644. [CrossRef]
49. Vaughan, L.; Corbin, A.L.; Goveas, J.S. Depression and frailty in later life: A systematic review. *Clin. Interv. Aging* **2015**, *10*, 1947–1958. [CrossRef]
50. Morin, V.; Hozer, F.; Costemale-Lacoste, J.F. The effects of ghrelin on sleep, appetite, and memory, and its possible role in depression: A review of the literature. *L'Encephale* **2018**, *44*, 256–263. [CrossRef]
51. Caroleo, M.; Carbone, E.A.; Primerano, A.; Foti, D.; Brunetti, A.; Segura-Garcia, C. The role of hormonal, metabolic and inflammatory biomarkers on sleep and appetite in drug free patients with major depression: A systematic review. *J. Affect. Disord.* **2019**, *250*, 249–259. [CrossRef]
52. Andréasson, A.; Arborelius, L.; Erlanson-Albertsson, C.; Lekander, M. A putative role for cytokines in the impaired appetite in depression. *Brain. Behav. Immun.* **2007**, *21*, 147–152. [CrossRef]
53. Gao, S.; Nguyen, J.T.; Hendrie, H.C.; Unverzagt, F.W.; Hake, A.; Smith-Gamble, V.; Hall, K. Accelerated weight loss and incident dementia in an elderly African-American cohort. *J. Am. Geriatr. Soc.* **2011**, *59*, 18–25. [CrossRef]
54. Soto, M.E.; Secher, M.; Gillette-Guyonnet, S.; van Kan, G.A.; Andrieu, S.; Nourhashemi, F.; Rolland, Y.; Vellas, B. Weight loss and rapid cognitive decline in community-dwelling patients with Alzheimer's disease. *J. Alzheimer's Dis. JAD* **2012**, *28*, 647–654. [CrossRef] [PubMed]
55. Jimenez, A.; Pegueroles, J.; Carmona-Iragui, M.; Vilaplana, E.; Montal, V.; Alcolea, D.; Videla, L.; Illán-Gala, I.; Pané, A.; Casajoana, A.; et al. Weight loss in the healthy elderly might be a non-cognitive sign of preclinical Alzheimer's disease. *Oncotarget* **2017**, *8*, 104706. [CrossRef] [PubMed]
56. López-Gambero, A.; Rosell-Valle, C.; Medina-Vera, D.; Navarro, J.; Vargas, A.; Rivera, P.; Sanjuan, C.; de Fonseca, F.R.; Suárez, J. A Negative Energy Balance Is Associated with Metabolic Dysfunctions in the Hypothalamus of a Humanized Preclinical Model of Alzheimer's Disease, the 5XFAD Mouse. *Int. J. Mol. Sci.* **2021**, *22*, 5365. [CrossRef] [PubMed]
57. Kim, S.H.; Won, Y.S.; Kim, M.G.; Baek, Y.J.; Oh, I.H.; Yeo, S.G. Relationship between obesity and hearing loss. *Acta Otolaryngol.* **2016**, *136*, 1046–1050. [CrossRef] [PubMed]
58. Jung, S.Y.; Kim, S.H.; Yeo, S.G. Association of Nutritional Factors with Hearing Loss. *Nutrients* **2019**, *11*, 307. [CrossRef]
59. Aytürk, D.G.; Castrucci, A.M.; Carr, D.E.; Keller, S.R.; Provencio, I. Lack of Melanopsin Is Associated with Extreme Weight Loss in Mice upon Dietary Challenge. *PLoS ONE* **2015**, *10*, e0127031. [CrossRef]
60. Lockhart, S.M.; Saudek, V.; O'Rahilly, S. GDF15: A Hormone Conveying Somatic Distress to the Brain. *Endocr. Rev.* **2020**, *41*, bnaa007. [CrossRef]
61. Niedźwiedzka, E.; Długosz, A.; Wądołowska, L. Validity of self-reported height and weight in elderly Poles. *Nutr. Res. Pract.* **2015**, *9*, 319–327. [CrossRef]
62. Sahyoun, N.R.; Maynard, L.M.; Zhang, X.L.; Serdula, M.K. Factors associated with errors in self-reported height and weight in older adults. *J. Nutr. Health Aging* **2008**, *12*, 108–115. [CrossRef]

MDPI AG
Grosspeteranlage 5
4052 Basel
Switzerland
Tel.: +41 61 683 77 34

Nutrients Editorial Office
E-mail: nutrients@mdpi.com
www.mdpi.com/journal/nutrients

www.ingramcontent.com/pod-product-compliance
Lightning Source LLC
Chambersburg PA
CBHW061440210926
51850CB00008B/1567
9783725850778